Hillslope and Watershed Hydrology

Special Issue Editors

Christopher J. Duffy

Xuan Yu

MDPI • Basel • Beijing • Wuhan • Barcelona • Belgrade

Special Issue Editors
Christopher J. Duffy
The Pennsylvania State University
USA

Xuan Yu
University of Delaware
USA

Editorial Office
MDPI
St. Alban-Anlage 66
Basel, Switzerland

This edition is a reprint of the Special Issue published online in the open access journal *Water* (ISSN 2073-4441) from 2015–2018 (available at: http://www.mdpi.com/journal/water/special issues/hillslope-watershed-hydrology).

For citation purposes, cite each article independently as indicated on the article page online and as indicated below:

Lastname, F.M.; Lastname, F.M. Article title. *Journal Name* **Year**, *Article number*, page range.

First Editon 2018

ISBN 978-3-03842-951-7 (Pbk)
ISBN 978-3-03842-952-4 (PDF)

Table of Contents

About the Special Issue Editors

Christopher J. Duffy, Professor Emeritus, Department of Civil and Environmental Engineering, The Pennsylvania State University, University Park, PA, USA. Professor Duffy held faculty appointments with Utah State University, Logan, UT, USA, 1981–1989; and visiting appointments with Los Alamos National Laboratory 1998–1999; Cornell University, Ithaca, NY, USA, 1987–1988; the Ècole Polytechnique Fédérale de Lausanne, Lausanne, Switzerland, 2006–2007. He was also a Visiting Scientist at the University of Bonn, Bonn, Germany, in 2015. He and his team have focused on developing the spatially distributed and physics-based computational code PIHM (The Penn State Integrated Hydrologic Model) for multiscale and multiprocess applications (http://www.pihm.psu.edu/) and an online national data service for access to geospatial watershed data (www.hydroterre.psu.edu) anywhere in the continental U.S. He was a Senior Fellow with the Smithsonian Institution while in residence at the Smithsonian Environmental Research Center, in 2007 and a Visiting Senior Fellow at the University of Bristol, 2014–2016.

Xuan Yu, Postdoctoral Researcher, Department of Geological Sciences, University of Delaware, Newark, DE, USA. Dr. Xu received his B.E. degree in water resources engineering from the China University of Geosciences, Beijing, China, in 2006; an M.E. degree from the China Institute of Water Resources and Hydropower Research, Beijing, China, in 2009; and a Ph.D. degree in civil and environmental engineering from The Pennsylvania State University, University Park, PA, USA, in 2014. His research interests include watershed models, coastal surface water-groundwater interactions, hydrology and climate change, and open science.

Preface to "Hillslope and Watershed Hydrology"

The goal of watershed hydrology is to better understand water movement and storage that can be managed and exploited for economic development and environmental sustainability. The hydrologists bring together information from topography, geology, land use, etc., in order to evaluate the watershed responses to different climate scenarios, ecological settings and human activities. The book attempts to present state-of-the-art methods of watershed hydrology, illustrated with worldwide case studies.

The book has three chapters: Chapter 1 presents advanced watershed models and understanding of processes, parameters, and uncertainty which is critical to the development of tools for the prediction of hydrologic state and flux variables, in order to manage water resources. Five watershed models are reported in different areas, which involve module development, model parameterization, sensitivity analysis, and uncertainty estimation.

Chapter 2 covers examples of watershed model applications for environmental assessment, management, and conservation. Application of watershed models requires integration of site-specific knowledge, regional calibration, environmental scenarios, and social problems, which increase the complexity of interpretation of model results for resource and environmental water management. This chapter collects case studies of watershed models from a variety of perspectives, including large watershed, urban watershed, intense human activities, and natural hazard vulnerability.

Chapter 3 lists several popular watershed models, summarizes the content of this book, and speculates on future research perspectives. It is anticipated that we will witness the introduction of more reliable models, comprehensive interpretations and broader implementation where water is driving human and ecosystem services. Such advances will be achieved by community efforts across disciplines and open science between researchers and the public.

Overall, the book covers different aspects of watershed hydrology that should be of interest for practitioners and academicians alike. We hope readers will glean a clear picture of how watershed models represent an important tool for environmental science and management.

Christopher J. Duffy and Xuan Yu
Special Issue Editors

Review

Watershed Hydrology: Scientific Advances and Environmental Assessments

Xuan Yu [1] **and Christopher J. Duffy** [2,*]

[1] Department of Geological Sciences, University of Delaware, Newark, DE 19716, USA; xuan@udel.edu
[2] Department of Civil and Environmental Engineering, Penn State University, University Park, PA 16802, USA
* Correspondence: cxd11@psu.edu; Tel.: +1-814-863-4384

Received: 9 February 2018; Accepted: 6 March 2018; Published: 8 March 2018

Abstract: The watershed is a fundamental concept in hydrology and is the basis for understanding hydrologic processes and for the planning and management of water resources. Storage and movement of water at a watershed scale is complicated due to the coupled processes which act over multiple spatial and temporal scales. In addition, climate change and human activities increase the complexity of these processes driving hydrologic change. Scientific advances in the field of watershed hydrology is now making use of the latest methods and technologies to achieve responsible management of water resources to meet the needs of rising populations and the protection of important ecosystems. The selected papers cover a wide range of issues that are relevant to watershed hydrology and have motivated model development, application, parameterization, uncertainty estimation, environment assessment, and management. Continued technological advances grounded in modern environmental science are necessary to meet these challenges. This will require a greater emphasis on disciplinary collaboration and integrated approaches to problem solving founded on science-driven innovations in technology, socio-economics, and public policy.

Keywords: watershed; catchment; models; climate change; ecosystem; management

---

1. Introduction

As water moves on, above, and below the Earth's surface, it forms the hydrologic cycle (i.e., hydrosphere), which lies between the atmosphere and the lithosphere and across the biosphere. Watershed is defined as the basic land unit for the hydrologic cycle description and resource management, because the water divide can be obtained from widely available topographic data and streamflow can be measured at the outlets [1]. Refining observations and modeling watershed hydrologic states and fluxes are the main components required to help improve the understanding of hydrologic processes and provide management support.

A modern approach to modeling watershed processes studies known coupled processes operating over a range of spatial and temporal scales. These processes include precipitation, overland flow, evapotranspiration, unsaturated flow, and groundwater flow, which describe the movement of water and simultaneous exchange between the various hydrological compartments, e.g., land surface, soil or vadose zone, and the underlying aquifers (i.e., phreatic zone). At the watershed scale, the hydrologic cycle also interacts with atmospheric processes, land surface, ecological processes, geological processes, and the pervasive effects of human activity. Therefore, the development of watershed models has been a strong objective of hydrologists [2,3] and remains challenging where gaps in our understanding of hydrologic processes and model capability are limited by data computational challenges [4].

Watershed analysis and models are important tools for environmental assessment, management, and conservation. In the United States, the Environmental Protection Agency, Army Corps of Engineers, and US Geological Survey provide repositories for tested watershed models for federal, state, and local water-resources planning, which have been applied on the national- and basin scales for water quality assessment (e.g., [5,6]). These model results have been widely applied for environmental management (e.g., best management practices (BMPs, [7]) and low impact development (LID, [8])). Technological advances in cyberinfrastructure have enabled the integrative hydrologic model and data development. A variety of watershed models have been archived or hosted by universities (e.g., The Pennsylvania State University (PIHM, http://www.pihm.psu.edu/), Texas A&M University (Hydrologic Modeling Inventory Website, http://hydrologicmodels.tamu.edu/)), research centers (e.g., Helmholtz Centre for Environmental Research (The Mesoscale Hydrologic Model, http://www.ufz.de/mhm/)), environmental companies (e.g., Aquanty (https://www.aquanty.com/hydrogeosphere), MIKE Powered by DHI (https://www.mikepoweredbydhi.com/products)), professional communities (e.g., CSDMS (Community surface dynamics modeling system, https://csdms.colorado.edu/wiki/Hydrological_Models/), CUAHSI (The Consortium of Universities for the Advancement of Hydrologic Science, Inc., https://www.cuahsi.org/data-models/)), and computer code repositories (e.g., ParFlow (https://github.com/parflow/parflow), UW Hydro (Computational Hydrology, http://uw-hydro.github.io/), GEOtop (GEOtop 2.x; http://geotopmodel.github.io/geotop/), RHESSys (The Regional Hydro-Ecologic Simulation System, https://github.com/RHESSys/RHESSys/)) to support broad applications in environmental management.

The main objective of the Special Issue is to assemble contributions of watershed models including model developments and environmental applications, which documents and inspires future directions in watershed hydrology.

2. Overview of This Special Issue

This Special Issue consists of 13 papers that cover diverse aspects of watershed hydrology. We summarize the articles using two main themes: (i) watershed models to advance scientific understanding of processes, parameters, and uncertainty; and (ii) model applications for environmental assessment, management, and conservation.

2.1. Advancing Process-Based Models in Watershed Hydrology

Begou et al. [9] conducted a sensitivity analysis and uncertainty estimation to compare the performance of catchment and sub-catchment calibration. The Soil and Water Assessment Tool (SWAT) was applied at Bani River, the major tributary of the Upper Niger River, Africa, and then calibrated using the Generalized Likelihood Uncertainty Estimation (GLUE) approach. The authors found that global parameter sets calibration was able to predict monthly and daily discharge with acceptable predictive uncertainty, which ensures transferability of the model parameters to ungauged sub-basins.

Cornelissen et al. [10] used the physics-based hydrological model HydroGeoSphere to test the role of distributed parameters in modeling hydrological processes. They investigated the sensitivity of discharge, water balance, and evapotranspiration patterns to spatial heterogeneity in land use, potential evapotranspiration, and precipitation. The results suggested that precipitation was the most sensitive input data set for discharge simulation, while spatially distributed land use parameterization had a much larger effect on evapotranspiration components and its pattern.

Son et al. [11] examined the impacts of fine-scale topography on ecohydrological processes by a spatially distributed model Regional Hydro-Ecological Simulation System (RHESSys). The results showed that the modeled streamflow was sensitive to digital elevation model (DEM) resolution, and coarser resolution models overestimated the climatic sensitivity of evapotranspiration and net primary productivity. These findings suggested that it is reliable to use at least 10-m DEM to simulate ecohydrological responses to climate change.

Muma et al. [12] modeled a coupled surface–subsurface flow process on an agricultural watershed. They applied a physics-based 3D hydrologic model CATHY (acronym for CATchment Hydrology) at Bras d'Henri River Watershed in Canada. Based on the calibrated model, subsurface drainage was found to increased baseflow and total flows, and decreased peak flows. In addition, the model can be applied to estimate the impacts on surface water quality of different agricultural practices.

Stern et al. [13] simulated streamflow and sediment transport using the Hydrological Simulation Program—Fortran (HSPF). The study area was a snow-dominated watershed contributing to the San Francisco Bay. The total sediment load has decreased by 50% during the last 50 years due to many reasons. The HSPF simulation matched the observed historical sediment reduction and highlighted the importance of climate as a main driving factor for sediment supply in this watershed. In particular, large storms associated with high peak flows are the most important driver of sediment transport.

Garee et al. [14] applied the SWAT model with a temperature index and elevation band algorithm in a glacier dominated watershed. The study area is one of the main sources of the Indus River, and upstream of Tarbela Dam, one of the largest dams in the world. The authors found that the combined effect of increased precipitation and warmer temperatures will increase streamflow by 10.63–43.70% by the year 2060. The projecting climate change impacts on the watershed will guide water resources management plans and dam construction and operation.

Li et al. [15] developed a stream network length estimation method based on flow recession dynamics. In headwater catchments, stream network length varies as the catchment wets and dries, both seasonally and in response to individual precipitation events, and direct observations of active stream network length (ASNL) is difficult. Based on flow recession rates, aquifer depth, and aquifer breath, the ASNL was estimated and agreed with GIS analysis results. This novel approach will bring more attention from both hydrologists and geomorphologists on stream network length estimation.

2.2. *Application of Watershed Models for Environmental Assessment*

Peng et al. [16] evaluated different soil water conservation methods on ecohydrological processes by RHESSys. The Loess Plateau is known for its highly erodible soil and fragile ecosystem. A variety of soil and water conservation methods have been applied since the 1950s, and a clear decline of streamflow has been observed. However, both climate change and water use could contribute to streamflow decreases as well. This study added modules of in-stream routing and reservoir operation to existing version of RHESSys, and evaluated soil and water conservation impacts on streamflow decline. The results suggested 78% of total impact on streamflow reduce is due to engineering construction of soil and water conservation.

Zhang et al. [17] combined several watershed analysis methods to evaluate the hydrologic impacts of dam construction. The study area was the Jiulong River Watershed (JRW), a medium-sized coastal watershed in Southeast China, which suffered from intensive human activities with over 13,500 hydraulic engineering facilities including over 120 small or medium dams along the mainstream and major tributaries. Flow duration curve analysis, hydrologic alteration, ranges of variability, and environmental flow were calculated to assess the impacts on daily and monthly streamflow regime. The approach is valuable for environmental impact assessment of dam construction.

Tang et al. [18] developed a flood frequency method to understand impacts of climate change on coastal watersheds. Many coastal areas are experiencing intensified flooding due to the combined impacts of the floods from the upstream watershed and the rising high tidal levels induced by sea-level rise (SLR). These climate change aspects have led to non-stationarity (i.e., trends) in flood records. The authors selected Pearl River Delta in South China due to its high density of river network and high frequency of extreme tides introduced by typhoons or storm surges. They combined a time-varying moments model and a hydrodynamic network model to estimate flood levels at 22 stations and found that when the non-stationarity was ignored, error up to 18% was found in the 100-year inflow floods and up to 14% in the 100-year tidal level.

Yu et al. [19] presented temporal variability of annual precipitation across a watershed to identify the spatial pattern of flood and drought frequency. The study area was the Huaihe River Basin with an area of 259,700 km^2, which typically suffered from both drought and flood hazards. The study found that the spatial distribution of precipitation varied remarkably, although a slight increasing trend could be observed over the entire basin, which provides important guidance for the water resources management in Huaihe River.

Li et al. [20] developed a classification method to assess catchment vulnerability to debris flow. They focused on the Wudongde Dam, one of the tallest hydroelectric dams in the world, and identified high, medium, and low vulnerability of debris flow susceptibility at 22 nearby watersheds. The approach will be applied to the whole Wudongde Dam area to assess debris flow hazard and secure dam stability.

Zhu and Chen [21] applied the storm water management model (SWMM) to assess the urban flooding reduction impacts of different engineering designs. In the late 20th century, low impact development (LID) and best management practices (BMPs) were proposed to control urban stormwater in United States. Gradually, these concepts and measures were introduced to China. The study modeled urban flooding in a typical residential area in Guangzhou, China to evaluate the effects of LID. The results showed that existing LID practices are able to control the flooding under the rainfall scenario of 2-year return period, 2-h rainfall duration, when the rainfall peak coefficient is 0.375. The control effects of LID practices are most affected by rainfall intensity compared to rainfall duration and rainfall peak.

3. Outlook

In this special issue, it is shown that advanced watershed model development and applications are a global enterprise, motivating many new technological innovations and interdisciplinary collaborations. As a result, we can expect in the future more reliable models, improved interpretations, and broader implementation where water is driving human and ecosystem services. Yet, there are still limitations in making significant advances, which we suggest follows four basic themes:

Design of a new generation of spatially resolved computational watershed models (e.g., [22]) driven by a new generation of field measurements including real time sensor networks and remote sensing ([23,24]);

Development of co-varying physical relationships for watershed processes that represent interactions between water, soil, vegetation, atmosphere, and geologic processes (e.g., [25,26]);

Enhance transparency, provisioning, and reproducibility of watershed models and analysis (e.g., [27,28]) to enable open communication between scientists; and

Improve the transfer and clarity of watershed knowledge among water resources scientists, engineers, policymakers, and the public to improve process understanding of a coupled, human–water system (e.g. [29,30]).

Acknowledgments: This work was supported by the U.S. National Science Foundation (EAR-0725019, EAR-1239285, EAR-1331726, and IIS-1344272) and U.S. Defense Advanced Projects Agency (World Modelers Program). The author of this paper and editor of this special issue would like to thank all authors for their notable contributions to this special issue, the reviewers for devoting their time and efforts to reviewing the manuscripts, and the Water Editorial team for their great support during the review of the submitted manuscripts.

Conflicts of Interest: The authors declare no conflicts of interest.

References

1. Edwards, P.J.; Williard, K.W.; Schoonover, J.E. Fundamentals of watershed hydrology. *J. Contemp. Water Res. Educ.* **2015**, *154*, 3–20. [CrossRef]
2. Clark, M.P.; Bierkens, M.F.; Samaniego, L.; Woods, R.A.; Uijlenhoet, R.; Bennett, K.E.; Pauwels, V.R.; Cai, X.; Wood, A.W.; Peters-Lidard, C.D. The evolution of process-based hydrologic models: Historical challenges and the collective quest for physical realism. *Hydrol. Earth Syst. Sci.* **2017**, *21*, 3427–3440. [CrossRef]
3. Ehret, U.; Gupta, H.V.; Sivapalan, M.; Weijs, S.V.; Schymanski, S.J.; Blöschl, G.; Gelfan, A.N.; Harman, C.; Kleidon, A.; Bogaard, T.A. Advancing catchment hydrology to deal with predictions under change. *Hydrol. Earth Syst. Sci.* **2014**, *18*, 649–671. [CrossRef]

4. Clark, M.P.; Fan, Y.; Lawrence, D.M.; Adam, J.C.; Bolster, D.; Gochis, D.J.; Hooper, R.P.; Kumar, M.; Leung, L.R.; Mackay, D.S. Improving the representation of hydrologic processes in Earth System Models. *Water Resour. Res.* **2015**, *51*, 5929–5956. [CrossRef]
5. Herrmann, M.; Najjar, R.G.; Kemp, W.M.; Alexander, R.B.; Boyer, E.W.; Cai, W.-J.; Griffith, P.C.; Kroeger, K.D.; McCallister, S.L.; Smith, R.A. Net ecosystem production and organic carbon balance of US East Coast estuaries: A synthesis approach. *Glob. Biogeochem. Cycles* **2015**, *29*, 96–111. [CrossRef]
6. Anning, D.W.; Flynn, M.E. *Dissolved-Solids Sources, Loads, Yields, and Concentrations in Streams of the Conterminous United States*; US Geological Survey: Reston, VA, USA, 2014.
7. Schueler, T.R.; Metropolitan Washington Water Resources Planning Board. *Controlling Urban Runoff: A Practical Manual for Planning and Designing Urban BMPs*; Metropolitan Information Center: Washington, DC, USA, 1987.
8. Prince George's County. *Low-Impact Development Design Strategies: An Integrated Design Approach*; Department of Environmental Resources: Largo, MD, USA, 1999.
9. Chaibou Begou, J.; Jomaa, S.; Benabdallah, S.; Bazie, P.; Afouda, A.; Rode, M. Multi-site validation of the SWAT model on the Bani catchment: Model performance and predictive uncertainty. *Water* **2016**, *8*, 178. [CrossRef]
10. Cornelissen, T.; Diekkrüger, B.; Bogena, H.R. Using high-resolution data to test parameter sensitivity of the distributed hydrological model HydroGeoSphere. *Water* **2016**, *8*, 202. [CrossRef]
11. Son, K.; Tague, C.; Hunsaker, C. Effects of model spatial resolution on ecohydrologic predictions and their sensitivity to inter-annual climate variability. *Water* **2016**, *8*, 321. [CrossRef]
12. Muma, M.; Rousseau, A.N.; Gumiere, S.J. Assessment of the impact of subsurface agricultural drainage on soil water storage and flows of a small watershed. *Water* **2016**, *8*, 326. [CrossRef]
13. Stern, M.; Flint, L.; Minear, J.; Flint, A.; Wright, S. Characterizing changes in streamflow and sediment supply in the Sacramento River Basin, California, using Hydrological Simulation Program—FORTRAN (HSPF). *Water* **2016**, *8*, 432. [CrossRef]
14. Garee, K.; Chen, X.; Bao, A.; Wang, Y.; Meng, F. Hydrological Modeling of the Upper Indus Basin: A Case Study from a High-Altitude Glacierized Catchment Hunza. *Water* **2017**, *9*, 17. [CrossRef]
15. Li, W.; Zhang, K.; Long, Y.; Feng, L. Estimation of Active Stream Network Length in a Hilly Headwater Catchment Using Recession Flow Analysis. *Water* **2017**, *9*, 348. [CrossRef]
16. Peng, H.; Jia, Y.; Tague, C.; Slaughter, P. An eco-hydrological model-based assessment of the impacts of soil and water conservation management in the Jinghe river basin, China. *Water* **2015**, *7*, 6301–6320. [CrossRef]
17. Zhang, Z.; Huang, Y.; Huang, J. Hydrologic alteration associated with dam construction in a medium-sized coastal watershed of southeast China. *Water* **2016**, *8*, 317. [CrossRef]
18. Tang, Y.; Guo, Q.; Su, C.; Chen, X. Flooding in Delta Areas under Changing Climate: Response of Design Flood Level to Non-Stationarity in Both Inflow Floods and High Tides in South China. *Water* **2017**, *9*, 471. [CrossRef]
19. Yu, Z.-L.; Yan, D.-H.; Ni, G.-H.; Do, P.; Yan, D.-M.; Cai, S.-Y.; Qin, T.-L.; Weng, B.-S.; Yang, M.-J. Variability of Spatially Grid-Distributed Precipitation over the Huaihe River Basin in China. *Water* **2017**, *9*, 489. [CrossRef]
20. Li, Y.; Wang, H.; Chen, J.; Shang, Y. Debris Flow Susceptibility Assessment in the Wudongde Dam Area, China Based on Rock Engineering System and Fuzzy C-Means Algorithm. *Water* **2017**, *9*, 669. [CrossRef]
21. Zhu, Z.; Chen, X. Evaluating the Effects of Low Impact Development Practices on Urban Flooding under Different Rainfall Intensities. *Water* **2017**, *9*, 548. [CrossRef]
22. Duffy, C.; Shi, Y.; Davis, K.; Slingerland, R.; Li, L.; Sullivan, P.L.; Goddéris, Y.; Brantley, S.L. Designing a Suite of Models to Explore Critical Zone Function. *Procedia Earth Planet. Sci.* **2014**, *10*, 7–15. [CrossRef]
23. Brantley, S.L.; DiBiase, R.A.; Russo, T.A.; Davis, K.J.; Eissenstat, D.M.; Dere, A.L.; Neal, A.L.; Brubaker, K.M.; Arthur, D.K. Designing a suite of measurements to understand the critical zone. *Earth Surf. Dyn.* **2016**, *4*, 211–235. [CrossRef]
24. Tauro, F.; Selker, J.; van de Giesen, N.; Abrate, T.; Uijlenhoet, R.; Porfiri, M.; Manfreda, S.; Caylor, K.; Moramarco, T.; Benveniste, J. Measurements and Observations in the XXI century (MOXXI): Innovation and multi-disciplinarity to sense the hydrological cycle. *Hydrol. Sci. J.* **2017**, *63*, 169–196. [CrossRef]
25. Paniconi, C.; Putti, M. Physically based modeling in catchment hydrology at 50: Survey and outlook. *Water Resour. Res.* **2015**, *51*, 7090–7129. [CrossRef]

26. Li, L.; Maher, K.; Navarre-Sitchler, A.; Druhan, J.; Meile, C.; Lawrence, C.; Moore, J.; Perdrial, J.; Sullivan, P.; Thompson, A. Expanding the role of reactive transport models in critical zone processes. *Earth-Sci. Rev.* **2017**, *165*, 280–301. [CrossRef]
27. Yu, X.; Duffy, C.J.; Rousseau, A.N.; Bhatt, G.; Pardo Álvarez, Á.; Charron, D. Open science in practice: Learning integrated modeling of coupled surface-subsurface flow processes from scratch. *Earth Space Sci.* **2016**, *3*, 190–206. [CrossRef]
28. Hutton, C.; Wagener, T.; Freer, J.; Han, D.; Duffy, C.; Arheimer, B. Most computational hydrology is not reproducible, so is it really science? *Water Resour. Res.* **2016**, *52*, 7548–7555. [CrossRef]
29. Sanderson, M.R.; Bergtold, J.S.; Heier Stamm, J.L.; Caldas, M.M.; Ramsey, S.M. Bringing the "social" into socio-hydrology: Conservation policy support in the Central Great Plains of Kansas, USA. *Water Resour. Res.* **2017**, *53*, 6725–6743. [CrossRef]
30. Chen, X.; Wang, D.; Tian, F.; Sivapalan, M. From channelization to restoration: Sociohydrologic modeling with changing community preferences in the Kissimmee River Basin, Florida. *Water Resour. Res.* **2016**, *52*, 1227–1244. [CrossRef]

Article

Multi-Site Validation of the SWAT Model on the Bani Catchment: Model Performance and Predictive Uncertainty

Jamilatou Chaibou Begou [1,2,*], Seifeddine Jomaa [3], Sihem Benabdallah [4], Pibgnina Bazie [2], Abel Afouda [1] and Michael Rode [3]

1 Graduate Research Program (GRP) Climate Change and Water Resources, West African Science Service Centre on Climate Change and Adapted Land Use (WASCAL), University of Abomey-Calavi, 01 BP 526 Cotonou, Benin; aafouda@yahoo.fr

2 Centre Regional AGRHYMET, PB 11011 Niamey, Niger; p.bazie@agrhymet.ne

3 Department of Aquatic Ecosystem Analysis and Management, Helmholtz Centre for Environmental Research—UFZ, Brueckstrasse 3a, 39114 Magdeburg, Germany; seifeddine.jomaa@ufz.de (S.J.); michael.rode@ufz.de (M.R.)

4 Centre de Recherche et des Technologies des Eaux (CERTE), BP 273, 8020 Soliman, Tunisia; sihem.benabdallah@certe.rnrt.tn

* Correspondence: jamilabegou@yahoo.fr; Tel.: +227-9136-7854

Academic Editor: Xuan Yu
Received: 14 March 2016; Accepted: 21 April 2016; Published: 30 April 2016

Abstract: The objective of this study was to assess the performance and predictive uncertainty of the Soil and Water Assessment Tool (SWAT) model on the Bani River Basin, at catchment and subcatchment levels. The SWAT model was calibrated using the Generalized Likelihood Uncertainty Estimation (GLUE) approach. Potential Evapotranspiration (PET) and biomass were considered in the verification of model outputs accuracy. Global Sensitivity Analysis (GSA) was used for identifying important model parameters. Results indicated a good performance of the global model at daily as well as monthly time steps with adequate predictive uncertainty. PET was found to be overestimated but biomass was better predicted in agricultural land and forest. Surface runoff represents the dominant process on streamflow generation in that region. Individual calibration at subcatchment scale yielded better performance than when the global parameter sets were applied. These results are very useful and provide a support to further studies on regionalization to make prediction in ungauged basins.

Keywords: SWAT; Bani catchment; West Africa; discharge; daily calibration; performance and predictive uncertainty

1. Introduction

Water resources managers are facing challenges in many river basins across the world due to limited data availability. Anthropogenic activities add more uncertainties to this task by inducing changes to land and climate at different scales [1,2]. This situation is more pronounced in developing countries, where in many river basins no runoff data are available [3–7] and the existing ones are of questionable quality or, at best, short or incomplete.

The Niger River basin is not an exception to that rule. The general situation of insufficient data is exacerbated by a deterioration of measurement networks. In the 80s and 90s, for instance, hydrometric stations were reduced to a minimum and many have been abandoned (e.g., [8]). To prevent the hydrologic observing system from more degradation, the Niger Basin Authority (NBA) has set the Niger-HYCOS project, which one of its specific objectives is to improve data quality of the Niger Basin.

For this purpose, the project identified and brings assistance in the installation and the management of 105 hydrometric stations shared by nine countries drained by the River, and contributes to the capacity building of national hydrological services.

In its fifth assessment report on regional aspects of climate change, the Inter-Governmental Panel on Climate Change [9] has shown that adaptation to climate change in Africa is confronted with a number of challenges among which is a significant data gap. Too many basins lack reliable data necessary to assess, in details, impacts of climate change on different components of the hydrological cycle and to develop strategies of adaptation related to each specific impact. Thus, it is germane to predict hydrological variables in ungauged basins for building high adaptive capacity by improving: (i) water resources knowledge, planning, and management; (ii) identification and implementation of strategies of adaptation to climate change in the sector of water, and (iii) ecological studies for a sustainable development.

The application of rainfall-runoff models and then, transferring model parameters from gauged to ungauged catchments is a long-standing method [10] for flow prediction in ungauged basins and has been highlighted during the decade of Prediction in Ungauged Basins (PUB) launched in 2003 by the International Association of Hydrological Sciences (IAHS) and concluded by the PUB Symposium held in 2012. This is the framework of the present study, in which the Soil and Water Assessment Tool (SWAT) model was calibrated on the Bani catchment (Niger River basin) and the most sensitive model parameters were estimated.

Many studies have successfully applied the SWAT model in West Africa, on different river basins. Examples include, among others: calibration of the SWAT model on the Niger basin [11–16], the Volta basin [12–15,17–19] and the Oueme catchment in Benin [15,20–22]. However there are few published papers on the application of the SWAT model on the Bani catchment. For instance, Schuol and Abbaspour [12] and Schuol *et al.* [14] applied the SWAT model to selected watersheds in West Africa including the Niger basin and modeled monthly values of river discharges (blue water) as well as the soil water (green water), and clearly showed the uncertainty of the model results. They developed and applied a daily weather generator algorithm [13] that uses 0.5 degree monthly weather statistics from the Climatic Research Unit (CRU) to obtain time series of daily precipitation as well as minimum and maximum temperatures for each sub-basin. These generated weather data were then used as input for model setup and the authors concluded that "discharge simulations using generated data were superior to the simulations using available measured data from local climate stations". Reported Nash-coefficient values obtained vary largely between sub-basins and were principally presented as average intervals limiting thus, our understanding of model performance at finer spatial (subbasin) and temporal (daily) scales.

Laurent and Ruelland [23] successfully calibrated SWAT on the Bani catchment using daily measured climate data. They interpolated precipitation data on a regular grid by the Inverse Distance Weighted (IDW) method, which has proven to yield better results than kriging, Thiessen and spline methods, especially when a hydrological model is used [24]. To show the model performance, Laurent and Ruelland [23] reported both discharge and biomass calibration results on an average annual basis, but did not assess model calibration uncertainty. Moreover, both above-mentioned studies performed interpolation of input data out of the model framework to obtain a time series of daily weather data for each sub-basin. However, the results of interpolation methods are strongly influenced by the density and spatial distribution of the measurement stations used in the interpolation [25]. Such a density of data is not always available in developing countries.

Against this background, the objective of this study was to assess the performance of the SWAT model and its predictive uncertainty on the Bani at catchment and subcatchment levels. More specifically, this meant to: (i) set up a hydrological model for the Bani catchment using the SWAT program; (ii) calibrate the model at the catchment outlet at daily and monthly time steps and assess the predictive performance and uncertainty; (iii) evaluate the spatial performance of the watershed-wide model within the catchment by validating it at two internal stations; and (iv) calibrate the model at the sub-catchments separately and provide a comparative assessment of the model performance at different spatial scales.

The originality of this study was the daily performance of the SWAT model at the whole catchment outlet and at two internal stations. Another important output of this paper was the involvement of evapotranspiration (the most important component of the water balance after rainfall especially under warm climate) in the verification of model outputs reasonability, a particular attention that has not been considered by any previous study in the region. In addition, we used in the current work point rain gauge data (as per SWAT's standard procedure) opposed to areal precipitation as used in previous studies [12–15,24,26,27] on the same basin in order to maintain the real data condition (limited in time and space) to the extent possible.

2. Material and Methods

2.1. The Study Area

The Bani is the major tributary of the Upper Niger River. Its drainage basin is principally located in Mali but spans in a lesser extent over Cote d'Ivoire and Burkina Faso and covers an area of about 100,000 km^2 at Douna gauging station (Figure 1). The Bani watershed was chosen for this study, on one hand, due to its relatively high-quality data availability compared to regional situation. It thus constitutes the appropriate gauged catchment in different hydro-climatic variables. On the other hand, this watershed has not been affected by important hydraulic structures able to significantly modify its flow regime, making the hydrological modeling of that catchment more convenient.

The catchment's topography (Figure 1) is characterized by a gentle elevation that ranges from 826 m in the South and the center-east to 249 m at the outlet in the North. According to FAO (2003) [28], major soil groups are mainly constituted by Luvisol, Acrisol, and Nitosol (Figure 2a). Based on the USGS Global Land Cover Characterization (GLCC) version 2.0 [29], agricultural land constitutes the dominant land use category followed by savannah and forest (Figure 2b). The Bani catchment is characterized by a Sudano-Sahelian climatic regime. The river flows from south to north along a high rainfall gradient. Annual precipitation varies from 1250 mm at Odienne to 615 mm at Segou (average of the period 1981–2000). The average annual discharge recorded at Douna gauging station between 1981 and 2000 was 184 $m^3 s^{-1}$, which is equivalent to 58 mm of surface runoff depth for an average annual precipitation of 1000 mm. The smallest runoff values were recorded during the years 1983, 1984, and 1987. Due to climate change, there was an abrupt decrease in rainfall in the period 1970–1971 and remained for two decades [27,30] with a more severe impact on water resources. A decrease of more than 60% in discharge at Douna [27,31] and lower contribution of baseflow to the annual flood [32,33] have been reported since the 70s. Concerning future climate change impacts, the Bani basin is projected to experience substantial decrease in rainfall and runoff especially in the long term behavior [27].

2.2. Model Description

SWAT is a river basin, or watershed, scale model developed to predict the impact of land management practices on water, sediment, and agricultural chemical yields in large, complex watersheds with varying soils, land use, and management conditions over long periods of time [34]. The model is semi-distributed, physically based and computationally efficient, uses readily available inputs and enables users to study long-term impacts [35]. For a detailed description of SWAT, see Soil and Water Assessment Tool input/output version 2012 [36] and the Theoretical Documentation, Version 2009 [37].

The ArcSWAT (ArcGIS extension) is a graphical user interface for the SWAT model. In the present study, the recent version, ArcSWAT2012, was used for building the hydrological model of the Bani catchment.

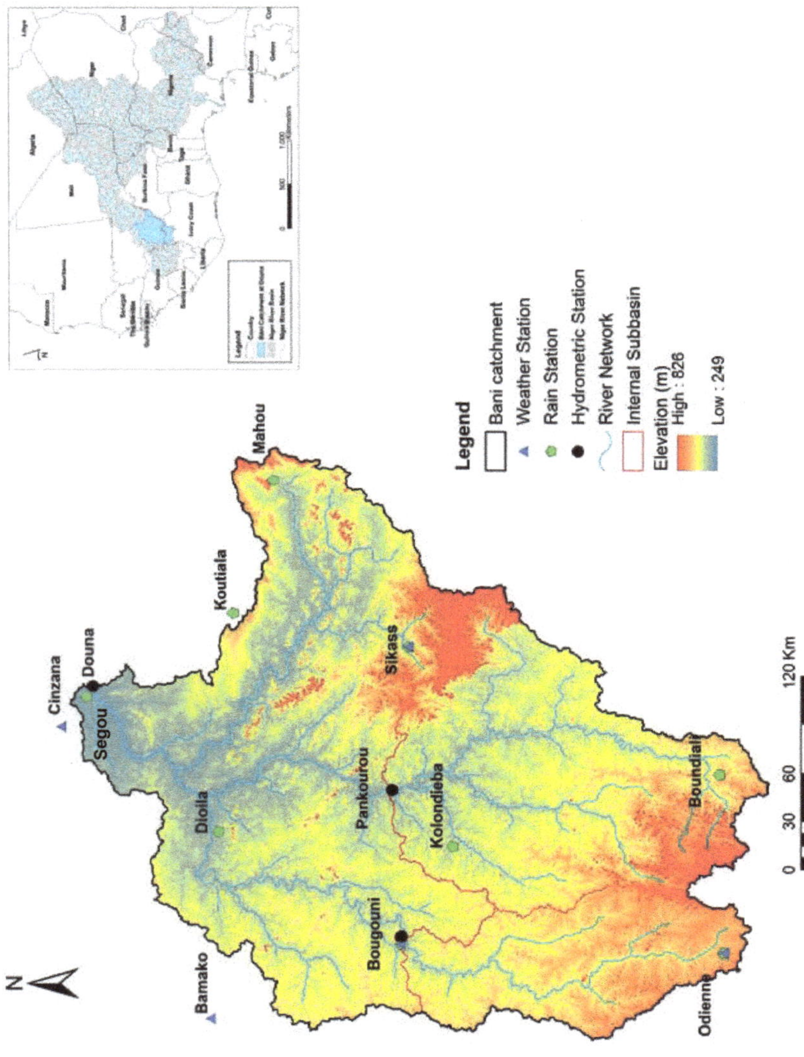

Figure 1. Localization of the Bani catchment at the Douna outlet. The altitude and the monitoring network of the catchment are also given.

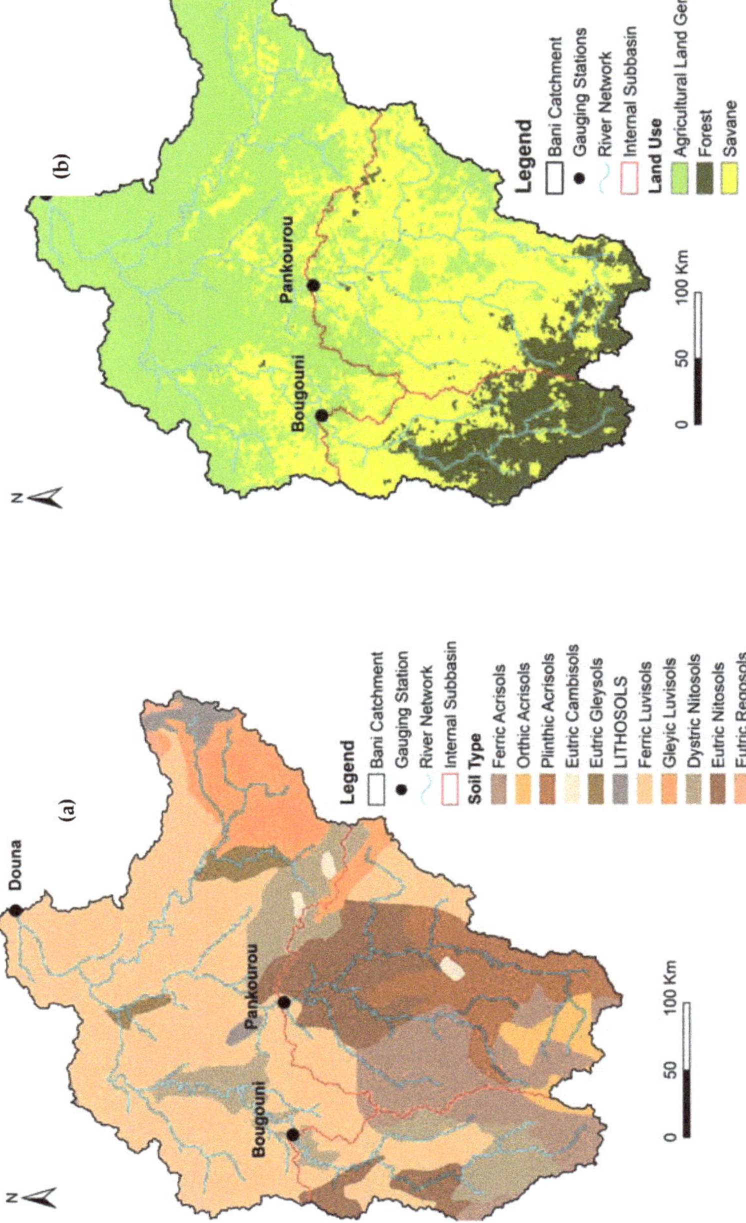

Figure 2. (a) Soil attributes and (b) land use categories of the Bani catchment.

The hydrologic cycle simulated by SWAT is based on the water balance equation:

$$SW_t = SW_0 + \sum_{i=1}^{t} \left(R_{day} - Q_{surf} - E_a - W_{seep} - Q_{gw} \right) \quad (1)$$

where, SW_t is the final soil water content (mm H_2O), SW_0 is the initial soil water content on day i (mm H_2O), t is the time (days), R_{day} is the amount of precipitation on day i (mm H_2O), Q_{surf} is the amount of surface runoff on day i (mm H_2O), E_a is the amount of evapotranspiration on day i (mm H_2O), W_{seep} is the amount of water entering the vadose zone from the soil profile on day i (mm H_2O) and Q_{gw} is the amount of groundwater exfiltration on day i (mm H_2O).

SWAT divides a basin into sub-basins which are further discretized into hydrologic response units (HRUs), based on unique soil-land use-slope combinations. The subdivision of the watershed enables the model to reflect differences in evapotranspiration for various crops and soils. Runoff is predicted separately for each HRU and routed to obtain the total runoff for the watershed. This increases accuracy and gives a much better physical description of the water balance [37].

Various hydrological models exist and there is no strict guideline on the selection of the model. The SWAT model uses a modified version of the Curve Number method, which was developed in the US for specifically calculating surface runoff generation. Therefore the model is especially suitable for regions with a high share of overland flow on total runoff. Other advantages of the SWAT model are that it allows a number of different physical processes (hydrologic, sediment, pollutants) to be simulated in a watershed. It has been previously validated for several large-scale watersheds throughout different climate contexts across the globe and has performed satisfactorily even in data poor and complex catchments (e.g., [38,39]). SWAT is also very flexible in terms of using specific and appropriate soil and land use information's of the watershed to be modeled by adding them to its database. However in this context, it is worth using a low cost or free model, which West African National Hydrological services could afford due to economic constraints.

2.3. *Input Data and Databases*

The SWAT model for the Bani was constructed using weather data and globally and freely available spatial information described in Table 1. Daily precipitation data from 11 rain gauges as well as daily maximum and minimum temperature from five weather stations located mainly on the catchment were used as input. The location and spatial distribution of input precipitation and temperature stations are represented in Figure 1.

It is worth noting the weak spatial density of the measuring network that is characterized by a rain gauge for more than 9000 km^2. Precipitation data are complete at the majority of the sites except for a few numbers of them, where the maximum missing data percentage varies between 8.5% and 100% in a year. Many more missing values are recorded in the temperature data. Collected climate data time series were of varying lengths. Thus, a common period of observation from 1981 to 2000 was first determined. Retained data then underwent a thorough quality control as recommended by the World Meteorological Organization (WMO) in the guide to climatological practices, third edition [40]. Three procedures were applied: (1) completeness check; (2) plausible value check; and (3) consistency check. The aim of the check is to detect erroneous data in order to correct and, if not possible, to delete it. Missing values were filled by the weather generator during the running time. For this purpose, the excel macro WGNmaker4 [41] was used to calculate weather stations statistics needed to generate representative daily climatic data.

Two different databases were used to set up the model. The SWAT database is composed by the crop database and the user soils database, both included in swat2012.mdb. They are named crop1 and soil1, respectively. Crop1 was kept default whereas soil1 was filled with soils transferred from mwswat2009.mdb (the database of the MapWindow interface for SWAT). The second database is composed by crop2 and soil2. Four land use categories define crop2: forest, savannah-bush, savannah, and steppe whereas six major soil groups are added to soil2: Acrisol, Cambisol, Gleysol, Lithosol, Luvisol ferrique, and Nitosol. Detailed description of this database can be found in [23].

Table 1. Input data of the SWAT model for the Bani catchment.

Data Type	Description	Resolution/Period	Source
		Simulation Data	
Topography	Conditioned DEM	90 m	USGS hydrosheds [42]
Land use/land cover	GLCC version 2	1 km	Waterbase [43]
Soil	FAO Soil Map	Scale 1:5000000	FAO [44]
River	River network map	500 m	USGS Hydrosheds [42]
Weather data	Rainfall, maximum and minimum temperature	Daily (1981-2000)	AGRHYMET
		Calibration/Verification Data	
Discharge	Discharge	Daily (1983–1997)	AGRHYMET/National hydrological service of Mali
PET	Potential evapotranspiration	10-day (1983–1998)	National Meteorological Agency of Mali
Epan	Pan evaporation	Monthly (1983–1997)	AGRHYMET

For calibration purpose, we used daily river discharge data at Douna, Bougouni and Pankourou stations covering the period 1981–2000, obtained from AGRHYMET and the National Hydraulic Direction of Mali. The period 1981–1997 was kept for calibration and validation processes as it exhibits few gaps. Small existing gaps were thus filled by a simple linear interpolation.

2.4. Model Setup

The catchment was delineated and divided into sub-catchments based on the DEM. A stream network was superimposed on the DEM in order to accurately delineate the location of the streams. The threshold drainage area was kept as default and additional outlets were considered at the location of stream gauging stations to enable comparison of measured discharge with SWAT results. The whole catchment was so discretized into 28 sub-catchments, which were further subdivided into 181 HRUs based on soil, land use, and slope combinations. Further parameters have been edited through the general watershed parameters and SWAT simulation menus and are reported in Table 2. Four simulations were performed based on land use and soil databases combinations: crop1soil1, crop1soil2, crop2soil1, and crop2soil2. A Nash-Sutcliffe Efficiency (*NSE*) [45] was thereafter calculated at Douna by comparing measured discharges against each default simulation and the one which will yield the highest *NSE* value will be kept for calibration and validation processes.

Table 2. Input methods for SWAT model simulation on the Bani catchment.

Code	Description	Method
	General Watershed Parameters	
IPET	Potential Evapotranspiration method	Hargreaves
IEVENT	Rainfall/runoff/routing option	Daily Rainfall/CN runoff/Daily routing
ICN	Daily Curve Number calculation method	Soil moisture (Plant ET at Bougouni)
IRTE	Channel water routing method	Variable storage
	SWAT Simulation	
Period of simulation	-	1981–2000
NYSKIP	Warm-up period	Two years (1981 and 1982)

2.5. Calibration and Validation Procedures

It is commonly accepted in hydrology to split the measured data either temporally or spatially for calibration and validation [36]. In addition to the split-sample method, a split-location calibration and validation approach has been performed because the global parameter set is not expected to be optimal for sub-catchments processes in view of the high heterogeneity in terms of climate, topography, soil, and land use characterizing such a large-area watershed. This approach is especially needed when prediction at data sparse sites is foreseen [46,47]. In the split-sample approach, the model was calibrated using discharge data solely measured at the catchment outlet by splitting the homogenous period mentioned in Section 2.3 into two datasets: two-thirds for calibration (1983–1992), and the other one for validation (1993–1997). To implement the split-location method, the model was calibrated at Douna and then validated at intermediate gauging stations (Bougouni and Pankourou) by turning the model on the same period (1983–1992), using the same behavioral parameter sets determined at the outlet.

Calibration was thereafter performed at Bougouni and Pankourou stations individually, and both modeling frameworks facilitated a comparative analysis of model performance and predictive uncertainty through scales. At this step, the calibration at Bougouni did not succeed within realistic range of the Curve Number (*CN*). Then, the daily *CN* calculation method was changed to Plant ET for simulation at Bougouni because soil moisture method is found to predict too much runoff in shallow soils [36]. An additional parameter (*CNCOEF*) was then necessary as required by the plant ET method and fixed to 0.5 in the Edit SWAT input menu.

Calibration/validation, uncertainty analysis, and sensitivity analysis were performed within the SWAT Calibration and Uncertainty Programs SWAT-CUP version 2012 [48] using Generalized Likelihood Uncertainty Estimation (GLUE) procedure [49]. GLUE is a Monte Carlo based method for model calibration and uncertainty analysis. It was constructed to partly account for non-uniqueness of model parameters. GLUE requires a large number of model runs with different combinations of parameter values chosen randomly and independently from the prior distribution in the parameter space. The prior distributions of the selected parameters are assumed to follow a uniform distribution over their respective range since the real distribution of the parameter is unknown. By comparing predicted and observed responses, each set of parameter values is assigned a likelihood value. The likelihood functions selected here is principally the *NSE* as it is very commonly used and included in SWAT-CUP for GLUE performance assessment. In this study, the number of model runs was set to 10,000 and the total sample of simulations were split into "behavioral" and "non-behavioral" based on a threshold value of 0.5, a minimum threshold for *NSE* recommended by [50] for streamflow simulation to be judged as satisfactory on a monthly time step. In that case, only simulations which yielded a $NSE \geqslant 0.5$ are considered behavioral and kept for further analysis.

In the calibration procedure, we included 12 parameters that govern the surface runoff and baseflow processes. The real approached baseflow alpha factor (*ALPHA_BF*) value has been determined by applying the baseflow filter program developed by [51] and modified by [52] to streamflow data measured at the three outlets. One novelty in this study was to involve the Manning's roughness coefficient for overland flow (*OV_N*) and the average slope length (*SLSUBBSN*) parameters that are not commonly used in calibration. The reason behind this choice was to correct the tendency of the model to delay the runoff as detected by graphical analysis. The remaining parameters were chosen based on the literature [53–55] and their adjusting ranges from the SWAT Input/Output version 2012 document (e.g., [56]).

2.6. Model Performance and Uncertainty Evaluation

To evaluate model performance, both statistical and graphical techniques were used as recommended by [50] based on previous published studies. The following quantitative statistics were chosen: *NSE* to quantify the relative magnitude of the residual variance ("noise") compared to the measured data variance, *PBIAS* for water balance error, and R^2 to describe the degree of collinearity between simulated and measured data, and were given for the best simulation. The *NSE*, R^2 and *PBIAS* were determined using the following equations:

$$NSE = 1 - \frac{\sum_{i=1}^{n}\left(Y_i^{obs} - Y_i^{sim}\right)^2}{\sum_{i=1}^{n}\left(Y_i^{obs} - \overline{Y^{obs}}\right)^2}, \quad (2)$$

$$R^2 = \left(\frac{\sum_{i=1}^{n}\left(Y_i^{obs} - \overline{Y^{obs}}\right)\left(Y_i^{sim} - \overline{Y^{sim}}\right)}{\sqrt{\sum_{i=1}^{n}\left(Y_i^{obs} - \overline{Y^{obs}}\right)^2}\sqrt{\sum_{i=1}^{n}\left(Y_i^{sim} - \overline{Y^{sim}}\right)^2}}\right)^2, \quad (3)$$

$$PBIAS = \frac{\sum_{i=1}^{n}\left(Y_i^{sim} - Y_i^{obs}\right) \times 100}{\sum_{i=1}^{n} Y_i^{obs}} \quad (4)$$

where Y_i^{sim} and Y_i^{obs} are the *i*th simulated and observed discharge, respectively, $\overline{Y^{sim}}$ and $\overline{Y^{obs}}$ the mean value of simulated and observed discharge, respectively and n the total number of observations.

The *NSE* varies between $-\infty$ and 1 (1 inclusive), with *NSE* = 1 being the optimal value. The optimal value of *PBIAS* is 0, with low *PBIAIS* in absolute values indicating accurate model simulation. Positive values indicate model overestimation bias, and negative values indicate model underestimation bias. R^2 ranges from 0 to 1, with higher values indicating less error variance, values greater than 0.5 are considered acceptable.

In the present study, model performance, for a monthly time step, will be judged as satisfactory if $NSE > 0.50$ and $PBIAS < \pm 25\%$ for discharge [50] and if the graphical analysis reveals a good agreement between predicted and measured hydrographs.

The GLUE prediction uncertainty was then quantified by two indices referred to as *P-factor* and *R-factor* [57]. The *P-factor* represents the percentage of observed data bracketed by the 95% predictive uncertainty (95PPU) band of the model calculated at the 2.5% and 97.5% levels of the cumulative distribution of an output variable obtained through Latin hypercube sampling. The *R-factor* is the ratio of the average width of the 95PPU band and the standard deviation of the measured variable. For uncertainty assessment, a value of *P-factor* > 0.5 (*i.e.*, more than half of the observed data should be enclosed within the 95PPU band) and *R-factor* < 1 (*i.e.*, the average width of the 95PPU band should be less than the standard deviation of the measured data) should be adequate for this study, especially considering limited data availability.

2.7. Sensitivity Analysis

A Global Sensitivity Analysis (GSA) was performed after 10,000 simulations on the 12 parameters included in the calibration process. Only GSA is allowed with GLUE in SWAT-CUP and can be performed after one iteration. A *t*-test is then used to identify the relative significance of each parameter. *T*-stat provides a measure of sensitivity and *p*-value determines the significance of the sensitivity. A larger *t*-stat in absolute value is more sensitive and a *p*-value close to zero has more significance [48].

2.8. Verification of Model Outputs

To evaluate the accuracy of the SWAT model to predict PET, we considered the model average annual basin output which was computed by the Hargreaves method [58] and compared it to PET values calculated with two other methods: the FAO-Penman Monteith method and the pan evaporation method. The estimates from those three methods are hereinafter referred to as PET_{har} (for average annual PET estimated by the Hargreaves method), PET_{pen} (for average annual PET estimated by the Penman-Monteith method) and PET_{pan} (for average annual PET estimated by the pan evaporation method). The modified Penman method is taken herein as the standard because it was considered to offer the best results with minimum possible error [59]. Average observed 10-day PET_{pen} were collected and computed to obtain average annual value on the calibration-validation period. Monthly observed pan evaporation data were used to estimate PET_{pan}. Doorenbos and Pruitt [60] related pan evaporation to reference evapotranspiration, ET_0 (or PET) using empirically derived coefficients. PET can be obtained by:

$$\mathrm{PET} = K_p \times E_{pan} \tag{5}$$

where, PET is the potential evapotranspiration in mm· day^{-1}, Epan represents the pan evaporation in mm· day^{-1}, and K_p is the pan coefficient, which is the adjustment factor that depends on mean relative humidity, wind speed, and ground cover.

As the pan factor in the Bani catchment could not be exactly determined due to lack of information about the pan environment and the climate, the average value of 0.7 [61] was used in this study. The *PBAIS* was again used as the evaluation criterion representing the deviation of the predicted PET compared to the one considered as the baseline.

3. Results

3.1. The Catchment Scale Model

3.1.1. Global Model Performance

In the preliminary analyses, we tested different land use and soil databases and kept for subsequent analysis the simulation of databases combination crop2soil2, which yielded the highest default, *i.e.*, before calibration, performance (*NSE* = 0.09). The impact of land use database was not so significant, but the type of soil database used to setup the model was very decisive in obtaining a simulation with the smallest overall error. SWAT-CUP output results are presented as 95PPU as well as the best simulation (Table 3).

Table 3. Model performance statistics for the Bani catchment at Douna, Pankourou, and Bougouni discharge gauging stations.

Time Step		Calibration (1983–1992)			Validation (1993–1997)		
	Criterion	Douna	Pankourou	Bougouni	Douna	Pankourou	Bougouni
Daily	*NSE*	0.76	0.73	0.66	0.85	0.77	0.37
	R^2	0.79	0.74	0.68	0.87	0.83	0.57
	PBIAS (%)	−12.23	6.08	−15.01	−23.26	−19.57	−59.53
Monthly	*NSE*	0.79	0.78	0.72	0.85	0.81	0.47
	R^2	0.82	0.78	0.76	0.88	0.91	0.68
	PBIAS (%)	−15.78	5.93	−13.14	−26.91	−19.54	−58.40

Overall, calibration and validation of the hydrological model SWAT on the Bani catchment at the Douna outlet yielded good results in terms of *NSE* and R^2 for both daily and monthly timesteps. 364 simulations for daily calibration against 588 for monthly calibration returned a $NSE \geqslant 0.5$ and were thus considered as behavioral. Very good *NSE* and R^2 values were obtained and were greater than 0.75 for the best simulations. Moreover, it can be noticed that the performance is slightly lower for daily calibration compared to monthly calibration, but always higher for the validation period. Only one year (1984) over 10 showed very low performance with a *NSE* of 0.23.

The water balance prediction can be considered as accurate at a daily time-step but becomes hardly satisfactory for monthly calibration, which is characterized by higher *PBIAIS* values showing increasing errors in the prediction. For example, the *PBIAIS* values increased from daily to monthly time intervals: from −12% to −16% in the calibration period and from −23% to −27% in the validation period (Figure 3). With regard to high flow events, visual analysis of simulated and observed hydrographs represented in Figure 3 came out with the following results: timing of peak is well reproduced although the simulation tends to underestimate peak flows especially during dry years (e.g., 1983, 1984, and 1987).

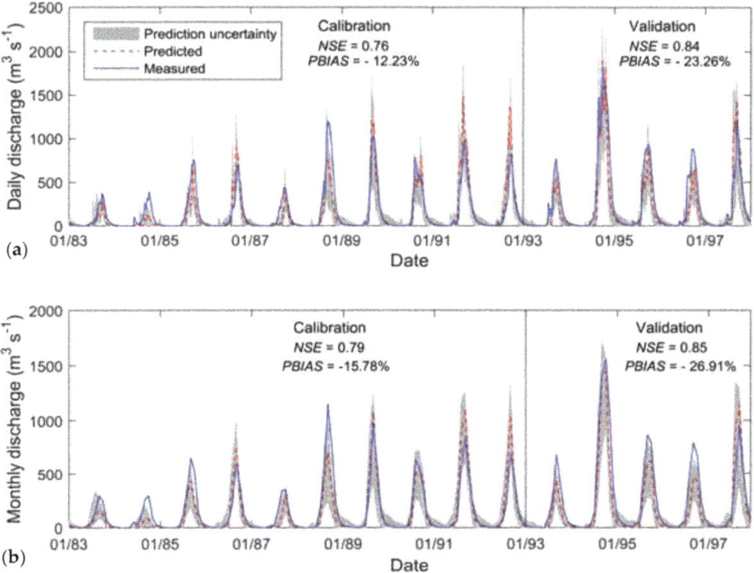

Figure 3. Simulated and observed hydrographs at Douna station at (**a**) daily and (**b**) monthly timesteps along with calculated statistics on calibration and validation periods.

3.1.2. Verification of Average Annual Basin Values

Table 4 reports the average annual values of the SWAT model simulated on the Bani catchment. However, there are not available data to enable a full verification of all model outputs at the watershed scale. In this case, we focused on available PET and biomass for which there exist regional values.

Table 4. Average annual basin values of precipitation (P), evapotranspiration (ET), potential evapotranspiration (PET), and biomass as SWAT outputs on the Bani catchment.

Period	P (mm)	ET (mm)	PET (mm) [a]	Biomass (ton ha^{-1})		
				Agricultural Land Generic	Savannah	Forest
Calibration (1983–1992)	960	895	1926	1.18	0.27	3.09
Validation (1993–1997)	1050	975	1925	1.72	0.53	5.51

[a] Average annual PET estimated by the Hargreaves method (herein used by SWAT).

Average annual basin values simulated by the model and described in Section 2.8 are shown in Table 4. The analysis of these values came out with several results. On average, PET_{har} presented a positive *PBIAS* of 11% compared with observed PET_{pen} herein equal to 1737 mm and the latter is very close to PET_{pan}, estimated to 1755 mm. These results give a clear indication of overestimation of PET by the SWAT model over the Bani catchment, an overestimation that can be attributed to the Hargreaves method used herein by the model to compute PET.

Table 5. Summary of the SWAT model parameters calibrated on the Bani catchment at Douna on a daily time interval.

Parameter	Description	Input Calibration Range	Calibrated Parameters: Best Parameter Value [Range]	Global Sensitivity Analysis	
				t-Stat	*p*-Value
CN2	SCS runoff curve number II (-)	±20%	−0.155 [−0.199; 0.102]	−54.03083	0.00000
OV_N	Manning's "*n*" value for overland flow (-)	0.01–30	23.153 [3.061; 29.915]	11.41603	0.00000
SLSUBBSN	Average slope length (m)	10–150	149.808 [12.677; 149.924]	8.87352	0.00000
ESCO	Soil evaporation compensation factor (-)	0.01–1	0.958 [0.768; 0.991]	−6.08880	0.00000
SOL_AWC	Available water capacity of the soil layer (mm H_2O/mm sol)	±20%	0.140 [−0.199; 0.197]	2.89864	0.00376
GW_DELAY	Groundwater delay (days)	0.0–50	4.938 [0.487; 49.823]	1.81341	0.06980
GWQMN	Threshold depth of water in the shallow aquifer required for return flow to occur (mm H_2O)	0.0–4000	3082.500 [0.043; 3995.710]	−1.51853	0.12891
REVAPMN	Threshold depth of water in the shallow aquifer for "revap" to occur (mm H_2O)	0–500	173.709 [0.636; 499.845]	−0.64939	0.51610
RCHRG_DP	Deep aquifer percolation fraction (-)	0–1	0.346 [0.001; 0.999]	0.46408	0.64260
GW_REVAP	Groundwater "revap" coefficient (-)	0.02–0.2	0.190 [0.021; 0.199]	−0.12613	0.89963
SURLAG	Surface runoff lag coefficient (-)	0.05–24	20.219 [0.076; 23.878]	−0.07433	0.94075
*ALPHA_BF**	Baseflow alpha factor (d^{-1})	0.034	0.034	ND	ND

* Determined on observed discharges by applying the baseflow filter program. ND: Not Determined.

To further investigate the model's accuracy, we evaluated predicted biomass values over the calibration/validation period (Table 4) against reported values for the study area. Simulated biomass was on average 4.3 ton· ha^{-1} for forest and 1.45 ton· ha^{-1} for agricultural land and both are in the ranges of observed values in the region (the observed biomass ranges between 2–4 and 2–3 ton· ha^{-1} for forest and cultivated land, respectively [23,62]). Nevertheless, this component is far underestimated for savannah with a simulated value of 0.4 ton ha^{-1} compared to the observed value which varies between 0.8 and 2 ton· ha^{-1} [62].

3.1.3. Sensitivity Analysis

There is a wide range of uses for which sensitivity analysis is performed. Based on the 12 selected SWAT parameters (*ALPHA_BF* being fixed), a GSA was used herein for identifying sensitive and important model parameters in order to better understand which hydrological processes are dominating the streamflow generation in the Bani catchment.

Sensitivity analysis results of 10,000 simulations are summarized in Table 5. The three most sensitive parameters (*CN2*, *OV_N*, and *SLSUBBSN*) are directly related to surface runoff, reflecting therefore the dominance of this process on the streamflow generation in the Bani catchment. Processes occurring at soil level followed at the second position as pointed out by the sensitivity of *ESCO* and *SOL_AWC*. Groundwater parameters happened in the last position demonstrating the low contribution of the latter to flows measured at the Douna outlet. The same sensitive parameters were identified by daily and monthly calibrations with only different ranks for soils parameters (*ESCO* and *SOL_AWC*).

3.1.4. Spatial Validation

The results of the spatial validation were divergent according to the location (Figure 4). For instance, at Pankourou, the same parameter sets determined at Douna produced a good simulation on a monthly basis (satisfactory for daily validation) whereas predictive uncertainty remained adequate and all met our requirements (*NSE* > 0.5, *P-factor* > 0.5 and *R-factor* < 1). In addition, the water balance was reasonably predicted at both time steps. In contrast, it has been recorded a complete loss of model performance at Bougouni with unsatisfactory *NSE* values and more uncertainty related to input discharge as expressed by a lower percentage of observed data (*P-factor* = 0.55 et 0.57 for daily and monthly validation) inside the 95PPU band (Figure 4). Accordingly, important uncertainty could be attributed to observed discharge at Bougouni.

3.2. The Subcatchment Model

Statistical evaluation results of the subcatchment calibration are presented in Table 3 and time series of observed and simulated hydrographs are shown in Figures 5 and 6. Good to very good performance was obtained at Pankourou with accurate predictive uncertainty. However, the validation period remained unsatisfactorily simulated at Bougouni. A comparative analysis of the catchment and subcatchment calibration performances came out with the following results:

- When calibrated separately, the prediction at Pankourou was slightly better, but greatly improved at Bougouni compared to when the catchment wide model was applied.
- The total uncertainty of the model is smaller at Pankourou (smaller *R-factor* and larger *P-factor*) than at the whole catchment, but larger at Bougouni.
- The water balance is better simulated at both internal stations compared to the watershed-wide water balance as depicted by smaller *PBIAIS* values, except always in the validation period at Bougouni.
- The model performance in terms of *NSE* and R^2 was higher at the watershed-wide level than at the sub-watershed level.

Overall, these results revealed that further calibration at the internal gauging stations was synonymous with gain of performance at the subcatchment level.

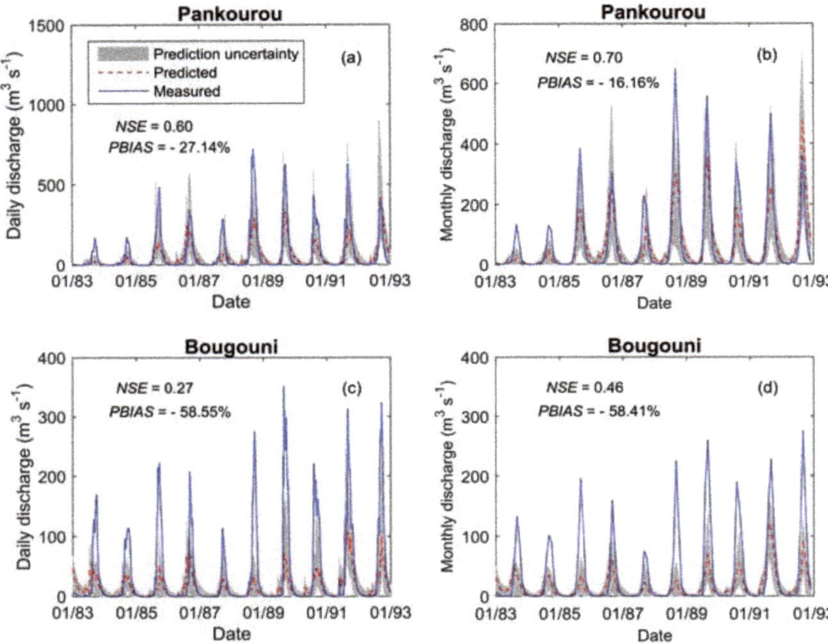

Figure 4. Spatial validation of the SWAT model on the Bani catchment. The model was turned at Pankourou ((**a**) daily and (**b**) monthly time steps) and at Bougouni ((**c**) daily and (**d**) monthly timesteps) by using the same behavioral parameter sets determined at the Douna outlet on the period 1983–1992.

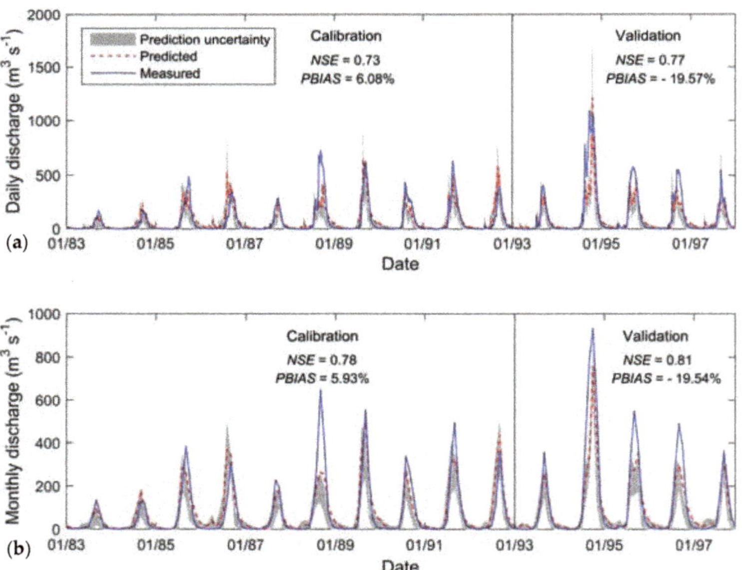

Figure 5. Simulated and observed hydrographs at Pankourou station at (**a**) daily and (**b**) monthly time steps along with calculated statistics on calibration and validation periods.

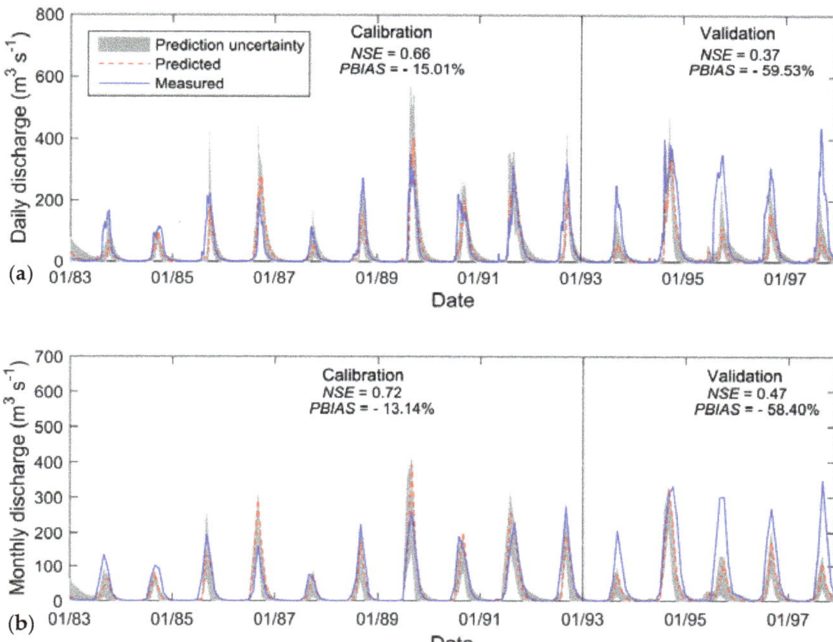

Figure 6. Predicted and measured discharges at Bougouni station at (**a**) daily and (**b**) monthly intervals during the calibration and validation periods with their corresponding statistics.

3.3. Model Predictive Uncertainty

In the global model, the predictive uncertainty, as indicated by the *P-factor* and *R-factor*, is adequate, though being larger during peak flow and recession periods (reflected by larger 95PPU band). On a daily basis, for instance, 61% of the observed discharge data are bracketed by a narrow 95PPU band depicted by the *R-factor* < 1 (Table 6). It has been noted that the entire uncertainty band is, however, very large during the year 1984 (Figure 3).

Table 6. Predictive uncertainty indices of the SWAT model for the Bani catchment at Douna, Pankourou, and Bougouni discharge gauging stations.

Time Step		Calibration (1983–1992)			Validation (1993–1997)		
	Criterion	Douna	Pankourou	Bougouni	Douna	Pankourou	Bougouni
Daily	*P-factor*	0.61	0.68	0.60	0.62	0.63	0.51
	R-factor	0.59	0.41	0.57	0.51	0.29	0.35
Monthly	*P-factor*	0.65	0.71	0.58	0.70	0.67	0.55
	R-factor	0.65	0.45	0.54	0.55	0.31	0.32

It is important to note the decrease of predictive uncertainty from Douna to Pankourou. In fact, the percentage of observed discharge bracket by 95PPU band has increased to 68%, while the width of the uncertainty band itself has decrease to 0.41 for the daily calibration (Table 6).

The same trend has been observed for the monthly calibration. At Bougouni, results showed a clear decrease of the uncertainty band (for daily and monthly calibration), but at the expense of bracketing less observed data. For instance, the *P-factor* and *R-factor* decreased from 0.65 to 0.58 and from 0.65 to 0.54, respectively, when moving from Douna to Bougouni during the monthly calibration.

Moreover, an increase of the uncertainty band with increasing time step (daily to monthly) has been recorded as depicted by higher *R-factor* values at Douna and Pankourou (from 0.59 to 0.65 and from 0.41 to 0.45, respectively). However, the uncertainty band was reduced during the validation period compared to the calibration period for all the stations (Table 6).

4. Discussion

4.1. Model Performance

In an effort to assess the performance of the SWAT model on the Bani catchment, we calibrated and validated the model at multiple sites on daily and monthly time steps by using measured climate data. There was no statistically significant difference in model performance among time intervals. Using guidelines given in Moriasi *et al.* [50], the overall performance of the SWAT model in terms of *NSE* and R^2 can be judged as very good, especially considering limited data conditions in the studied area. On a monthly basis, we obtained at the Douna outlet a *NSE* value equal to 0.79 for the calibration period (0.85 for the validation period). These results are greater than the ones of the studies by Schuol and Abbaspour [12], and Schuol *et al.* [14] at the same outlet. Schuol and Abbaspour [12] reported indeed a negative *NSE* (between −1 and 0) for the monthly calibration and a value ranging between 0 and 0.7 for monthly validation, while Schuol *et al.* [14] obtained a *NSE* between 0 and 0.70 for both monthly calibration and validation. However, Laurent and Ruelland [23] reported a greater performance (*NSE* values varying between 0.81 and 0.91 for calibration and validation period, respectively) but on a coarser time step (average annual basis). The water balance is less well simulated, especially for monthly time step with a *PBIAIS* greater than 25% in absolute value.

The quantified prediction uncertainty is surprisingly satisfactory (Table 6). At the end of the daily calibration, the model was able to account for 61% of observed discharge data (65% for monthly calibration) in a narrow uncertainty band. These results are close to the result of Schuol *et al.* [14] who estimated the observed discharge data bracketed by the 95PPU between 60% and 80% for monthly calibration (40% and 60% for monthly validation). However, one explanation that could be attributed to the small uncertainty band we obtained is that model predictive uncertainty derived by GLUE depends largely on the threshold value to separate "behavioral" from "non-behavioral" parameter sets [63,64].

This means, a high threshold value (as in this case) will generally lead to a narrower uncertainty band [65–67] but this will be achieved at the cost of bracketing less observed data within the 95PPU band. In addition, GLUE accounts partly for uncertainty due to the possible non-uniqueness (or equifinality) of parameter sets during calibration and could therefore underestimate total model uncertainty [68]. For instance, Sellami *et al.* [69] showed that the GLUE predictive uncertainty band was larger and surrounded more observation data when uncertainty in the discharge data was explicitly considered. Engeland and Gottschalk [70] demonstrated that the conceptual water balance model structural uncertainty was larger than parameter uncertainty. In spite of all the aforementioned limitations of GLUE, we succeeded in enclosing interestingly most of the observed data within a narrow uncertainty band (the sought adequate balance between the two indices) hence increasing confidence in model results. These are encouraging results showing, on one hand, the good performance of the SWAT model on a large Soudano-Sahelian catchment under limited data and varying climate conditions and, on the other hand, the capability of observed climate and hydrological input data of this catchment, even though contested, to provide reliable information about hydro-meteorological systems prevailing in the region.

It has been also noted that the model did not perform well during the year 1984 particularly (lower performance and larger uncertainty). This loss of performance can be attributed to the disruption in rainfall-runoff relationship consequence of consecutive years of drought, which has prevailed in the beginning of the 80s. The over-predicted PET on the Bani catchment could be attributed to the Hargreaves method, which could give a greater estimate of PET than it actually is. Ruelland *et al.* [28] applied a temperature-based method given by Oudin *et al.* [71] and provided a similar estimate of PET (1723 mm) than the values calculated herein by the Penman and pan evaporation methods hence corroborating our results. These results demonstrated the valuable of pan evaporation measurements for estimating PET and that the simple pan evaporation method appears to be suited for application in the study area and can be used when all the climatic data required by the Penman method are missing.

As far as biomass is concerned, the underestimation of this component in savannah could be explained by inappropriate specification of all categories in the land use map grid to be modeled by SWAT as savannah or inaccurate savannah characteristics added in the SWAT database, directly affecting biomass production such as *BIO_E* and *LAI* parameters, among others.

4.2. Impact of Spatial and Temporal Scales on the Model Uncertainty

Results showed that transferring the model parameters from the catchment outlet (Douna) to the internal gauging stations performs reasonably well only in the case of similarity between donor and target catchments. The case of catchments controlled by Douna and Pankourou gives a clear example of such physical proximity where precipitation, soil and land use vary smoothly between both catchments. However, the SWAT model parameters determined at the outlet could not reproduce well the measured discharge at Bougouni mainly due to more significant spatial dissimilarities. Bougouni is indeed situated in a more humid zone and dominated by forest whereas Douna is more arid. Moreover, it has been demonstrated that the individual calibration at subcatchment scale has led to a narrower uncertainty band and more observed discharge data enclosed in it, which is the sought adequate balance between the two indices. Hence, predictive uncertainty was found to decrease with decreasing spatial scale. This finding can be attributed to the presence of less heterogeneity in hydrological variables in smaller catchments. These results showed the importance of the calibration of hydrological models at finer spatial scale to ensure that predominant processes in each subcatchment are captured, and this is particularly relevant in case of large-area global catchments. Concerning the effect of temporal scale, we demonstrated that the validation period is characterized by less predictive uncertainty as opposed to the calibration period. One explanation that can be given is the fact that 1993-1997 constitutes a more humid period than 1983-1992 and is characterized, therefore, by less variability in precipitation. In contrast, when moving from daily to monthly calibration, the uncertainty of the model, in terms of uncertainty band width, increased. This could be attributed to the cumulative effect of uncertainty in daily discharge data used to compute monthly discharge, resulting therefore in larger monthly uncertainty. Overall, due to decreasing prediction uncertainty with decreasing spatial and temporal scales, it is germane to develop on the basin a more efficient system of hydro-meteorological data collection to account for spatial and temporal variabilities in hydro-meteorological systems prevailing in the region, especially under changing climate and land use conditions.

4.3. Advance in Understanding of Hydrological Processes

The GSA confirms what has already been reported on and around the Bani catchment about the contribution of hydrological processes to streamflow generation. In order to better understand the origin of flows at Kolondieba (a tributary of the Bani River), Dao *et al.* [72] showed that Groundwater contribution to the hydrodynamic equilibrium at the outlet of watershed Kolondieba is small and the direct flow from the soil surface governs the runoff process. This fact can be explained by the double impact of a general impoverishment of shallow aquifers due to reduction in precipitation in West Africa in general since the great drought of the 70s as well as a concurrent increase of the recession coefficient of the Bani river as demonstrated by Bamba *et al.* [32] and Mahé [73] with a decrease of baseflow contribution to total flow in absolute and relative values as corollary.

4.4. Spatial Performance

The results of different calibration and validation techniques showed varying predictive abilities of the SWAT model through scales. Firstly, it can be derived from these findings that model performance in terms of *NSE* and R^2 was higher on the watershed-wide level than on the sub-watershed level. However, this could be attributed to compensation between positive and negative errors of processes occurring at a larger scale [74,75]. This suggests that calibrating a model only at the basin outlet leads to an overconfidence in its performance than at the sub-basin scale. Secondly, individual calibration of subcatchment processes expectedly improved model accuracy in predicting flows at the internal gauging stations, due to reducing heterogeneities with downsizing space [76], and is especially beneficent while the donor and receiver catchments are substantially different. Finally, predictive uncertainty appears to decrease with reducing spatial scale, but increases with humidity as shown by the lower performance recorded at Bougouni. The inability of the model to perform during the validation period at Bougouni could be attributed to the structure of the validation period which is substantially different to that of calibration, and is solely composed by average to wet years while in contrast, the occurrence of dry, average, and wet years during the calibration period is noted.

These results have an important role to play in the calibration and validation approaches of large-area watershed models and constitutes a first step to model parameter regionalization for prediction in ungauged basins.

Generally speaking, it is well known that in recent decades the Niger River basin has suffered from a serious degradation of its natural resources, which in turn lead to severe environmental issues. To this end, different agreements and collaborations on water and climate data sharing have been established between the 9 countries sharing the basin through different national and international programs. Thus, the need to reinforce the existing framework of integrated, coordinated, and sustainable water management strategies in the Bani basin and therefore the Niger River Basin become more urgent than ever.

Therefore, this study is a step in that long-term direction, where an integrated water management tool has been developed and validated spatially on the Bani catchment, which allows investigation of future effects of land use and climate change scenarios on water resources.

5. Conclusions

In this study, the performance of the widely-used SWAT model was evaluated on the Bani catchment using both split-sample and split-location calibration and validation techniques on daily and monthly intervals. The model was calibrated at the Douna outlet and at two internal stations. Freely available global data and daily observed climate and discharge data were used as inputs for model simulation and calibration. Calibration, validation, uncertainty, and sensitivity analyses were performed with GLUE within SWAT-CUP. Both graphical and statistical techniques were used for hydrologic calibration results evaluation. Evapotranspiration and biomass production outputs were verified and compared to regional values to make sure these components were reasonably predicted. Sensitivity analysis contributed to a better understanding of the hydrological processes occurring at the study area.

Final results showed a good SWAT model performance to predict daily as well as monthly discharge at Douna with acceptable predictive uncertainty despite the poor data density and the high gradient of climate and land use characterizing the study catchment. However, the daily calibration resulted in less predictive uncertainty than the monthly calibration. The performance of the model is somehow lower at an internal sub-catchments level when the global parameter sets are applied, especially at the one with higher humidity and dominated by forest. However, subcatchment calibration induced an increase of model performance at intermediate gauging stations as well as a decrease of total uncertainty. With regard to predicted PET, this component is overestimated by the model when the Hargreaves method is applied in that specific region while biomass production remained low in the savannah land use category. The GSA revealed the predominance of surface and subsurface processes in the streamflow generation of the Bani River.

Overall, this study has shown the validity of the SWAT model for representing globally hydrological processes of a large-scale Soudano-Sahelian catchment in West Africa. Given the high spatial variability of climate, soil, and land use characterizing the catchment, additional calibration is however needed at subcatchment level to ensure that predominant processes are captured in each subcatchment. Accordingly, the importance of spatially distributed hydrological measurements is demonstrated and constitutes the backbone of any type of progress in hydrological process understanding and modeling. The calibrated SWAT model for the Bani can be used to assess the current and future impacts of climate and land use change on water resources of the catchment, increasingly necessary information awaited by water resources managers. Knowing this information, a strategy of adaptation in response to the current and future impacts can be clearly proposed and the vulnerability of the population can therefore be reduced. More widely, this impact study can increase the transferability of the model parameters from the Bani subcatchment to another ungauged basin with some similarities, and then predicting discharge without the need of any measurement. These findings are very useful, especially in West Africa, where many river basins are ungauged or poorly gauged.

Acknowledgments: Authors are extremely grateful to AGRHYMET for providing the climatic database. We would like also to thank Kone Soungalo for climate data on the Ivorian part of the Bani basin and Francois Laurent for soil and land use databases.

Funding: This study was funded by the German Ministry of Education and Research (BMBF) through the West African Science Service Centre on Climate Change and Adapted Land Use (**WASCAL**; www.wascal.org) that supports the Graduate Research Program Climate Change and Water Resources at the University of Abomey-Calavi. Additional funds were provided by **AGRHYMET** through the French Global Environment Facility (FFEM/CC) project.

Author Contributions: All authors designed the objectives and methods of the study. J. C. Begou, P. Bazie, and A. Afouda collected the climatological and hydrological datasets required for the model. J. C. Begou, S. Jomaa, S. Benabdallah, and M. Rode conducted the model setup and performed the simulations. All authors analyzed the obtained data. J. C. Begou prepared the manuscript with contributions from all co-authors. All authors read and approved the final manuscript.

Conflicts of Interest: The authors declare no conflict of interest.

References

1. Pomeroy, J.W.; Spence, C.; Whitfield, P.H. Putting prediction in ungauged basins into practice. In *Putting Prediction in Ungauged Basins into Practice*; Canadian Water Resources Association: Ottawa, ON, Canada, 2013; pp. 1–12.
2. Sivapalan, M.; Takeuchi, K.; Franks, S.W.; Gupta, V.K.; Karambiri, H.; Lakshmi, V.; Liang, X.; McDonnell, J.J.; Mendiondo, E.M.; O'Connell, P.E.; *et al.* Iahs decade on predictions in ungauged basins (PUB), 2003–2012: Shaping an exciting future for the hydrological sciences. *Hydrol. Sci. J.* **2003**, *48*, 857–880. [CrossRef]
3. Bormann, H.; Diekkrüger, B. Possibilities and limitations of regional hydrological models applied within an environmental change study in Benin (West Africa). *Phys. Chem. Earth Parts A/B/C* **2003**, *28*, 1323–1332. [CrossRef]
4. Kapangaziwiri, E.; Hughes, D.A.; Wagener, T. Incorporating uncertainty in hydrological predictions for gauged and ungauged basins in Southern Africa. *Hydrol. Sci. J.* **2012**, *57*, 1000–1019. [CrossRef]
5. Mazvimavi, D.; Meijerink, A.M.J.; Savenije, H.H.G.; Stein, A. Prediction of flow characteristics using multiple regression and neural networks: A case study in Zimbabwe. *Phys. Chem. Earth Parts A/B/C* **2005**, *30*, 639–647. [CrossRef]
6. Minihane, M. Estimating mean monthly streamflow in the Lugenda River, Northern Mozambique. In *Putting Prediction in Ungauged Basins into Practice*; Canadian Water Resources Association: Ottawa, ON, Canada, 2013; pp. 185–196.
7. Ndomba, P.; Mtalo, F.; Killingtveit, A. SWAT model application in a data scarce tropical complex catchment in Tanzania. *Phys. Chem. Earth Parts A/B/C* **2008**, *33*, 626–632. [CrossRef]

8. Nkamdjou, L.S.; Bedimo, J.B. *Critique des Donnees Hydrometriques du Bassin du Niger: Rapport Preliminaire*; Niger Basin Authority: Niamey, Niger, 2008; p. 28.
9. IPCC. *Climate Change 2014: Impacts, Adaptation, and Vulnerability. Part B: Regional Aspects*; Barros, V.R., Field, C.B., Dokken, D.J., Mastrandrea, M.D., Mach, K.J., Bilir, T.E., Chatterjee, M., Ebi, K.L., Estrada, Y.O., Genova, R.C., *et al.*, Eds.; Cambridge University Press: Cambridge, UK; New York, NY, USA, 2014; p. 688.
10. Wagener, T.; Wheater, H.S.; Gupta, H.V. *Rainfall-Runoff Modelling in Gauged and Ungauged Catchments*; Imperial College Press: London, UK, 2004; p. 332.
11. Amadou, A.; Djibo Abdouramane, G.; Ousmane, S.; Sanda Ibrah, S.; Sittichok, K. Changes to flow regime on the Niger River at Koulikoro under a changing climate. *Hydrol. Sci. J.* **2015**, *60*, 1709–1723.
12. Schuol, J.; Abbaspour, K.C. Calibration and uncertainty issues of a hydrological model (SWAT) applied to West Africa. *Adv. Geosci.* **2006**, *9*, 137–143. [CrossRef]
13. Schuol, J.; Abbaspour, K.C. Using monthly weather statistics to generate daily data in a SWAT model application to West Africa. *Ecol. Model.* **2007**, *201*, 301–311. [CrossRef]
14. Schuol, J.; Abbaspour, K.C.; Srinivasan, R.; Yang, H. Estimation of freshwater availability in the West African sub-continent using the SWAT hydrologic model. *J. Hydrol.* **2008**, *352*, 30–49. [CrossRef]
15. Schuol, J.; Abbaspour, K.C.; Yang, H.; Srinivasan, R.; Zehnder, A.J.B. Modeling blue and green water availability in Africa. *Water Resour. Res.* **2008**, *44*, 1–18. [CrossRef]
16. Sittichok, K.; Gado Djibo, A.; Seidou, O.; Moussa Saley, H.; Karambiri, H.; Paturel, J. Statistical seasonal rainfall and streamflow forecasting for the Sirba watershed, West Africa, using sea surface temperatures. *Hydrol. Sci. J.* **2014**, *61*, 1700–1712. [CrossRef]
17. Awotwi, A.; Yeboah, F.; Kumi, M. Assessing the impact of land cover changes on water balance components of White Volta Basin in West Africa. *Water Environ. J.* **2015**, *29*, 259–267. [CrossRef]
18. Guzinski, R.; Kass, S.; Huber, S.; Bauer-Gottwein, P.; Jensen, I.; Naeimi, V.; Doubkova, M.; Walli, A.; Tottrup, C. Enabling the use of earth observation data for integrated water resource management in Africa with the water observation and information system. *Remote Sens.* **2014**, *6*, 7819–7839. [CrossRef]
19. Sood, A.; Muthuwatta, L.; McCartney, M. A SWAT evaluation of the effect of climate change on the hydrology of the Volta River Basin. *Water Int.* **2013**, *38*, 297–311. [CrossRef]
20. Bossa, A.Y.; Diekkrüger, B.; Giertz, S.; Steup, G.; Sintondji, L.O.; Agbossou, E.K.; Hiepe, C. Modeling the effects of crop patterns and management scenarios on N and P loads to surface water and groundwater in a semi-humid catchment (West Africa). *Agric. Water Manag.* **2012**, *115*, 20–37. [CrossRef]
21. Bossa, A.Y.; Diekkrüger, B.; Igué, A.M.; Gaiser, T. Analyzing the effects of different soil databases on modeling of hydrological processes and sediment yield in Benin (West Africa). *Geoderma* **2012**, *173–174*, 61–74.
22. Sintondji, L.O.; Barnabé, Z.; Ahouansou, D.M.; Vissin, W.E.; Agbossou, K.E. Modelling the water balance of ouémé catchment at the savè outlet in Benin: Contribution to the sustainable water resource management. *Int. J. AgriSci.* **2014**, *4*, 74–88.
23. Laurent, F.; Ruelland, D. Modélisation à base physique de la variabilité hydroclimatique à l'échelle d'un grand bassin versant tropical. In Global Change: Facing Risks and Threats to Water Resources, Proceedings of the sixth world FRIEND conference, Fez, Morroco, 25–29 October 2010; International Association of Hydrological Sciences (IAHS): Wallingford, UK, 2010; pp. 474–484.
24. Ruelland, D.; Ardoin-Bardin, S.; Billen, G.; Servat, E. Sensitivity of a lumped and semi-distributed hydrological model to several methods of rainfall interpolation on a large basin in West Africa. *J. Hydrol.* **2008**, *361*, 96–117.
25. Masih, I.; Maskey, S.; Uhlenbrook, S.; Smakhtin, V. Assessing the impact of areal precipitation input on streamflow simulations using the SWAT model. *J. Am. Water Resour. Assoc.* **2011**, *47*, 179–195. [CrossRef]
26. Paturel, J.E.; Ouedraogo, M.; Mahé, G.; Servat, E.; Dezetter, A.; Ardoin, S. The influence of distributed input data on the hydrological modelling of monthly river flow regimes in West Africa. *Hydrol. Sci. J.* **2003**, *48*, 881–890.
27. Food and Agriculture Organization of the United Nations (FAO). *Digital Soil Map of the World and Derived Soil Properties*; Version 3.6; Land and Water Development Division, FAO: Rome, Italy, 2003.
28. Ruelland, D.; Ardoin-Bardin, S.; Collet, L.; Roucou, P. Simulating future trends in hydrological regime of a large sudano-sahelian catchment under climate change. *J. Hydrol.* **2012**, *424–425*, 207–216.

29. Loveland, T.R.; Reed, B.C.; Brown, J.F.; Ohlen, D.O.; Zhu, Z.; Yang, L.; Merchant, J.W. Development of a global land cover characteristics database and IGBP DISCover from 1 km AVHRR data. *Int. J. Remote Sens.* **2000**, *21*, 1303–1330.
30. L'Hote, Y.; Mahé, G.; Some, B.; Triboulet, J.P. Analysis of a sahelian annual rainfall index from 1896 to 2000; the drought continues. *Hydrol. Sci. J.* **2002**, *47*, 563–572. [CrossRef]
31. Mahé, G.; Olivry, J.-C.; Dessouassi, R.; Orange, D.; Bamba, F.; Servat, E. Relations eaux de surface-eaux souterraines d'une rivière tropicale au Mali. *Comptes Rendus de l'Académie des Sciences Series IIA Earth Planet. Sci.* **2000**, *330*, 689–692. [CrossRef]
32. Bamba, F.; Mahe, G.; Bricquet, J.P.; et Olivry, J.C. Changements climatiques et variabilite des ressources en eau des bassins du haut niger et de la cuvette lacustre. *XIIèmes journees hydrologiques de l'orstom* **1996**, *10–11*, 1–27.
33. Ruelland, D.; Guinot, V.; Levavasseur, F.; Cappelaere, B. Modelling the longterm impact of climate change on rainfall-runoff processes over a large sudano-sahelian catchment. In New Approaches to Hydrological Prediction in Data Sparse Regions, Proceedings of the Symposium HS.2 at the Joint Convention of The International Association of Hydrological Sciences (IAHS) and The International Association of Hydrogeologists (IAH), Hyderabad, India, 6–12 September 2009; IAHS: Wallingford, UK, 2009; pp. 59–68.
34. Arnold, J.G.; Srinivasan, R.; Muttiah, R.S.; Williams, J.R. Large area hydrologic modeling and assessment part i: Model development. *J. Am. Water Resour. Assoc.* **1998**, *34*, 73–89. [CrossRef]
35. Winchell, M.; Srinivasan, R.; di Luzio, M.; Arnold, J.G. *Arcswat Interface for SWAT. 2012 User's Guide*; Soil and Water Research Laboratory, USDA Agricultural Research Service: Temple, TX, USA, 2013; p. 459.
36. Arnold, J.G.; Kiniry, J.R.; Srinivasan, R.; Williams, J.R.; Haney, E.B.; Neitsch, S.L. *Soil and Water Assessment Tool Input/Output Documentation: Version 2012*; Texas Water Resources Institute Technical Report: College Station, TX, USA, 2012; Volume TR-439.
37. Neitsch, S.L.; Arnold, J.G.; Kiniry, J.R.; Williams, J.R. *Soil and Water Assessment Tool Theoretical Documentation: Version 2009*; Technical Report No. 406; Texas Water Resources Institute Texas: College Station, TX, USA, 2011.
38. Bouraoui, F.; Benabdallah, S.; Jrad, A.; Bidoglio, G. Application of the SWAT model on the Medjerda River Basin (Tunisia). *Phys. Chem. Earth Parts A/B/C* **2005**, *30*, 497–507. [CrossRef]
39. Ouessar, M.; Bruggeman, A.; Abdelli, F.; Mohtar, R.H.; Gabriels, D.; Cornelis, W.M. Modelling water-harvesting systems in the arid south of Tunisia using SWAT. *Hydrol. Earth Syst. Sci.* **2009**, *13*, 2003–2021. [CrossRef]
40. World Meteorological Organization (WMO). *WMO-No.100. Guide to Climatological Practices*, 3rd ed.; WMO: Geneva, Switzerland, 2011.
41. Boisramé, G. WGN Excel Macro. Available online: http://swat.Tamu.Edu/software/links/ (accessed on 25 April 2016).
42. USGS Hydrosheds. Available online: http://hydrosheds.cr.usgs.gov/dataavail.php (accessed on 26 April 2016).
43. Waterbase. Available online: http://www.waterbase.org/resources.html (accessed on 26 April 2016).
44. FAO. Available online: http://www.fao.org/geonetwork/srv/en/main.home (accessed on 26 April 2016).
45. Nash, J.E.; Sutcliffe, J.V. River flow forecasting through conceptual models part I—A discussion of principles. *J. Hydrol.* **1970**, *10*, 282–290. [CrossRef]
46. Moussa, R.; Chahinian, N.; Bocquillon, C. Distributed hydrological modelling of a mediterranean mountainous catchment—Model construction and multi-site validation. *J. Hydrol.* **2007**, *337*, 35–51. [CrossRef]
47. Robson, B.J.; Dourdet, V. Prediction of sediment, particulate nutrient and dissolved nutrient concentrations in a dry tropical river to provide input to a mechanistic coastal water quality model. *Environ. Model. Softw.* **2015**, *63*, 97–108. [CrossRef]
48. Abbaspour, K.C. *SWAT-Cup 2012: SWAT Calibration and Uncertainty Programs—A User Manual*; Eawag, Swiss Federal Institute of Aquatic Science and Technology: Duebendorf, Switzerland, 2014; p. 106.
49. Beven, K.; Binley, A. The future of distributed models: Model calibration and uncertainty prediction. *Hydrol. Process.* **1992**, *6*, 279–298. [CrossRef]

50. Moriasi, D.N.; Arnold, J.G.; van Liew, M.W.; Bingner, R.L.; Harmel, R.D.; Veith, T.L. Model evaluation guidelines for systematic quantification of accuracy in watershed simulations. *Trans. ASABE* **2007**, *50*, 885–900. [CrossRef]
51. Arnold, J.G.; Allen, P.M.; Muttiah, R.; Bernhardt, G. Automated base flow separation and recession analysis techniques. *Ground Water* **1995**, *33*, 1010–1018. [CrossRef]
52. Arnold, J.G.; Allen, P.M. Automated methods for estimating baseflow and ground water recharge from streamflow records. *J. Am. Water Resour. Assoc.* **1999**, *35*, 411–424. [CrossRef]
53. Betrie, G.D.; Mohamed, Y.A.; van Griensven, A.; Srinivasan, R. Sediment management modelling in the Blue Nile Basin using SWAT model. *Hydrol. Earth Syst. Sci.* **2011**, *15*, 807–818. [CrossRef]
54. Van Griensven, A.; Meixner, T.; Grunwald, S.; Bishop, T.; Diluzio, M.; Srinivasan, R. A global sensitivity analysis tool for the parameters of multi-variable catchment models. *J. Hydrol.* **2006**, *324*, 10–23. [CrossRef]
55. Zhang, X.; Srinivasan, R.; Liew, M.V. Multi-site calibration of the SWAT model for hydrologic modeling. *Trans. ASABE* **2008**, *51*, 2039–2049. [CrossRef]
56. Arnold, J.G.; Moriasi, D.N.; Gassman, P.W.; Abbaspour, K.C.; White, M.J.; Srinivasan, R.; Santhi, C.; Harmel, R.D.; Griensven, A.V.; Liew, M.W.V.; *et al.* SWAT: Model use, calibration, and validation. *Trans. ASABE* **2012**, *55*, 1491–1508. [CrossRef]
57. Abbaspour, K.C.; Johnson, C.A.; van Genuchten, M.T. Estimating uncertain flow and transport parameters using a sequential uncertainty fitting procedure. *Vadose Zone J.* **2004**, *3*, 1340–1352. [CrossRef]
58. Hargreaves, G.L.; Hargreaves, G.H.; Riley, J.P. Agricultural benefits for Senegal River Basin. *J. Irrig. Drain. Eng.* **1985**, *111*, 113–124. [CrossRef]
59. Allen, R.G.; Pereira, L.S.; Raes, D.; Smith, M. *Crop Evapotranspiration-Guidelines for Computing Crop Water Requirements-FAO Irrigation and Drainage Paper 56*; Food and Agriculture Organization of the United Nations (FAO): Rome, Italy, 1998.
60. Doorenbos, J.; Pruitt, W.O. *Guidelines for Predicting Crop Water Requirements, Irrigation and Drainage Paper 24*; Land and Water Development Division, FAO: Rome, Italy, 1975; p. 179.
61. Brouwer, C.; Heibloem, M. *Irrigation Water Management: Training Manual No. 3*; Food and Agriculture Organization of the United Nations (FAO): Rome, Italy, 1986.
62. Coulibaly, A. *Profil Fourrager (Mali)*; Food and Agriculture Organization of the United Nations (FAO): Rome, Italy, 2003.
63. Mantovan, P.; Todini, E. Hydrological forecasting uncertainty assessment: Incoherence of the GLUE methodology. *J. Hydrol.* **2006**, *330*, 368–381. [CrossRef]
64. Montanari, A. Large sample behaviors of the generalized likelihood uncertainty estimation (GLUE) in assessing the uncertainty of rainfall-runoff simulations. *Water Resour. Res.* **2005**, *41*, 1–13. [CrossRef]
65. Blasone, R.-S.; Madsen, H.; Rosbjerg, D. Uncertainty assessment of integrated distributed hydrological models using GLUE with Markov chain Monte Carlo sampling. *J. Hydrol.* **2008**, *353*, 18–32. [CrossRef]
66. Viola, F.; Noto, L.V.; Cannarozzo, M.; La Loggia, G. Daily streamflow prediction with uncertainty in ephemeral catchments using the GLUE methodology. *Phys. Chem. Earth Parts A/B/C* **2009**, *34*, 701–706. [CrossRef]
67. Xiong, L.; O'Connor, K.M. An empirical method to improve the prediction limits of the GLUE methodology in rainfall—Runoff modeling. *J. Hydrol.* **2008**, *349*, 115–124. [CrossRef]
68. Yen, H.; Wang, X.; Fontane, D.G.; Harmel, R.D.; Arabi, M. A framework for propagation of uncertainty contributed by parameterization, input data, model structure, and calibration/validation data in watershed modelling. *Environ. Model. Softw.* **2014**, *54*, 211–221. [CrossRef]
69. Sellami, H.; Vanclooster, M.; Benabdallah, S.; La Jeunesse, I. Assessment of the SWAT model prediction uncertainty using the GLUE approach a case study of the Chiba catchment (Tunisia). In Modeling, Simulation and Applied Optimization (ICMSAO), Proceedings of the 5th International Conference, Hammamet, Tunisia, 28–30 April 2013; Institute of Electrical and Electronics Engineers (IEEE): New York, NY, USA, 2013; pp. 1–6.
70. Engeland, K.; Gottschalk, L. Bayesian estimation of parameters in a regional hydrological model. *Hydrol. Earth Syst. Sci.* **2002**, *6*, 883–898. [CrossRef]
71. Oudin, L.; Hervieu, F.; Michel, C.; Perrin, C.; Andréassian, V.; Anctil, F.; Loumagne, C. Which potential evapotranspiration input for a lumped rainfall-runoff model? Part 2 towards a simple and efficient potential evapotranspiration model for rainfall-runoff modelling. *J. Hydrol.* **2005**, *303*, 290–306. [CrossRef]

72. Dao, A.; Kamagate, B.; Mariko, A.; Seguis, L.; Maiga, H.B.; Goula Bi, T.A.; Savane, I. Deconvolution of the flood hydrograph at the outlet of watershed kolondieba in the south of Mali. *Int. J. Eng. Res. Appl.* **2012**, *2*, 1174–1181.
73. Mahé, G. Surface/groundwater interactions in the Bani and Nakambe rivers, tributaries of the Niger and Volta Basins, West Africa. *Hydrol. Sci. J.* **2009**, *54*, 704–712. [CrossRef]
74. Cao, W.; Bowden, W.B.; Davie, T.; Fenemor, A. Multi-variable and multi-site calibration and validation of SWAT in a large mountainous catchment with high spatial variability. *Hydrol. Process.* **2006**, *20*, 1057–1073. [CrossRef]
75. Wellen, C.; Kamran-Disfani, A.-R.; Arhonditsis, G.B. Evaluation of the current state of distributed watershed nutrient water quality modeling. *Environ. Sci. Technol.* **2015**, *49*, 3278–3290. [CrossRef] [PubMed]
76. Daggupati, P.; Yen, H.; White, M.J.; Srinivasan, R.; Arnold, J.G.; Keitzer, C.S.; Sowa, S.P. Impact of model development, calibration and validation decisions on hydrological simulations in West Lake Erie Basin. *Hydrol. Process.* **2015**, *29*, 5307–5320. [CrossRef]

Article

Using High-Resolution Data to Test Parameter Sensitivity of the Distributed Hydrological Model HydroGeoSphere

Thomas Cornelissen [1,*], Bernd Diekkrüger [1] and Heye R. Bogena [2]

[1] Department of Geography, University of Bonn, Bonn 53115, Germany; b.diekkrueger@uni-bonn.de

[2] Agrosphere Institute (IBG-3), Forschungszentrum Jülich, Jülich 52425, Germany; h.bogena@fz-juelich.de

* Correspondence: thomas.cornelissen@gmx.net; Tel.: +49-228-732-401

Academic Editor: Xuan Yu
Received: 5 March 2016; Accepted: 10 May 2016; Published: 16 May 2016

Abstract: Parameterization of physically based and distributed hydrological models for mesoscale catchments remains challenging because the commonly available data base is insufficient for calibration. In this paper, we parameterize a mesoscale catchment for the distributed model HydroGeoSphere by transferring evapotranspiration parameters calibrated at a highly-equipped headwater catchment in addition to literature data. Based on this parameterization, the sensitivity of the mesoscale catchment to spatial variability in land use, potential evapotranspiration and precipitation and of the headwater catchment to mesoscale soil and land use data was conducted. Simulations of the mesoscale catchment with transferred parameters reproduced daily discharge dynamics and monthly evapotranspiration of grassland, deciduous and coniferous vegetation in a satisfactory manner. Precipitation was the most sensitive input data with respect to total runoff and peak flow rates, while simulated evapotranspiration components and patterns were most sensitive to spatially distributed land use parameterization. At the headwater catchment, coarse soil data resulted in a change in runoff generating processes based on the interplay between higher wetness prior to a rainfall event, enhanced groundwater level rise and accordingly, lower transpiration rates. Our results indicate that the direct transfer of parameters is a promising method to benefit highly equipped simulations of the headwater catchments.

Keywords: parameter transfer; distributed hydrological modeling; mesoscale catchment; headwater catchment sensitivity; HydroGeoSphere

1. Introduction

The usage of integrated and distributed hydrological models (e.g., HydroGeoSphere [1], ParFlow-CLM [2], MIKE-SHE [3] and Cathy [4]) has considerably increased during the last decade alongside advances in computer and measurement technology. It is widely acknowledged that hydrological models integrating the surface and subsurface flow systems have on the one hand, a great potential to give insights into temporal and spatial patterns of fluxes, state variables and feedback [5–9]. On the other hand, the complexity of these models causes overparameterization (e.g., [10]) and hinders transferability of achieved simulation as well as parameterization results to other spatio-temporal scales (e.g., [11,12]).

Distributed observation networks, necessary for a reliable calibration and validation of spatial patterns simulated by 3D-models—for example soil moisture sensor networks—are typically only available for small test sites [13], such as the Wüstebach catchment in Germany [14] or the Little Washita catchment in the United States [2]. Due to their high data demand, distributed hydrological 3D-models are currently predominantly used for small-scale applications [6,8,15]. Rare examples at large scales include the study of Goderniaux *et al.* [16] who estimated climate change effects on groundwater reserves in a 480 km^2 mesoscale catchment with HydroGeoSphere and the study of Rahman *et al.* [17] who applied ParFlow-CLM to a 2364 km^2 macroscale catchment to investigate spatio-temporal patterns of land surface mass and energy fluxes.

The rarity of simulations at meso- (>10 km^2) or macroscale (>1000 km^2) catchments is in sharp contrast to potential feedback with boundary processes at these scales. For example, Hauck *et al.* [18] stated that the simulation of convection could be improved with more detailed information on spatial and vertical distribution of soil moisture.

Modeling catchments larger than headwater catchments is closely connected to a decrease in quantity and quality of available calibration data, especially concerning their spatial distribution. In addition, mesoscale catchments can exhibit stronger variability in land use and climate variables. Thus, assembling a data base which is sufficient for a reliable calibration of distributed and process-based models is a challenging task. Given the large uncertainties inherent in mesoscale catchment modeling with distributed hydrological models, it is necessary to facilitate the model setup and to investigate the sensitivity of model parameters and input data. The model setup can be facilitated by using prior knowledge of parameters and their spatial distribution. For example, parameters can be regionalized with different methods from other catchments or parameters can be directly transferred from subcatchments, as described in Bogena and Diekkrüger [19]. According to Moriasi *et al.* [20], a sensitivity analysis investigates the reaction of the model output to changes in parameters or input data and is a requirement for a successful calibration. A sensitivity analysis of a highly heterogeneous mesoscale catchment must include the investigation of spatial heterogeneity in climate input data and of model parameters. Similar to the transfer of parameters, information about the sensitivity of model parameters can potentially be inferred from a subcatchment, if the spatial and temporal discretization does not change.

In this study we use the distributed hydrological 3D-model HydroGeoSphere in a nested simulation approach to conduct a sensitivity analysis across scales. The first research aspect investigates the sensitivity of discharge, water balance and evapotranspiration patterns to spatial heterogeneity in land use, potential evapotranspiration and precipitation at the mesoscale. Evapotranspiration parameters for the mesoscale catchment setup are taken from existing research and are directly transferred from calibrated and validated simulations of a well-equipped sub-catchment. As these transferred evapotranspiration parameters originate from a homogeneously covered spruce forest, the question arises if they can be applied for evapotranspiration simulation of different land use types. Thus, the second research aspect gives a validation of simulated monthly evapotranspiration for the different land use types of the mesoscale catchment. Apart from transferred parameters, the setup of the mesoscale catchment involves the incorporation of land use data and soil parameters that are different to the subcatchment due to change in resolution and heterogeneity. Thus, the third research aspect of this study is a sensitivity analysis of the subcatchment to land use and soil parameters used for the setup of the mesoscale catchment.

The simulation results at the mesoscale catchment, in terms of discharge dynamics and monthly evapotranspiration, highlight the potential of heavily instrumented test sites to deliver reliable estimates of evapotranspiration parameters for the simulation of different land use conditions.

2. Materials and Methods

2.1. Description of the Study Area

The Erkensruhr catchment is located in western Germany close to the Belgian border (Figure 1A). It is 41.9 km^2 large and its elevation increases from 286 m.a.s.l. to 631 m.a.s.l. in eastern and southern directions. The slope varies between 0° and 7° in flatter areas in the central and south-eastern catchment part. It rapidly increases with proximity to the river bed with slopes above 25° in the northern catchment part. Mean annual temperatures range between 7.6 °C at high and 10 °C at low altitudes. The catchment is characterized by a strong west-east gradient in precipitation with a mean annual precipitation of 1150 mm over areas west to the catchment and 740 mm over areas east to the catchment.

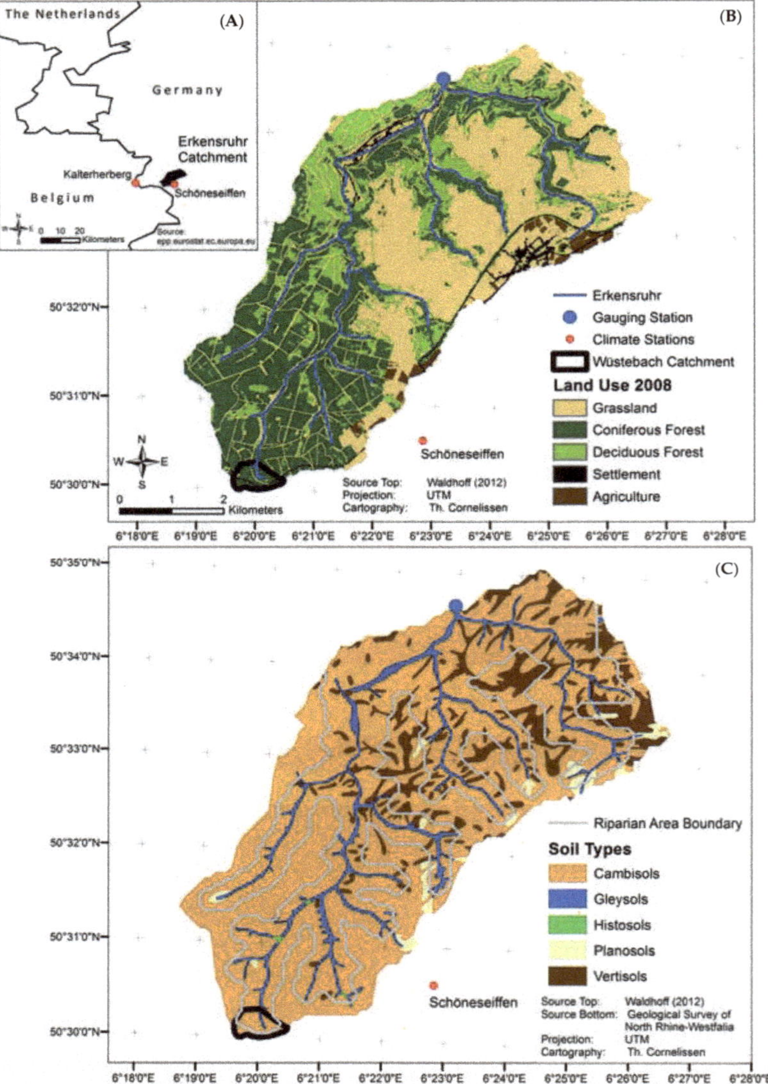

Figure 1. Location of the Erkensruhr catchment and climate stations (**A**); land use (**B**) and soil type distribution in the Erkensruhr catchment (**C**). The bottom map also illustrates the border between the slope of the hill and the riparian area (refer to Chapter 2.3).

Cambisols are the dominant soil type in the Erkensruhr catchment, whereas river valleys are dominated by Gleysols and Planosols (Figure 1). Silt loam is the dominant soil texture in the first soil layer (mean depth: 0.8 m) and clay in the second layer (mean depth: 1.8 m). In the second layer, Cambisols, Gleysols and Vertisols exhibit an high skeleton content of at least 66% with a maximum skeleton content of 90%, whereas Planosols have a mean skeleton content of only 10%. Native rocks are Devonian clayshales with sandstone intrusions and filled fractures [21].

The Erkensruhr catchment is dominated by coniferous forest (mainly Picea abies) in the southern part; and deciduous forest (mainly Fagus sylvatica) in the north-western part. Grassland and pasture predominantly occur in the eastern and central parts of the catchment (Figure 1). Arable land and urban areas cover only less than 3% of the catchment area.

The Wüstebach headwater catchment is located at the southern border of the Erkensruhr catchment. It is 0.385 km^2 large [22] with a mean annual precipitation of 1220 mm (1979–1999 [23]) and has been completely covered with Norwegian spruce (Picea Abies [24]) since 1950. The catchment is heavily monitored due to hydrological fluxes and states, as well as transportation of matter [25].

2.2. Model Description

In our study we applied the fully coupled surface-subsurface flow model HydroGeoSphere (HGS; [1,16,26]) in the parallel mode [27]. HGS solves the 3D Richards equation for subsurface flow and the 2D wave approximations of the Saint Venant equation for surface flow. The simulation of interception and evapotranspiration follows the approach of Kristensen and Jensen [28]. Interception is modeled with a bucket approach, where precipitation reaches the ground when the precipitation rate exceeds the maximum interception storage and its evaporation. Interception storage is emptied prior to other evapotranspiration processes.

The transpiration rate depends (1) on LAI according to a linear correlation function; (2) on a root distribution function which distributes root extraction among the root zone confined by the maximum root depth; (3) on the difference between potential and canopy evapotranspiration and (4) nonlinearly on the current soil moisture (see [26] for a detailed description).

The impact of the LAI on transpiration depends on two fitting parameters. With the chosen LAI and fitting parameters for the Wüstebach catchment, the function is compatible.

The maximum root depth and the root extraction function are sensitive parameters in term of soil moisture simulation. Quadratic and cubic root distribution functions are used in the context of this study.

Transpiration nonlinearly depends on soil moisture according to the following rules: the transpiration is zero for soil moistures (θ) below the wilting point (θ_{wp}) and beyond the anoxic limit (θ_{an}). Between the wilting point (θ_{wp}) and the field capacity (θ_{fc}), as between the oxic (θ_o) and anoxic limits (θ_{an}), the transpiration increases to the potential rate depending on a dimensionless fitting parameter. Between the field capacity (θ_{fc}) and the oxic limit (θ_o), the actual transpiration occurs at the potential rate. The oxic (θ_o) and anoxic limits (θ_{an}) were the most sensitive parameters in the Wüstebach study in terms of simulated transpiration amount [15].

To distinguish between different runoff components, the hydraulic mixing cell method was applied to HGS [29]. This method extracts discharge components from flux and storage information of the model for rectangular cells. Currently, the method distinguishes between (1) baseflow to the stream and to overland areas (return flow) and (2) direct rainfall input into the stream and onto overland areas. As the baseflow filter does currently not support gridded input data, it was only applied for the simulations of the Wüstebach headwater catchment.

In this study, the dual node approach for the implementation of the surface domain was used. This means that the 2D surface flow domain follows the uppermost node layer of the subsurface domain. Flow equations of both domains are coupled via an interconnection term describing leakage through an artificial skin layer on top of the uppermost 3D subsurface nodes. The subsurface domain was discretized using 3D-prisms.

2.3. Conceptual Model

In order to capture the variation in catchment slope (refer to Chapter 2.1), the Erkensruhr mesh consisted of two zones with different grid spacing. The riparian zone contained those catchment parts that either have a slope of more than 15° or are in a maximum distance of 200 m to the river. The riparian zone and the slope of the hill zone were discretized using 100 m and 200 m spacing, respectively. The Wüstebach catchment was fully discretized using 100 m spacing to facilitate the comparison with the independent Wüstebach simulation of Cornelissen *et al.* [15] who used the same spacing. The subsurface domain was 2 meters deep and was resolved in 28 numerical layers with increasing thickness in increasing distance to the infiltration zone.

Simulations of the Erkensruhr and the Wüstebach catchment started with a spin-up period of half a year. Initial conditions were set equal to the results of a 20 year warm-up run for the Wüstebach and a 10 year warm-up run for the Erkensruhr. At the catchment outlet, the critical depth boundary condition was assigned while all other boundaries were no flow boundaries.

2.4. Data Base

Table 1 summarizes spatial and temporal input data used in both simulations and data only used in either the Wüstebach or the Erkensruhr simulations.

The term mesoscale is used in the context of this study to refer to catchments with a size larger than 10 km^2 and to distinguish the low resolution data set of the Erkensruhr catchment (e.g., soil data at 1:50,000) from the high resolution data set of the headwater catchment Wüstebach (e.g., soil data at 1:2500).

For the calculation of **potential evapotranspiration** (PET hereafter), the climate station at Schöneseiffen (at 610 m.a.s.l.; refer to Figure 1 for its location) was used for all simulations of both catchments. Due to the high correlation between altitude and temperature (R^2: 0.95) with a mean temperature gradient of 0.695 °C/100 m calculated with measured data from 5 stations near the catchment, it was necessary to account for a spatial variability in PET in the Erkensruhr catchment. The climate data of the reference climate station Schöneseiffen were arbitrarily defined to be valid for the 50 m above and below the station height, leading to the definition of the following altitude layers: ⩽360, 360–460, 460–560, >560 m.a.s.l. To compute the FAO-Penman-Monteith PET [30], a new weighted albedo value according to the fraction of land use was defined for each altitude layer. The albedo values were taken from Breuer *et al.* [31]. This method resulted in an increase in PET (between the lowest and the highest altitude class) of 113 mm in 2010 and 126 mm in 2011. The total amount of PET did not change compared to the homogeneous PET because 80% of the catchment has an altitude above 460 m.

Precipitation data for simulations with homogeneous precipitation were taken from the station of Kalterherberg (9.6 km west of the catchment; refer to Figure 1 for its location) and corrected according to the method of Richter [32]. The simulation of the Erkensruhr catchment with distributed precipitation used ground-truth corrected precipitation radar data with a spatial resolution of 1 × 1 km^2 and a temporal resolution of 5 min provided by the local water-authority (Wasserverband Eifel-Rur). A snow model following the degree-day method [33] was applied to both precipitation data sets at hourly time steps and subsequently aggregated to daily time steps.

Table 1. Data sources, availability, spatial and temporal resolution and measurement location of input data used in the Wüstebach and Erkensruhr simulations.

Data Type	Source	Spatial Resolution	Temporal Resolution	Availability	Measurement Location
		Wüstebach and Erkensruhr simulations			
Digital Elevation	Land Surveying Office of North Rhine-Westphalia	10 × 10 m^2	-	-	-
Climate	TERENO Observation Network	1 Station	Hourly	Since 2009	Schöneseiffen (3.4 km east to Wüstebach)
		Wüstebach simulations			
Soil	Geological Survey of North Rhine-Westphalia	1:2500	-	-	-
Precipitation	German Weather Service	1 Station	Hourly	Since 2001	Kalterherberg (9.6 km west to Wüstebach)
		Erkensruhr simulations			
Soil	Geological Survey of North Rhine-Westphalia	1:50,000	-	-	-
Precipitation (Radar Data)	Wasserverband Eifel-Rur	1 × 1 km^2	5 min	Since 2002	-
Land Use	[34]	15 × 15 m^2	-	Since 2008	-

2.5. Parameterization and Calibration

In HGS the nonlinear relationship between soil suction and soil moisture is described by the van-Genuchten-Mualem (VGM) model [35]. The VGM parameters were derived from soil texture and bulk density using the pedotransfer function of Rawls and Brakensiek [36] and if applicable, the function of Brakensiek and Rawls [37] to account for skeleton content. We assumed a litter layer of 5 cm thickness using VGM parameters as suggested by Bogena *et al.* [38] at both catchments. In the context of this study, the litter layer is defined as the layer overlaying the actual soil horizons with a mixture of differently decomposed foliage. In the case of the Erkensruhr catchment, saturated conductivity values were directly taken from the soil map. At the Wüstebach, they were calculated with the pedotransfer function of Brakensiek *et al.* 1986 [39] including the influence of rock fragments on conductivity. At the Erkensruhr, the soil map does not show any skeleton content in the first layer. These differences in methodology resulted in large deviations in conductivity values (refer to Tables S1 and S2). For the simulations of the Wüstebach catchment (independently of the Erkensruhr), the VGM parameters were calculated for a model resolution of 25 m. To investigate the effects of spatial aggregation of soil parameter [15,40], mean VGM parameters were calculated for the 100 m setups used in this study. Parameters used for these setups are given in Tables S1 and S2 as area averaged mean values for the 5 soil types of the Erkensruhr catchment.

Mean monthly LAI values for agriculture, grassland and deciduous broadleaf forests were computed as an arithmetic mean of the 8-day LAI values between 2003 and 2013 from the Moderate Resolution Imaging Spectroradiometer (MODIS) with a spatial resolution of 1 km^2.

Meinen *et al.* [41] and Dannowski and Wurbs [42] report the distribution of root biomass with soil depth for Fagus sylvatica and the extensive grassland. HGS offers 4 different root distribution functions: constant, linear, quadratic and cubic decay. A quadratic decay function was fitted (R^2: 0.99) to the data for deciduous forest and a cubic decay function was fitted (R^2: 0.90) to the data for extensive grassland.

A **calibration** to discharge measurements of the Erkensruhr catchment was not performed in this study because calibration of a distributed model to an aggregated value (in this case discharge) leads to equifinality in the model parameters [43]. In addition, the process-based model HGS does not contain any empirical values related to discharge calculation. Instead, calibrated evapotranspiration parameters from the Wüstebach were used. The oxic and anoxic transpiration limits of HGS which influence the dependency between transpiration rate and soil moisture have been calibrated in Cornelissen *et al.* [15] to match soil moisture dynamics and an actual evapotranspiration of 40% of catchment rainfall. Soil moisture [44] and actual evapotranspiration [14] was measured at the coniferous forest of the Wüstebach catchment. As measured actual evapotranspiration data from a nearby grassland site (Marius Schmidt, personal communication) suggest an actual evapotranspiration of 60% of local precipitation rates, the oxic and anoxic transpiration limits for grassland areas were set to a value which allows transpiration to be unlimited, if the saturation exceeds the field capacity.

Table 2 lists all vegetation parameters used in the simulation together with their sources. The term "transferred" used in the table means that the parameter was equal to the calibrated parameters for coniferous forest of the Wüstebach catchment.

Table 2. Land use parameters used in the Erkensruhr simulation study.

Parameter \ Land Use Class	Coniferous	Deciduous	Grassland	Agriculture	Urban
Fraction of land use type (%)	40	19	38	2	1
Mean annual LAI (-)	6.7 [1]	1.93 [2]	1.51 [2]	1.16 [2]	25.5 [2]
Evaporation depth (m)	0.2 [1]	Transferred			
Root depth (m)	0.5 [1]	1.8 [3]	0.35 [4]	1.0 [5]	Deactivation
Root and evaporation distribution function (-)	Quadratic [1]	Quadratic [6]	Cubic [4]	Quadratic [5]	Deactivation
Transpiration fitting parameters (-)	0.3 [1],0.2 [1], 1.0 [1]	Transferred			Deactivation
Transpiration limiting saturations (Wilting point, Field capacity, Oxic, Anoxic) (-)	0.3 [1], 0.4 [1], 0.89 [7], 0.97 [7]	Transferred	0.3 [1], 0.4 [1], 1.0, 1.0	Transferred	
Canopy storage (mm)	0.8	0.83 [8]	1.0 [8]	2.5 [8]	15.0 [5]
Evaporation limiting saturations (min, max) (-)	0.3 [1], 0.4 [1]	Transferred			

[1]: [40]; [2]: MODIS data; [3]: [31] (Fagus Sylvatica on deep loam in Germany); [4]: Values for extensive grassland by Dannowski and Wurbs [42]; [5]: Assumption; [6]: [41]; [7]: [15]; [8]: Mean interception capacities for grassland (1.5 mm), agriculture (2.9 mm) and Fagus sylvatica (1.6 mm) according to Breuer *et al.* [31] and Mendel [45] were divided by corresponding mean LAI (1.51 for grassland, 1.16 for agriculture and 1.93 for Fagus sylvatica) according to MODIS data.

2.6. Modeling Procedure

In this study we compare ten model scenarios with different parameterizations: six scenarios of the Wüstebach headwater catchment and four scenarios of the Erkensruhr catchment with unique combinations of soil and land use parameters (refer to Table 3 for an overview).

For the first set of model setups, the coniferous forest of the Wüstebach reference setup (Wbach) was changed to deciduous forest (WbachDeci) and grassland (WbachGrass) while keeping all other inputs constant. For the second set of model setups, the mesoscale soil parameters of the Erkensruhr were applied to the Wüstebach simulation using the three different land use parameterizations for coniferous (WbachEsoilConi), deciduous (WbachEsoilDeci) and grassland (WbachEsoilGrass).

For the base setup of the Erkensruhr catchment (Erk) we used distributed soil data but assumed homogeneous land use, PET and precipitation. Land use, PET and precipitation were set equal to the Wüstebach catchment. Spatial heterogeneity of land use, PET and precipitation (in the form of radar data) was introduced step-wise into the Erkensruhr setup leading to three additional simulations: Erk_LN, Erk_LN_PET and Erk_LN_PET_P. This procedure means that the simulation Erk_LN still uses PET and precipitation equal to that of the Wüstebach catchment, but soil and land use data originated from the Erkensruhr catchment. In the Erk_LN_PET simulation, only precipitation is equal to the Wüstebach. In the Erk_LN_PET_P simulation, all inputs originate from the Erkensruhr. Additionally, a simulation run using spatially mean radar data was accomplished to separate possible effects of the precipitation pattern from the effects of changes in precipitation amount (results not shown).

Table 3. Summary of abbreviations of conducted simulations with applied soil and land use data.

Simulation Scenarios	Soil Data	Land Use	Additional Information
		Wüstebach Catchment	
Wbach	Wüstebach	Coniferous	Reference scenario
WbachDeci	Wüstebach	Deciduous	
WbachGrass	Wüstebach	Grassland	
WbachEsoilConi	Erkensruhr	Coniferous	
WbachEsoilDeci	Erkensruhr	Deciduous	
WbachEsoilGrass	Erkensruhr	Grassland	
		Erkensruhr Catchment	
Erk	Erkensruhr	Coniferous	
Erk_LN	Erkensruhr	All	Distributed land use
Erk_LN_PET	Erkensruhr	All	Distributed land use and potential evapotranspiration
Erk_LN_PET_P	Erkensruhr	All	Distributed land use, potential evapotranspiration and precipitation

The reference model setup for the Wüstebach simulations corresponds to calibrated and validated 100 m setup without bedrock inclusion used in Cornelissen *et al.* [15]. Please note that differences between model results reported in Cornelissen *et al.* [15] and those reported in this study resulted from the usage of a new version of HGS that corrected a bug in the interception module and differences in residual saturation and porosities. In Cornelissen *et al.* [15] residual saturations and porosities were calibrated to measured soil moistures. The calibration resulted in an increase in residual saturations and porosities approximately equal to the correction for skeleton content according to Brakensiek and Rawls [37]. Due to the fact that no skeleton content was reported in the first layer in the Erkensruhr soil data, we ran the simulations Wbach, WbachDeci and WbachGrass (refer to Table 3) with uncalibrated residual saturations and porosities. We acknowledge that the usage of different residual saturations and porosities slightly changed the impact of transpiration parameters on actual evapotranspiration and soil moisture simulation.

2.7. Measures of Model Performance

We applied three statistical measures to quantify the quality of discharge simulations (also refer to Equations (1) and (2)): (1) the bias, defined as the ratio between simulated and observed mean discharge of a given time period; (2) the coefficient of variation measuring the agreement between observed and simulated distributions of discharge values; (3) the R^2 (squared form of Pearson's correlation coefficient) as an indicator of linear correlation. All values were calculated separately for the hydrological summer and winter periods due to substantial performance differences. As all measures reached their optimum value at unity, numerals larger or smaller than 1.0 given in Figures 3 and 8 can be interpreted as percentage deviations from the optimum.

$$Bias = \frac{\mu\left(Q_{sim}\right)}{\mu\left(Q_{mes}\right)} \tag{1}$$

$$Coefficient\ of\ Variation = \frac{\frac{\sigma(Q_{sim})}{\mu(Q_{sim})}}{\frac{\sigma(Q_{mes})}{\mu(Q_{mes})}} \tag{2}$$

with $\mu\left(Q_{sim}\right)$ and $\mu\left(Q_{mes}\right)$ being simulated and observed mean discharge and $\sigma\left(Q_{sim}\right)$ and $\sigma\left(Q_{mes}\right)$ being simulated and observed standard deviation. Following Moriasi *et al.* [20], we rate values of all three measures larger than 0.9 and smaller than 1.1 as "very good", values between 0.9 (1.1) and 0.85 (1.15) as "good", values between 0.85 (1.15) and 0.75 (1.25) as "satisfactory" and all other values as unsatisfactory.

3. Results

3.1. Influence of Mesoscale Soil and Land Use Parameterization on the Simulation of the Headwater Catchment

Figure 2A shows measured and simulated discharge rates for the two simulations Wbach and WbachEsoilConi for the years 2010 and 2011. Observed discharge was characterized by a strong seasonality with a pronounced low flow period during the summer and high variability during snow dominated periods in the winter. Generally, both simulation scenarios reproduced the discharge dynamics well but overestimated peaks during the winter (due to an overestimation of snow melt by the snow model) and omitted some peaks during the summer. The usage of mesoscale soil data from the Erkensruhr (model scenario WbachEsoilConi) intensified the tendency to overestimate peak discharge rates.

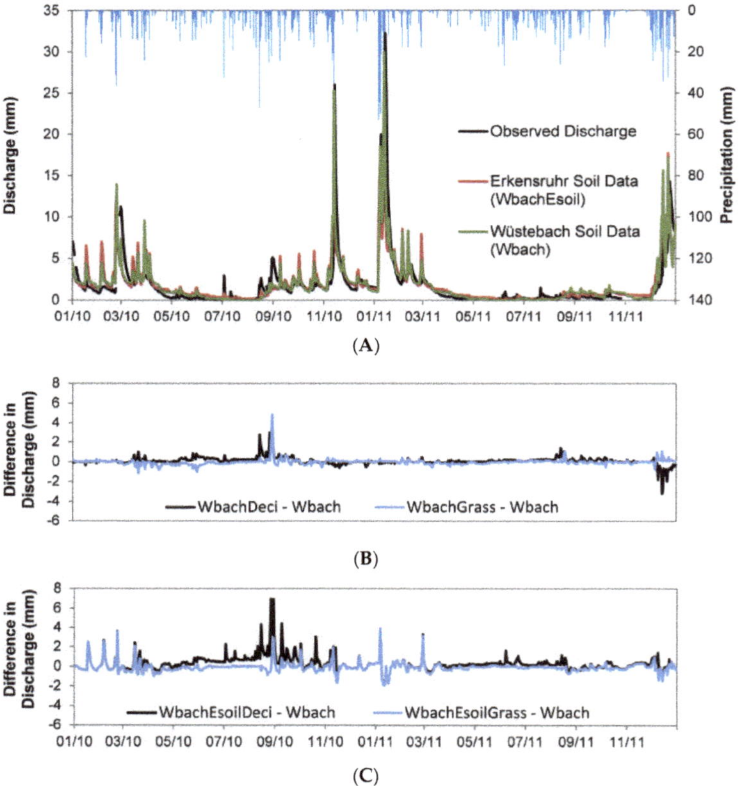

Figure 2. (**A**) Comparison of observed and simulated discharge of the Wüstebach for simulations with high-resolution soil data (Wbach) and low-resolution soil data (WbachEsoilConi); (**B**) Discharge difference between simulations with changing land use; (**C**) Discharge difference between simulations with changing land use and soil data.

Differences in discharge between the reference simulation Wbach and the simulations WbachDeci and WbachGrass (Figure 2B) were smaller than ± 0.5 mm for more than 90% of the simulation period indicating a weak sensitivity of discharge to changes in land use parameterization. Higher discharge rates of WbachDeci and WbachGrass in late summer 2010 resulted from differences in LAI development and corresponding changes in interception. At the end of 2011, differences in discharge resulted from differences in soil moisture. The WbachDeci simulation had lower soil moistures in all depths than the Wbach simulation and therefore rainfall was primarily replenishing the water storage. The WbachGrass simulation had highest soil moisture at the same time and accordingly, highest discharge rates. In Figure 2C, differences between Wbach and the WbachEsoilDeci and WbachEsoilGrass simulations are shown. Both simulations produced higher discharge rates during both years with an extreme overestimation during 2010 of the WbachEsoilDeci model scenario.

Figure 3 summarizes statistical measures of model performance for the hydrological winter 2010/2011 and—as a mean value—for the hydrological summer periods in 2010 and 2011.

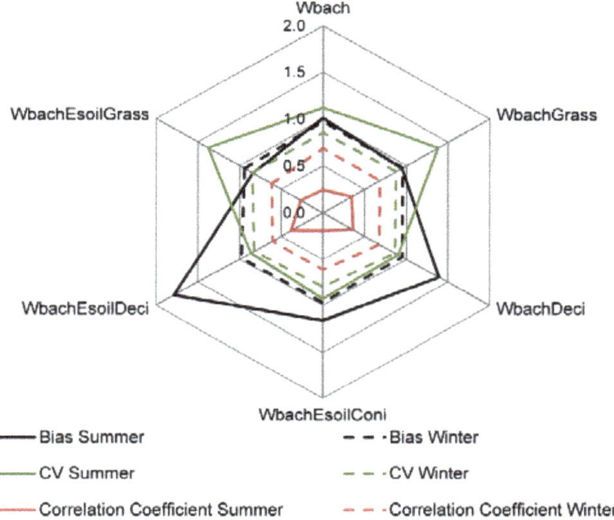

Figure 3. Bias (black line), coefficient of variation (CV; green line) and correlation coefficient (red line) in hydrological summer (solid lines) and winter (dashed lines) for the Wüstebach discharge simulations.

All statistical measures varied more strongly between simulations during summer than during winter because (1) differences in evapotranspiration simulation only became apparent during summer and (2) small changes in discharge amount and timing had a high impact on statistical measures during the low flow period.

During winter, all model scenarios produced unsatisfactory R^2 values (0.61–0.68) but very good bias values (0.94–0.98); the coefficient of variation was lower than unity for all simulations due to the underestimation of discharge variability during winter.

Changing land use primarily affected the coefficient of variation during the hydrological summer with increases for grassland and decreases for deciduous forest. In contrast, a change in soil data heavily influenced the bias and the R^2. The unique behavior of the simulation WbachEsoilDeci in terms of very high increases in bias and R^2 compared to WbachDeci has already been mentioned in the previous paragraph. The reason for this increase in bias will be further analyzed in this chapter and in the discussion section.

The water balance of the Wüstebach simulations (Table 4) showed some interesting features concerning evapotranspiration components and infiltration sums. The total amount of actual evapotranspiration significantly changed between different land uses with highest values for WbachGrass due to the changes in transpiration parameters. In 2010, the amount of actual evapotranspiration for the WbachDeci simulation equaled that of Wbach but in 2011 the evapotranspiration was higher by

50 mm. Infiltration sums and fractions of subsurface flow varied between the years but not between simulation variants using the same soil data.

Table 4. Water balance components for Wüstebach simulations.

Water Balance Setup Component \ Simulation	Wbach	Wbach Deci	Wbach Grass	Wbach Esoil Coni	Wbach Esoil Deci	Wbach Esoil Grass
				2010		
Rainfall (mm)				1226		
Potential ET (mm)				694		
Measured Discharge (mm)				608		
Transpiration (mm)	232	227	279	195	99	282
Evaporation (mm)	247	254	289	247	256	293
Actual Evapotranspiration (mm)	479	481	568	442	355	575
Discharge [1] (mm)	611	657	591	647	764	587
Baseflow (%)	76	76	75	64	62	63
Infiltration (mm)	968	992	1011	891	879	954
				2011		
Rainfall (mm)				1348		
Potential ET (mm)				756		
Measured Discharge (mm)				630		
Transpiration (mm)	272	312	290	247	250	306
Evaporation (mm)	273	283	314	273	289	325
Actual Evapotranspiration (mm)	545	595	604	520	539	631
Discharge [1] (mm)	637	640	626	652	673	594
Baseflow (%)	62	64	60	56	58	53
Infiltration (mm)	894	959	960	832	870	896

[1]: Yearly sums of simulated discharge exclude time steps with gaps in measured discharge data.

Comparing simulations with high-resolution soil data of the Wüstebach to those with mesoscale Erkensruhr soil data, significant differences in the water balance components and in the fractions of subsurface flow became apparent. For both forested land uses, actual evapotranspiration decreased by 37 mm (2010) and 25 mm (2011) for coniferous and by 126 mm (2010) and 56 mm (2011) for the deciduous forest. The decrease in evapotranspiration resulted from a decrease in infiltration sums by 77 mm (2010) and 62 mm (2011) for coniferous and by 113 mm (2010) and 89 mm (2011) for the deciduous forest. Despite the decrease in infiltration sums, discharge sums were much higher and as a result, the fraction of subsurface flow decreased by 12%–14% in 2010 and 6%–7% in 2011. The effect described above was stronger in 2010 than in 2011. In April and May 2010 precipitation rates were larger than PET rates but in April and May 2011 precipitation rates were lower. The surplus in PET significantly reduced soil moisture in 2011 thus dampening the effect of mesoscale soil data on runoff generation processes.

In contrast to the forest land uses, the WbachEsoilGrass scenario showed small changes in total evapotranspiration ($\leqslant$27 mm) and correspondingly lowest variations in infiltration sums. The deviations in the water balance during 2010 and 2011 arose from intense rainfall rates during December in both years.

Comparing water balance results of the setups Wbach and WbachEsoilConi to measured water balance components mentioned in Cornelissen *et al.* [15], the following observations can be stated: (1) Simulated discharge amounts match well with measured discharge rates for the Wbach scenario in both years; the WbachEsoilConi overestimated discharge by 20 mm (2010) and 40 mm (2011); (2) The estimated fraction of evaporation (20% of precipitation) matches very well to simulated fractions; (3) The amounts of actual evapotranspiration are largely underestimated in 2011 by 50 mm (Wbach) and 80 mm (WbachEsoilConi) due to low transpiration rates.

In the context of this paper, **soil moisture simulation results** are compared between simulations but not to measurements. For a detailed comparison between simulated and measured soil moisture of the Wüstebach catchment, the reader is referred to Cornelissen *et al.* (2014) [15].

We noted pronounced differences in simulated soil moisture dynamics between land use types at all depths. At 5 cm depth, differences were most pronounced during August and July 2010 when the WbachDeci simulation maintained soil moisture values above 0.5 while soil moisture for both the Wbach and the WbachGrass simulations dropped below 0.3. In August and July 2011, the WbachDeci simulation was again the wettest but differences to Wbach and WbachGrass were smaller. The Wbach and WbachGrass scenarios showed small differences at 5 cm depth because their root depth (refer to Table 2) was comparably small with 0.5 m and 0.35 m respectively. At 20 cm depth (Figure 4) the WbachDeci scenario produced the lowest soil moisture in both years. During July and August of both years, WbachGrass and Wbach maintained soil moisture values of about 0.6 while WbachDeci dropped below 0.4 in 2011. In both years, the WbachGrass scenario produced the highest soil moisture. At 50 cm depth a clear hierarchy following root depths was found in both years with the highest moistures for WbachGrass (featuring the lowest root depth) and lowest values for WbachDeci (featuring the highest root depth).

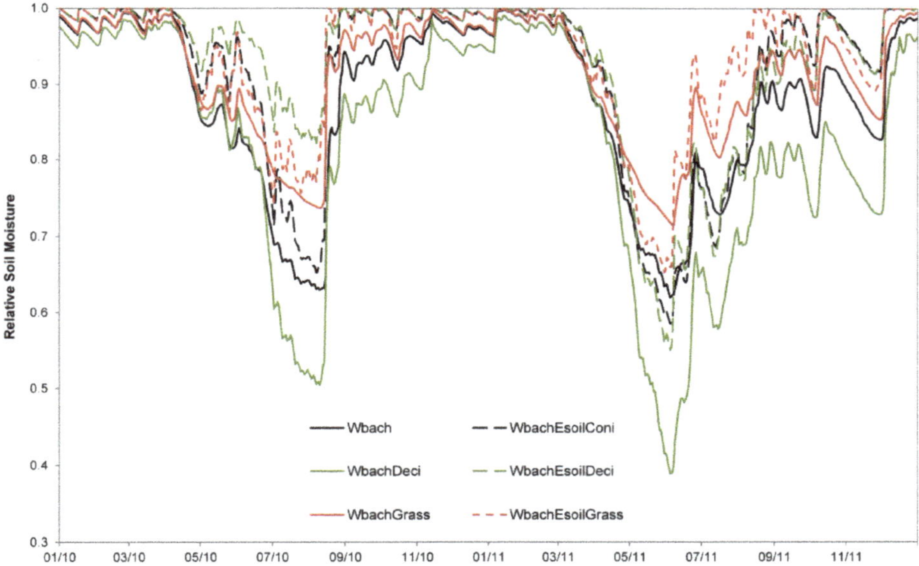

Figure 4. Soil moisture dynamics of the Wüstebach simulations at 20 cm depth.

The usage of mesoscale soil data generally increased soil wetness and intensified short term soil moisture dynamics down to 50 cm depth. Differences were again most pronounced for the simulation with deciduous land use. The increased soil moisture dynamic with coarser soil data led to a decrease in infiltration and transpiration with a corresponding increase in discharge.

The relationship between mean soil moisture and its standard deviation ($\sigma_\theta(\langle\theta\rangle)$) showed little variations between different land use types at 5 cm depth (Figure 5). Simulations with Erkensruhr soil data produced a steeper slope with higher standard deviations at the same moisture. This is attributed to the fact that the VGM parameters of the model setups used in this study were aggregated from a model resolution of 25 m ([15]; also refer to Chapter 2.5). As demonstrated recently by Qu *et al.* [46] the shape of $\sigma_\theta(\langle\theta\rangle)$ can be explained to a large extent by the spatial variance of soil hydraulic properties.

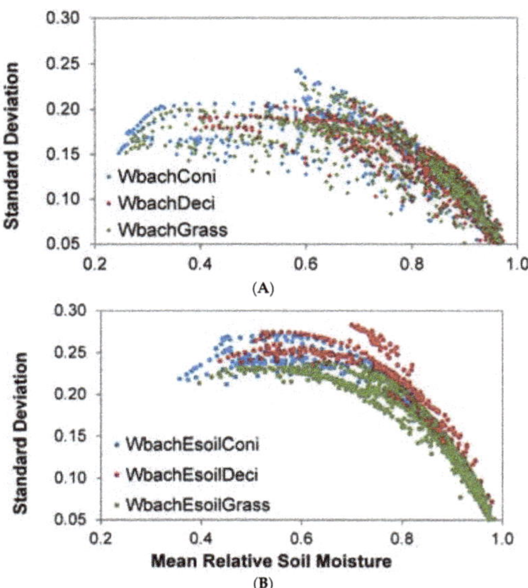

Figure 5. Relationship between mean soil moisture and its standard deviation for Wüstebach simulations at 5 cm depth with (**A**) Wüstebach and (**B**) Erkensruhr soil data.

3.2. Influence of Parameter Regionalization and Spatially Distributed Input Data on the Simulation of the Mesoscale Catchment

In the following, the results of the four Erkensruhr simulations are analyzed separately for the whole Erkensruhr catchment and for the Wüstebach sub-catchment. Water balance results were only available for the Erkensruhr as HGS does not enable the export of water balance results for sub-catchments.

The three Erkensruhr simulations with homogeneous rainfall (Figure 6) heavily overestimated discharge amounts, especially during autumn because the applied rainfall originated from a climate station located in the southwestern—and thus wettest—part of the catchment. The usage of distributed precipitation substantially improved the discharge simulation of the Erkensruhr in terms of total sum, rising and falling limbs and low flows (Figure 7). However the discharge peaks were underestimated, possibly because the same interception and transpiration parameterization was used for a different precipitation input data set. We further found that using spatially aggregated instead of distributed radar precipitation data, produced equal simulation results in terms of water balance and discharge (not shown).

The overestimation of simulated discharge amounts at the Erkensruhr outlets caused bias values around 1.6 for the Erk model scenario during summer (Figure 8).

The R^2 values of the Erkensruhr simulations were considerably higher during winter (0.86) than during summer (0.22). As the R^2 during winter was higher for the Erkensruhr simulations than for the independent Wüstebach simulations (refer to Figure 3), we assumed that the snow model used in both simulations performed better for the smoother discharge curve of the mesoscale catchment. A plausible reason for this is a smoothing effect on winter discharges due to the larger catchment size.

The usage of distributed precipitation data mainly improved the bias during winter. During summer, distributed precipitation rates caused the bias to change from a 50% overestimation (Erk_LN_PET) to a 50% underestimation.

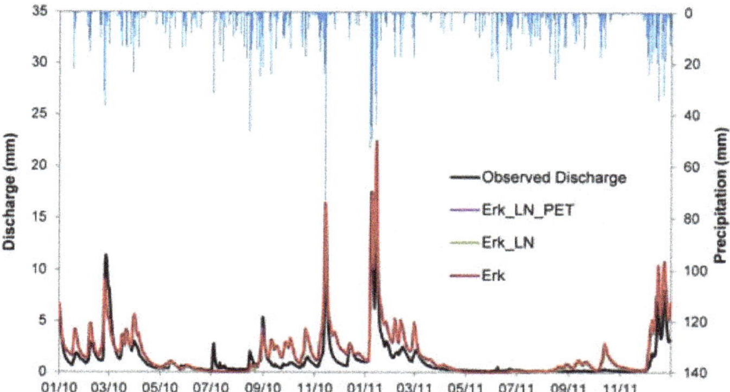

Figure 6. Observed and simulated discharge of the Erkensruhr for simulations with heterogeneous soil data (Erk), heterogeneous soil and land use data (Erk_LN), heterogeneous soil, land use and potential evapotranspiration data (Erk_LN_PET).

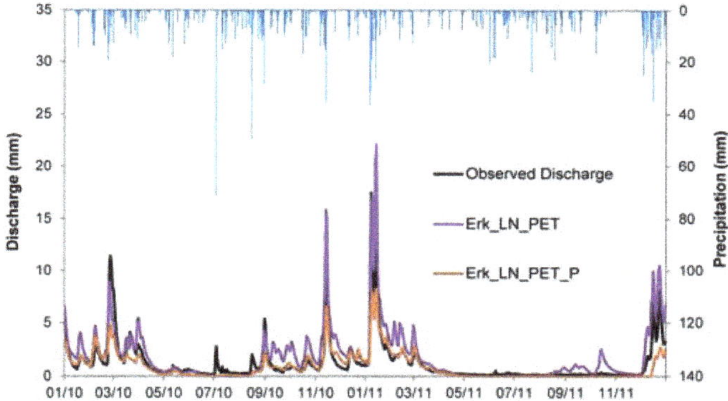

Figure 7. Observed and simulated discharge of the Erkensruhr for simulations with homogeneous (Erk_LN_PET) and distributed precipitation (Erk_LN_PET_P).

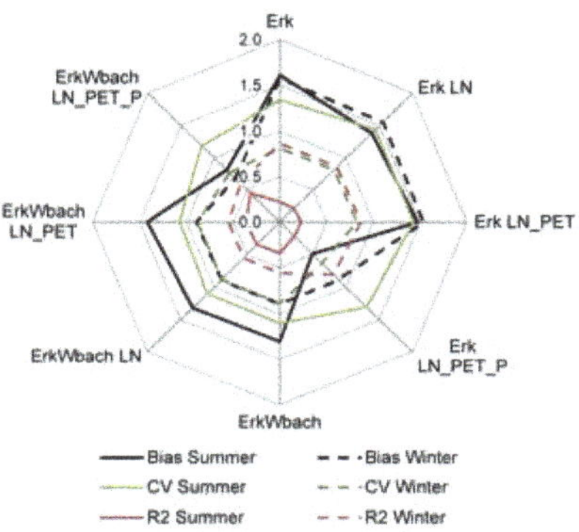

Figure 8. Bias (black line), coefficient of variation (CV; green line) and correlation coefficient (red line) in hydrological summer (solid lines) and winter (dashed lines) for the Erkensruhr discharge simulations.

Evaporation amounts of the Erk simulation which considered spatially homogeneous coniferous land use throughout the catchment were slightly lower (by 15 mm) than that of Wbach and WbachEsoilConi, with the same land use type (refer to Tables 4 and 5). The consideration of heterogeneous land use in the Erk_LN scenario slightly increased evaporation by 13 mm in 2010 and 17 mm in 2011. As already mentioned, the mesoscale soil data decreased simulated transpiration and infiltration amounts in the Wüstebach simulations independent of land use type. However transpiration of the Erk scenario was equal to that of the Wbach scenario and infiltration slightly increased.

Table 5. Water balance components for simulations of the Erkensruhr catchment.

Water Balance Setup Component \ Simulation	Erk		Erk_LN		Erk_LN_PET		Erk_LN_PET_P	
	2010	2011	2010	2011	2010	2011	2010	2011
Rainfall (mm)	1226	1348	1226	1348	1226	1348	956	902
Potential ET (mm)	694	756	694	756	694	757	694	757
Measured Discharge (mm)	524	396	524	396	524	396	524	396
Transpiration (mm)	226	272	268	286	260	305	283	332
Evaporation (mm)	265	289	278	306	288	312	265	267
Actual Evapotranspiration (mm)	491	561	546	592	548	617	548	599
Discharge (mm)	721	654	696	623	692	619	391	245
Infiltration (mm)	996	954	1024	980	1016	976	771	683

The total evapotranspiration amount increased when heterogeneous land use information was used compared to the Erk setup. The consideration of distributed PET increased total evapotranspiration in 2011 by 25 mm with increases in both evapotranspiration components.

Reasons for the deviations in the water balance of the simulations Erk, Erk_LN, Erk_LN_PET are comparable to the Wüstebach simulations that have already been explained in Chapter 4.1.

Figure 9 shows measured and simulated monthly deviations from mean annual evapotranspiration rates for coniferous (Figure 9A), grassland (Figure 9B) and deciduous (Figure 9C) vegetation. The simulated values were compared with measured eddy-covariance data in the case of coniferous and grassland vegetation; and to literature values from Mendel [45] in the case of deciduous vegetation.

For coniferous and grassland vegetation, the trend in mean monthly evapotranspiration was well simulated with R^2 values larger than 0.9. In the case of coniferous vegetation, the monthly evapotranspiration was overestimated between April and July and in December while it was underestimated in August, September and February. Distributed precipitation rates improved the simulation in July while the simulation Erk_LN_PET was the most unfavorable simulation meaning that PET was slightly underestimated for coniferous land use when the distribution method was used. Simulation of the evapotranspiration of the grassland vegetation was best during the winter and worst during March and April. For deciduous vegetation, Figure 9C reveals largest deviations between simulated and measured data taken from literature [45].

Figure 10 shows the pattern of simulated mean actual evapotranspiration given as a relative value of the sum of evapotranspiration for 2010 and 2011. The pattern of the Erk scenario (Figure 10A), shows a clearly defined riparian and stream area with very high relative evapotranspiration values close to unity. Driest conditions were found at the ridge of hills at the eastern, western and southern borders of the catchment. The pattern shown in Figure 10B for the Erk_LN scenario, illustrates that the incorporation of heterogeneous land use enhanced evapotranspiration in the central part of the catchment covered with grassland. Distributed PET decreased actual evapotranspiration in higher parts of the catchments (e.g., south-western border). The incorporation of distributed precipitation generally decreased the contribution of grassland areas.

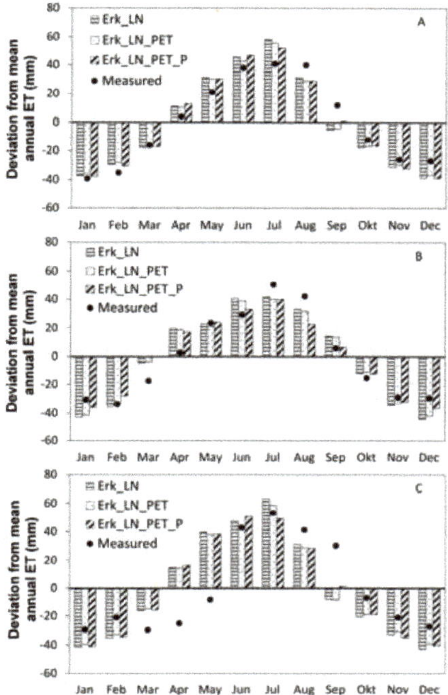

Figure 9. Measured and simulated monthly deviation from mean annual evapotranspiration rates for (**A**) coniferous; (**B**) grassland and (**C**) deciduous vegetation. Measured data refer to (**A**) eddy-covariance data from [14]; (**B**) eddy-covariance data from Schmidt (personal communication) and; (**C**) mean monthly data from a low mountain catchment in northern Germany between 1969 and 1972 with a mean rainfall of 1066 mm [45].

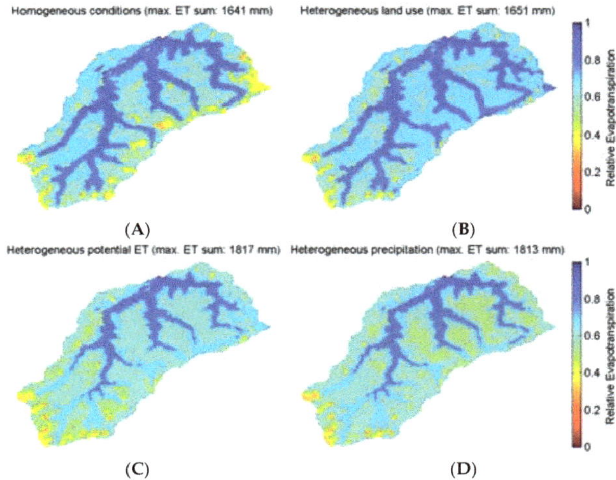

Figure 10. Pattern of actual evapotranspiration for simulations Erk (**A**), Erk_LN (**B**), Erk_LN_PET (**C**) and Erk_LN_PET_P (**D**). Evapotranspiration is given relative to the maximum of the actual evapotranspiration sums of 2010 and 2011 as specified in the brackets.

4. Discussion

4.1. Influence of Mesoscale Soil and Land Use Parameterization on the Simulation of the Headwater Catchment

Results of the Wüstebach simulations revealed a strong influence of (1) soil data on runoff generating processes and of (2) land use parameterization on evapotranspiration components.

The observed increase in fast runoff components due to an increase in saturated conductivity, led to a unique interplay between antecedent wetness, infiltration, groundwater level rise and transpiration parameterization which strongly depended on the difference between precipitation and PET during spring.

This observed high sensitivity of fast runoff sums and runoff generation mechanisms to changes in soil properties agreed well with the findings of many studies; but the result that higher saturated conductivity led to an increase in fast runoff components contradicted to results reported in the literature for simulations with distributed hydrological models [47–49].

The applicability of mesoscale land use parameterization was validated with a comparison between evapotranspiration amounts simulated with Wüstebach soil data (Table 4) and values reported in the literature.

Simulated evapotranspiration of grassland amounted to 46% of total catchment rainfall but literature values ranged between 36% of 800 mm precipitation for a measurement site in Germany [50] and 60% of total rainfall at a grassland site near to the Erkensruhr catchment (Schmidt, personal communication). Data cited in Mendel [45] ranged between 55% of 800 mm precipitation and 75% of only 581 mm precipitation).

Despite the lack of calibration, simulated interception fractions for the deciduous forest (17% of total precipitation) corresponded well to the value reported by Oishi *et al.* [51] for a hardwood forest in the United States with a precipitation of 1091 mm. Mendel [45] reported interception values between 5% and 48% of precipitation for a beech forest. Simulated fractions of evapotranspiration (without interception) amounted to 25% but literature values ranged between 26% [45] and 40% [51] of catchment rainfall.

The broad range of reported evapotranspiration components for both land uses pointed to considerable uncertainty in evapotranspiration validation thus expressing the need for additional land use specific evapotranspiration measurements.

4.2. Influence of Parameter Regionalization and Spatially Distributed Input Data on the Simulation of the Mesoscale Catchment

Erkensruhr simulations revealed that the quality of the discharge simulation in terms of dynamics, amount and peak flow rates, was most sensitive to differences in precipitation data. Spatially distributed land use parameterization only affected discharge amounts while spatially distributed PET had a weak effect on discharge but a significant effect on the pattern of actual evapotranspiration during winter (not shown).

An influence of spatial precipitation patterns on hydrological simulations has been long under debate. For example, Schuurmans and Bierkens [52] and Arnaud *et al.* [53] compared simulation results of distributed models using spatially distributed and spatially aggregated precipitation input. Schuurmans and Bierkens [52] found that spatial variability of rainfall is necessary to simulate spatial variability in daily discharge, groundwater level and soil moisture content but not required for aggregated measures like catchment mean groundwater table or water balance. Arnaud *et al.* [53] showed at two mesoscale catchments (103 and 22 km^2), that differences in simulated discharge amount and peak flow rates decreased with increasing peak flow rate. In our study, differences in discharge rates between simulations with aggregated and spatially distributed precipitation were therefore marginally supporting the results from Schuurmans and Bierkens [52], that spatially distributed precipitation is not required for the correct simulation of aggregated measures of hydrologic functioning, like the water balance. In contrast to Schuurmans and Bierkens [52], simulated evapotranspiration patterns were similar between simulations with aggregated and distributed precipitation, indicating that the spatial evapotranspiration simulation in HGS is more sensitive to parameterization than precipitation data. The high sensitivity of peak flow rates to precipitation sum and distribution highlights the extraordinary importance of meteorological forcing data in comparison to parameterization efforts.

Mean monthly evapotranspiration was well simulated at the mesoscale catchment for coniferous forest and grassland (R^2: 0.9) and to a lesser degree for deciduous vegetation (R^2: 0.58) as the data used were book values and therefore the most unreliable [45]. Interestingly, the simulated trend improved when values were re-shifted in positive direction by one month, giving an R^2 of 0.88 revealing a systematic error in MODIS LAI data and/or measured evapotranspiration data from Mendel [45]. The results indicated that usage of the same model-specific transpiration parameters for different land uses can be sufficient to reproduce monthly dynamics of evapotranspiration.

We acknowledge that the presented results originated from a study of two catchments with similar catchment properties. To allow for a more general evaluation of our parameter transfer approach, applications to further catchments with different catchment properties, e.g., in terms of topography, soil properties, geology, climate *etc.* will need to be made. Such an analysis would also help to define the limits of our model parameter transfer approach.

5. Conclusions

In this study, we dealt with one of the major limitations to the application of distributed hydrological models at mesoscale catchments: the commonly available data base at mesoscale catchments is insufficient for a thorough calibration and validation. We parameterized the distributed hydrological model HydroGeoSphere for a mesoscale catchment by transferring model-specific parameters calibrated for water balance and soil moisture dynamics at a small headwater catchment. To account for the spatial variability in land use at the mesoscale catchment, parameters that are not model-specific (e.g., LAI, root depth and distribution) were taken from the literature.

At the mesoscale, the effect of transferred parameterization was differentiated from the influence of spatial variability in soil, land use, potential evapotranspiration and precipitation by a step-wise introduction of their spatial distributions into the model setup. In addition, the possible range of model behavior was exploited by using mesoscale soil and land use data for the simulation of the headwater catchment.

Using mesoscale soil data from the mesoscale catchment in the headwater simulation revealed a model specific behavior, as higher saturated conductivities enhanced faster runoff components instead of reducing them, as reported in the literature [47–49]. We attributed this finding to a unique interplay between wetness prior to rainfall events during the summer, groundwater level rise and transpiration parameterization.

The mesoscale simulations with regionalized model-specific parameters reproduced discharge dynamics and evapotranspiration in a satisfying manner. Precipitation was the most sensitive input data set for discharge simulation, while spatially distributed land use parameterization had a much larger effect on evapotranspiration components and its pattern, than distributed precipitation or potential evapotranspiration.

Supplementary Materials: The following are available online at http://www.mdpi.com/2073-4441/8/5/202/s1, Table S1: Area weighted soil parameters at the Erkensruhr catchment. θs, θr, α and n are the porosity, residual saturation and the Van-Genuchten-Mualem shape parameters; Ks is the saturated conductivity, Table S2: Area weighted soil parameters at the Wüstebach catchment for the soil types of the Erkensruhr catchment. θs, θr, α and n are the porosity, residual saturation and the Van-Genuchten-Mualem shape parameters; Ks is the saturated conductivity.

Acknowledgments: The authors thank the Deutsche Forschungsgemeinschaft (DFG) for financial support of sub-project C1 of the Transregional Collaborative Research Centre 32 "Patterns in Soil-Vegetation-Atmosphere Systems" and TERENO (Terrestrial Environmental Observatories) funded by the Helmholtz-Gemeinschaft. We gratefully thank Daniel Partington of the University of Adelaide for providing us with the baseflow filter for HGS and for his intensive support. We also thank Christof Homann from the Wasserverband Eifel-Rur for providing the radar data. MODIS LAI data were taken from: Land Processes Distributed Active Archive Center (LP DAAC), 2003–2013, Leaf Area Index—Fraction of Photosynthetically Active Radiation 8-Day L4 Global 1 km (MOD15A2): NASA EOSDIS Land Processes DAAC, USGS Earth Resources Observation and Science (EROS) Center, Sioux Falls, South Dakota (https://lpdaac.usgs.gov), accessed on 17 April 2014. Measurement and processing of eddy covariance and precipitation data of the Rollesbroich test site were provided by the central service project Z3 of the DFG TR32.

Author Contributions: Thomas Cornelissen conducted the data analysis, performed model setup and simulations and wrote the paper. Bernd Diekkrüger and Heye Bogena substantially contributed to the study design, supported the model setup and did proofreading of the manuscript.

Conflicts of Interest: The authors declare no conflict of interest.

Abbreviations

The following abbreviations are used in this manuscript:

FAO	Food and Agriculture Organization of the United Nations
HGS	HydroGeoSphere
LAI	Leaf Area Index
MODIS	Moderate Resolution Imaging Spectroradiometer
PET	Potential Evapotranspiration
TERENO	Terrestrial Environmental Observatories
VGM	Van-Genuchten-Mualem

References

1. Panday, S.; Huyakorn, P.S. A fully coupled physically-based spatially-distributed model for evaluating surface/subsurface flow. *Adv. Water Resour.* **2004**, *27*, 361–382. [CrossRef]
2. Kollet, S.J.; Maxwell, R.M. Capturing the influence of groundwater dynamics on land surface processes using an integrated, distributed watershed model. *Water Resour. Res.* **2008**, *44*, W02402. [CrossRef]
3. Graham, D.N.; Butts, M.B. Flexible, integrated watershed modelling with MIKE SHE. In *Watershed Models*; Singh, V.P., Frevert, D.K., Eds.; CRC Press: Boca Raton, FL, USA, 2005; pp. 245–271.
4. Camporese, M.; Paniconi, C.; Putti, M.; Orlandini, S. Surface-subsurface flow modeling with path-based runoff routing, boundary condition-based coupling, and assimilation of multisource observation data. *Water Resour. Res.* **2010**, *46*, W02512. [CrossRef]
5. Ala-aho, P.; Rossi, P.M.; Isokangas, E.; Kløve, B. Fully integrated surface-subsurface flow modelling of groundwater–lake interaction in an esker aquifer: Model verification with stable isotopes and airborne thermal imaging. *J. Hydrol.* **2015**, *522*, 391–406. [CrossRef]
6. Frei, S.; Fleckenstein, J.H. Representing effects of micro-topography on runoff generation and sub-surface flow patterns by using superficial rill/depression storage height variations. *Environ. Model. Softw.* **2014**, *52*, 5–18. [CrossRef]
7. Li, Q.; Unger, A.J.A.; Sudicky, E.A.; Kassenaar, D.; Wexler, E.J.; Shikaze, S. Simulating the multi-seasonal response of a large-scale watershed with a 3D physically-based hydrologic model. *J. Hydrol.* **2008**, *357*, 317–336. [CrossRef]
8. Voeckler, H.M.; Allen, D.M.; Alila, Y. Modeling coupled surface water—Groundwater processes in a small mountainous headwater catchment. *J. Hydrol.* **2014**, *517*, 1089–1106. [CrossRef]
9. Weill, S.; Altissimo, M.; Cassiani, G.; Deiana, R.; Marani, M.; Putti, M. Saturated area dynamics and streamflow generation from coupled surface-subsurface simulations and field observations. *Adv. Water Resour.* **2013**, *59*, 196–208. [CrossRef]
10. Kirchner, J.W. Getting the right answers for the right reasons: Linking measurements, analyses, and models to advance the science of hydrology. *Water Resour. Res.* **2006**, *42*, W03S04. [CrossRef]

11. Liang, X.; Lettenmaier, D.P.; Wood, E.F.; Burges, S.J. A simple hydrologically based model of land surface water and energy fluxes for general circulation models. *J. Geophys. Res. Atmos.* **1994**, *99*, 14415–14428. [CrossRef]
12. Troy, T.J.; Wood, E.F.; Sheffield, J. An efficient calibration method for continental-scale land surface modeling. *Water Resour. Res.* **2008**, *44*, W09411. [CrossRef]
13. Romano, N. Soil moisture at local scale: Measurements and simulations. *J. Hydrol.* **2014**, *516*, 6–20. [CrossRef]
14. Graf, A.; Bogena, H.R.; Drüe, C.; Hardelauf, H.; Pütz, T.; Heinemann, G.; Vereecken, H. Spatiotemporal relations between water budget components and soil water content in a forested tributary catchment. *Water Resour. Res.* **2014**, *50*, 4837–4857. [CrossRef]
15. Cornelissen, T.; Diekkrüger, B.; Bogena, H.R. Significance of scale and lower boundary condition in the 3D simulation of hydrological processes and soil moisture variability in a forested headwater catchment. *J. Hydrol.* **2014**, *516*, 140–153. [CrossRef]
16. Goderniaux, P.; Brouyère, S.; Fowler, H.J.; Blenkinsop, S.; Therrien, R.; Orban, P.; Dassargues, A. Large scale surface-subsurface hydrological model to assess climate change impacts on groundwater reserves. *J. Hydrol.* **2009**, *373*, 122–138. [CrossRef]
17. Rahman, M.; Sulis, M.; Kollet, S.J. The concept of dual-boundary forcing in land surface-subsurface interactions of the terrestrial hydrologic and energy cycles. *Water Resour. Res.* **2014**, *50*, 8531–8548. [CrossRef]
18. Hauck, C.; Barthlott, C.; Krauss, L.; Kalthoff, N. Soil moisture variability and its influence on convective precipitation over complex terrain. *Q. J. R. Meteorol. Soc.* **2011**, *137*, 42–56. [CrossRef]
19. Bogena, H.R.; Diekkrüger, B. Modelling solute and sediment transport at different spatial and temporal scales. *Earth Surf. Process. Landf.* **2002**, *27*, 1475–1489. [CrossRef]
20. Moriasi, D.N.; Arnold, J.G.; Van Liew, M.W.; Bingner, R.L.; Harmel, R.D.; Veith, T.L. Model evaluation guidelines for systematic quantification of accuracy in watershed simulations. *Trans. ASABE* **2007**, *50*, 885–900. [CrossRef]
21. Stoltidis, I.; Krapp, L. *Hydrological Map NRW. 1:25.000, Sheet 5404*; State Agency for Water and Waste of North Rhine-Westfalia: Düsseldorf, Germany, 1980.
22. Stockinger, M.P.; Bogena, H.R.; Lücke, A.; Diekkrüger, B.; Weiler, M.; Vereecken, H. Seasonal soil moisture patterns: Controlling transit time distributions in a forested headwater catchment. *Water Resour. Res.* **2014**, *50*, 5270–5289. [CrossRef]
23. Bogena, H.R.; Herbst, M.; Huisman, J.A.; Rosenbaum, U.; Weuthen, A.; Vereecken, H. Potential of Wireless Sensor Networks for Measuring Soil Water Content Variability. *Vadose Zone J.* **2010**, *9*, 1002–1013. [CrossRef]
24. Lehmkuhl, F.; Loibl, D.; Borchardt, H. Geomorphological map of the Wüstebach (Nationalpark Eifel, Germany)—An example of human impact on mid-European mountain areas. *J. Maps* **2010**, *6*, 520–530. [CrossRef]
25. Bogena, H.R.; Bol, R.; Borchard, N.; Brüggemann, N.; Diekkrüger, B.; Drüe, C.; Groh, J.; Gottselig, N.; Huisman, J.A.; Lücke, A.; *et al.* A terrestrial observatory approach to the integrated investigation of the effects of deforestation on water, energy, and matter fluxes. *Sci. China Earth Sci.* **2015**, *58*, 61–75. [CrossRef]
26. Aquanty. *HGS 2013: HydroGeoSphere—User Manual*; Aquanty: Waterloo, ON, Canada, 2013.
27. Hwang, H.-T.; Park, Y.-J.; Sudicky, E.A.; Forsyth, P.A. A parallel computational framework to solve flow and transport in integrated surface–subsurface hydrologic systems. *Environ. Model. Softw.* **2014**, *61*, 39–58. [CrossRef]
28. Kristensen, K.J.; Jensen, S.E. A model for estimating actual evapotranspiration form potential evapotranspiration. *Nord. Hydrol.* **1975**, *6*, 170–188.
29. Partington, D.; Brunner, P.; Frei, S.; Simmons, C.T.; Werner, A.D.; Therrien, R.; Maier, H.R.; Dandy, G.C.; Fleckenstein, J.H. Interpreting streamflow generation mechanisms from integrated surface-subsurface flow models of a riparian wetland and catchment. *Water Resour. Res.* **2013**, *49*, 5501–5519. [CrossRef]
30. Allen, R.G.; Pereira, L.S.; Raes, D.; Smith, M. *FAO Irrigation and Drainage Paper No. 56*; FAO: Rome, Italy, 1998.
31. Breuer, L.; Eckhardt, K.; Frede, H.-G. Plant parameter values for models in temperate climates. *Ecol. Model.* **2003**, *169*, 237–293. [CrossRef]
32. Richter, D. *Ergebnisse Methodischer Untersuchungen zur Korrektur des Systematischen Meßfehlers des Hellmann-Niederschlagsmessers. Berichte des Deutschen Wetterdienstes*; German Weather Service: Offenbach am Main, Germany, 1995.
33. Maidment, D. *Handbook of Hydrology*; McGraw-Hill: New York, NY, USA, 1993.

34. Waldhoff, G. *Enhanced Land Use Classification of 2008 for the Rur Catchment*; CRC/TR32 Database (TR32DB), doi:10.5880/TR32DB.1; University of Cologne: Cologne, Germany, 2012.
35. Van Genuchten, M.T. A closed-form equation for predicting the hydraulic conductivity of unsaturated soils. *Soil Sci. Soc. Am. J.* **1980**, *44*, 892–898. [CrossRef]
36. Rawls, W.J.; Brakensiek, D.L. Prediction of soil water properties for hydrologic modeling. In Proceedings of the Symposium Watershed Management in the Eighties, Denver, CO, USA, 30 April–1 May 1985; pp. 293–399.
37. Brakensiek, D.L.; Rawls, W.J. Soil containing rock fragments: Effects on infiltration. *Catena* **1994**, *23*, 99–110. [CrossRef]
38. Bogena, H.R.; Huisman, J.A.; Baatz, R.; Hendricks Franssen, H.-J.; Vereecken, H. Accuracy of the cosmic-ray soil water content probe in humid forest ecosystems: The worst case scenario: Cosmic-Ray Probe in Humid Forested Ecosystems. *Water Resour. Res.* **2013**, *49*, 5778–5791. [CrossRef]
39. Brakensiek, D.L.; Rawls, W.J.; Stephenson, G.R. Determining the Saturated Hydraulic Conductivity of a Soil Containing Rock Fragments. *Soil Sci. Soc. Am. J.* **1986**, *50*, 834–835. [CrossRef]
40. Sciuto, G.; Diekkrüger, B. Influence of soil heterogeneity and spatial discretization on catchment water balance modeling. *Vadose Zone J.* **2010**, *9*, 955–969. [CrossRef]
41. Meinen, C.; Hertel, D.; Leuschner, C. Biomass and morphology of fine roots in temperate broad-leaved forests differing in tree species diversity: Is there evidence of below-ground overyielding? *Oecologia* **2009**, *161*, 99–111. [CrossRef] [PubMed]
42. Dannowski, M.; Wurbs, A. Spatial differentiated representation of maximum rooting depths of different plant communities on a field wood-area of the Northeast German Lowland. *Bodenkultur* **2003**, *54*, 93–108.
43. Beven, K.J. *Rainfall-Runoff Modelling—The Primer*; Wiley: Chichester, UK, 2001.
44. Rosenbaum, U.; Bogena, H.R.; Herbst, M.; Huisman, J.A.; Peterson, T.J.; Weuthen, A.; Western, A.W.; Vereecken, H. Seasonal and event dynamics of spatial soil moisture patterns at the small catchment scale. *Water Resour. Res.* **2012**, *48*, W10544. [CrossRef]
45. Mendel, H. *Elemente des Wasserkreislaufs: Eine Kommentierte Bibliographie zur Abflußbildung*; Analytica: Berlin, Germany, 2000.
46. Qu, W.; Bogena, H.R.; Huisman, J.A.; Vanderborght, J.; Schuh, M.; Priesack, E.; Vereecken, H. Predicting subgrid variability of soil water content from basic soil information: Predict soil water content variability. *Geophys. Res. Lett.* **2015**, *42*, 789–796. [CrossRef]
47. Bormann, H.; Breuer, L.; Gräff, T.; Huisman, J.A. Analysing the effects of soil properties changes associated with land use changes on the simulated water balance: A comparison of three hydrological catchment models for scenario analysis. *Ecol. Model.* **2007**, *209*, 29–40. [CrossRef]
48. Herbst, M.; Diekkrüger, B.; Vanderborght, J. Numerical experiments on the sensitivity of runoff generation to the spatial variation of soil hydraulic properties. *J. Hydrol.* **2006**, *326*, 43–58. [CrossRef]
49. Kværnø, S.H.; Stolte, J. Effects of soil physical data sources on discharge and soil loss simulated by the LISEM model. *CATENA* **2012**, *97*, 137–149. [CrossRef]
50. Harsch, N.; Brandenburg, M.; Klemm, O. Large-scale lysimeter site St. Arnold, Germany: Analysis of 40 years of precipitation, leachate and evapotranspiration. *Hydrol. Earth Syst. Sci.* **2009**, *13*, 305–317. [CrossRef]
51. Oishi, A.C.; Oren, R.; Stoy, P.C. Estimating components of forest evapotranspiration: A footprint approach for scaling sap flux measurements. *Agric. For. Meteorol.* **2008**, *148*, 1719–1732. [CrossRef]
52. Schuurmans, J.M.; Bierkens, M.F.P. Effect of spatial distribution of daily rainfall on interior catchment response of a distributed hydrological model. *Hydrol. Earth Syst. Sci. Discuss.* **2007**, *11*, 677–693. [CrossRef]
53. Arnaud, P.; Bouvier, C.; Cisneros, L.; Dominguez, R. Influence of rainfall spatial variability on flood prediction. *J. Hydrol.* **2002**, *260*, 216–230. [CrossRef]

Article

Effects of Model Spatial Resolution on Ecohydrologic Predictions and Their Sensitivity to Inter-Annual Climate Variability

Kyongho Son [1,*], Christina Tague [2] and Carolyn Hunsaker [3]

1 Research Foundation of the City University of New York, New York, NY 10036, USA
2 Bren School of Environmental Science & Management, University of California Santa Barbara, Santa Barbara, CA 93117, USA; ctague@bren.ucsb.edu
3 Pacific Southwest Research Station, USDA Forest Service, Fresno, CA 93710, USA; chunsaker@fs.fed.us
* Correspondence: kkyong77@hotmail.com; Tel.: +1-805-570-8553

Academic Editors: Christopher J. Duffy and Xuan Yu
Received: 9 May 2016; Accepted: 15 July 2016; Published: 29 July 2016

Abstract: The effect of fine-scale topographic variability on model estimates of ecohydrologic responses to climate variability in California's Sierra Nevada watersheds has not been adequately quantified and may be important for supporting reliable climate-impact assessments. This study tested the effect of digital elevation model (DEM) resolution on model accuracy and estimates of the sensitivity of ecohydrologic responses to inter-annual climate variability. The Regional Hydro-Ecologic Simulation System (RHESSys) was applied to eight headwater, high-elevation watersheds located in the Kings River drainage basin. Each watershed was calibrated with measured snow depth (or snow water equivalent) and daily streamflow. Modeled streamflow estimates were sensitive to DEM resolution, even with resolution-specific calibration of soil drainage parameters. For model resolutions coarser than 10 m, the accuracy of streamflow estimates largely decreased. Reduced model accuracy was related to the reduction in spatial variance of a topographic wetness index with coarser DEM resolutions. This study also found that among the long-term average ecohydrologic estimates, summer flow estimates were the most sensitive to DEM resolution, and coarser resolution models overestimated the climatic sensitivity for evapotranspiration and net primary productivity. Therefore, accounting for fine-scale topographic variability in ecohydrologic modeling may be necessary for reliably assessing climate change effects on lower-order Sierra Nevada watersheds ($\leqslant$2.3 km^2).

Keywords: DEM resolution; ecohydrologic modeling; climate change effects; RHESSys; California's Sierra

1. Introduction

In recent decades, warmer temperatures in the western United States have led to a reduction of snow accumulation as well as earlier melt and streamflow [1,2]. Changing snowmelt input has also altered the timing and magnitude of soil moisture, vegetation water use and productivity [3]. A variety of hydrological models have been used to assess the effect of climate change on the ecohydrologic response at various watershed scales [1,4,5]. However, spatial units in these models tend to be defined at relatively coarse spatial resolutions (>100 m) and thus ignore the fine-scale variation of topography. Particularly in mountain environments, substantial variation in topographic properties over relatively short spatial scales is observed, and the distribution of atmospheric forcing variables (radiation, temperature and precipitation), and local and lateral moisture are often related to this fine-scale variation in topography. Therefore, ignoring the fine-scale variation of mountain topography may result in poor predictions of ecohydrologic responses to climate change for small watersheds.

Previous studies have emphasized the importance of detailed topographic information for characterizing hydrologic and geomorphic properties of watersheds and obtaining accurate hydrologic and ecologic predictions [6–8]. Cline et al. [7] showed that the mean snow water equivalent (SWE) predictions using a 90 m digital elevation model (DEM) are different from the predictions obtained using a 30 m DEM in the Emerald Lake watershed in California. Zhang and Montgomery [6] showed that TOPMODEL [9] using a DEM with 10 m resolution improved streamflow predictions compared to simulations using coarser DEM (30 m and 90 m) for two small catchments in the western United States. Lassueur et al. [8] demonstrated the usefulness of a fine-resolution DEM to estimate plant species richness in an alpine landscape.

Studies evaluating the response of model performance to DEM resolution show that sensitivity varies across sites. Kuo et al. [10] showed that model estimates for slowly undulating landscapes tend to be less degraded with increasing grid size than those for landscapes with steep valleys. Their research also found that runoff does not change with grid size in wet years, but does change in dry years. Model predictions for snow-dominated watersheds may be more sensitive to DEM resolution than those for rain-dominated watersheds because topographic parameters (elevation, aspect and slope) determine the energy input, thereby controlling the snow melt patterns [11,12]. DEM resolution also affects the snow accumulation estimates because many snow models use a simple lapse rate based on air temperature and elevation to partition the total precipitation into snow and rain.

The effect of DEM resolution on model predictions also varied with the variable of interest [13]. Using Soil and Water Assessment Tool (SWAT) modeling, coarsened DEM resolution was found to reduce the accuracy of both streamflow and NO_3-N load prediction, but not the accuracy of total P load predictions [13]. A distributed hydrologic model used to predict average soil moisture and streamflow at the hillslope scale showed that using a coarser DEM did not reduce model accuracy, but the spatial pattern of soil moisture was distorted [14]. Estimates from a distributed ecohydrologic model showed that the grid-size effect on net primary productivity (NPP) estimates is more significant than on evapotranspiration (ET) estimates [15].

These previous studies have focused on the effect of DEM resolution on model predictions in general. However, the importance of fine-scale topographic variation in hydrologic modeling for climate change studies and other issues is not well understood. The declines in accuracy with coarsening resolution noted above may or may not be critical for using models to make inferences about climate change effects. Vegetation water and productivity are important variables for assessing the effect of climate change on ecosystem productivity, but previous hydrologic studies do not integrate the effect of DEM resolution on changes in water availability and the related impacts on modeled ET and NPP.

This study evaluated the effect of DEM resolution on the accuracy of modeled streamflow, specifically for rain-snow transition watersheds and snow-dominated watersheds that are expected to be particularly sensitive to climate change. This study also explicitly tested how DEM resolution influences the sensitivity of modeled ecohydrologic responses (annual streamflow, summer flow, annual ET and annual NPP) to inter-annual climate variability. Investigation of the influences of DEM resolution on the estimates of ecohydrologic responses to historic climate variability serves as an indicator of the likely importance of DEM resolution for future predictions.

The Regional Hydrologic-Ecologic Simulation System (RHESSys) [16] was applied to eight small Sierra Nevada watersheds. The watersheds have different dominant precipitation phases (snow vs. rain), topographic properties (elevation, slope and aspect), and vegetation properties (leaf area index, rooting depths). This study answers three questions: (1) does the total precipitation phase (snow vs. rain) control the sensitivity of model estimates to DEM resolution; (2) which topographic parameters determine the sensitivity of model estimates to DEM resolution; and (3) which variable of interest among model estimates is the most sensitive to DEM resolution? Model estimates consider both annual means and inter-annual variation in ecohydrologic variables. To answer these questions, this study follows the framework outlined in Figure 1. First, this study investigates the effect of DEM resolution on topographic parameters (elevation, slope, aspect and wetness index) in the eight watersheds. Second, this study identifies the watershed sensitivity based on the difference in estimates of the snow water equivalent (SWE), and the accuracy of modeled daily streamflow among various resolution models. Finally, this study estimates the sensitivity of the model estimates of the four ecohydrologic variables (annual streamflow, summer streamflow, annual ET and annual NPP) to DEM resolution. These tests provide a guideline for determining the appropriate DEM resolution in ecohydrologic modeling for climate effect assessment for the Sierra Nevada watersheds.

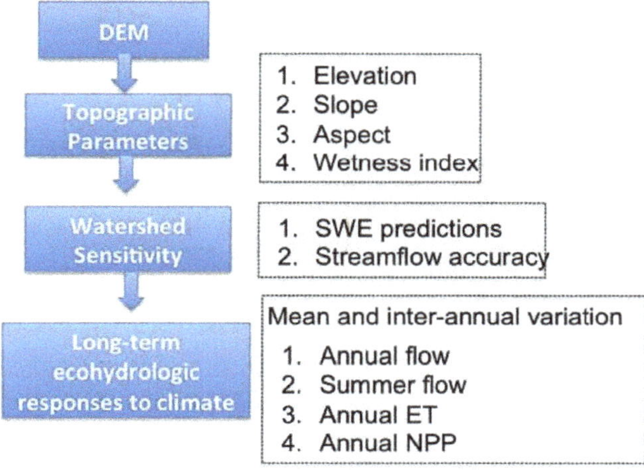

Figure 1. Framework for studying the effect of DEM resolution on the topographic parameters, the watershed sensitivity and the long-term ecohydrologic responses to climate in the eight Sierra Nevada watersheds.

2. Research Sites

This study site is located at the Forest Service's Kings River Experimental Watersheds (KREW) in California (Figure 2). RHESSys was implemented for gauged watersheds within the KREW Providence site (P301, P303, P304 and D102) and the Bull site (B201, B203, B204 and T003). Detailed descriptions of each watershed are provided below.

2.1. Providence Sites

The Providence sites include P301 (0.99 km^2), P303 (1.32 km^2), P304 (0.49 km^2) and D102 (1.2 km^2). Elevations range from 1485 m to 2115 m. The average annual precipitation (from the year 2002 to the year 2006) is 1350 mm. Precipitation occurs primarily in the winter as a mixture of snow and rain, and with little contribution of storm rainfall during the summer. In Providence sites, 20% to 50% of the annual precipitation falls as snow [17]. Following the snow regime classification developed by Jefferson [18], P301, P303, P304 and D102 are transient snow watersheds (TSWs). The major soil types are Shaver soil and Gerle-Cagwin soil [17]. The runoff ratio for the Providence sites ranges from 0.23 to 0.36. P304 has the largest runoff ratio at 0.36, and P303 has the lowest value at 0.23. The dominant forest type is Sierran mixed-conifer forest with some mixed chaparral and barren land cover. Sierran mixed-conifer vegetation in this location consists largely of white fir (*Abies concolor*), ponderosa pine (*Pinus ponderosa*), black oak (*Quercus kelloggii*), sugar pine (*Pinus lambertiana*) and incense cedar (*Calocedrus decurrens*). Two climate stations are located near or in the P303 watershed. A station is located near the outlet of the P303 watershed, while the other station is at the top of the P303 watershed. At the two stations, hourly precipitation, minimum and maximum air temperature, relative humidity, solar radiation, wind speed and direction and snow depth have been measured since 2002. The snow depth data are collected using acoustic snow-depth sensors (Judd Communications TM LLC). At the upper climate station, SWE is measured with snow pillows (Mendenhall Manufacturing, McClellan, CA, USA). Each watershed has two flumes at the outlet to measure low and high flows.

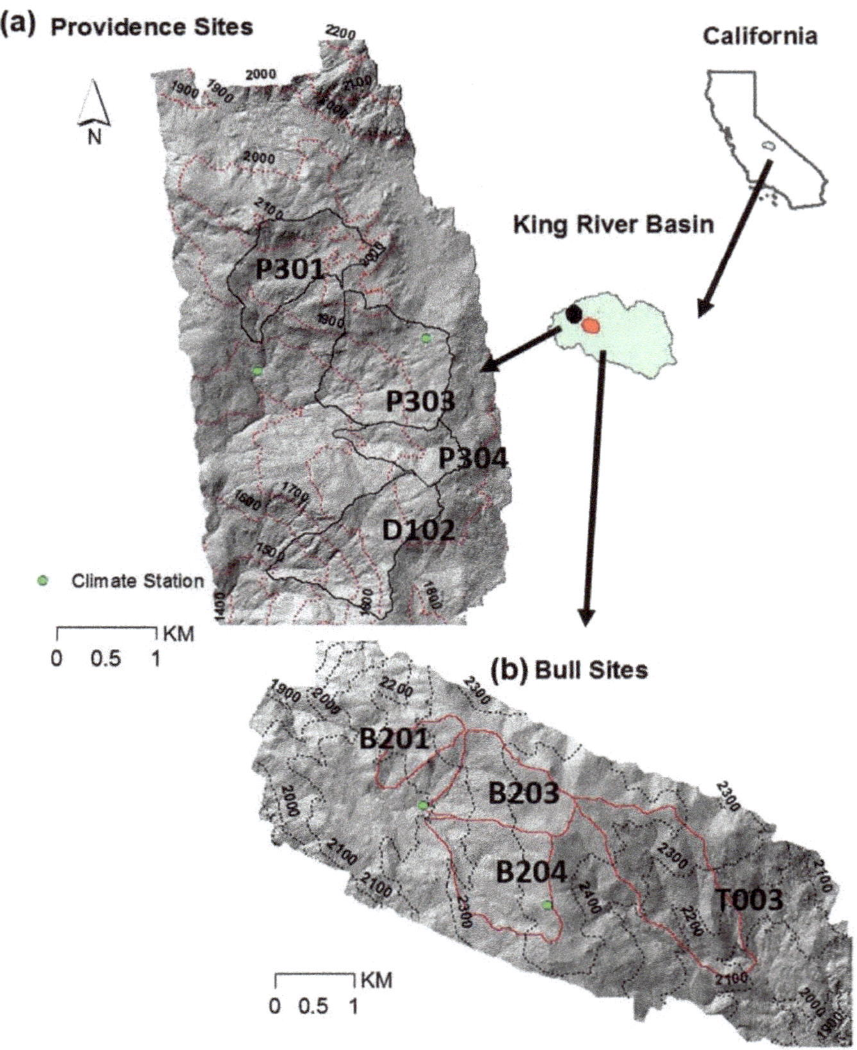

Figure 2. The location of climate stations, streamflow gauge stations and elevation gradient of study sites: (**a**) Providence sites and (**b**) Bull sites.

2.2. Bull Sites

The Bull sites include B201 (0.53 km^2), B203 (1.4 km^2), B204 (1.7 km^2) and T003 (2.3 km^2). Elevations range from 2050 m to 2490 m. The average annual precipitation (from the year 2003 to the year 2007) is 1300 mm. The Bull site is more snow-dominated (75% to 95% of precipitation falls as snow) than the Providence site. B201, B203, B204 and T003 are classified as snow-dominated watersheds (SDWs). The runoff ratio for the Bull sites ranges from 0.36 to 0.53. The runoff ratio values for the Bull sites are larger than those for the Providence sites. B203 has the largest runoff ratio at 0.53, and T003 and B201 have the lowest value at 0.36. The major soil type is Cagwin soil [18]. Similar to the Providence site, the dominant forest type in the Bull site is Sierra mixed conifer; however, red fir (*Aibes magnifica*) is more dominant at this elevation than white fir. Two climate stations are located near or in the B204 watershed. The lower climate station is located near the outlet of the B204 watershed, while the upper station is located at the top of the watershed. The station measures the same meteorological variables as at the Providence meteorological stations, and the meteorological data has been collected since year 2003. Each watershed has two flumes at the outlet of the watershed to measure high and low flow, and streamflow has been measured since 2003. A detailed description of measurements and instruments is provided by Hunsaker et al. [17].

3. Methodology

3.1. Effect of DEM Resolution on Topographic Parameters

This study tested the effect of the DEM resolution on topographic parameters including elevation, aspect, slope and topographic wetness index for the eight Sierra Nevada watersheds. The topographic wetness index [9] is defined as $ln(\frac{a}{tan\beta})$, where a is the local upslope contributing area per unit contour length and $tan\beta$ is the slope angle of the ground surface. Each DEM product was derived from a 1 m LIDAR DEM (available at https://eng.ucmerced.edu/snsjho/files/MHWG/LiDAR) with a bilinear interpolation algorithm. Providence and Bull site boundaries were derived from 5 m LIDAR DEM in order to minimize the effect of the DEM resolution on deriving the watershed area. However, due to the irregular edges, small differences in estimating the watershed area are unavoidable. Other topographic parameters including elevation, slope, aspect and wetness index were derived with five different resolutions (5 m, 10 m, 30 m, 90 m and 150 m).

The Wilcox rank-sum test was used to quantify the difference of the topographic parameters between the finest DEM (5 m) and other coarser DEM resolutions (10 m, 30 m, 90 m and 150 m) (Table 1). Values for topographic parameters were taken from each grid cell within the watershed boundaries. For all watersheds and for all DEM resolutions, the watershed mean values of slope and wetness index are significantly different (p-value < 0.01) from those computed using the 5 m resolution. However, coarsening the DEM generally does not influence the mean values of elevation and aspect except in few cases. For example, for two of Providence's transient snow watersheds (D102 and P304), the mean values of elevation using 10 m are significantly different (p-value < 0.05) than those computed using 5 m. Among the snow-dominated watersheds, B204's mean aspect values using 90 m and 150 m are significantly different (p-value < 0.1) from those computed using 5 m. T003 also has significantly different (p-value < 0.1, p-value < 0.01) mean values of aspect using 5 m compared with those computed using 10 m, and the resolution greater than 10 m, respectively.

Density plots were used to qualitatively compare the overall distributions of the topographic parameters. The density plots for only slope and wetness index parameters are presented in Figures 3 and 4 because these two parameters have the most significant change with coarsening DEM. In general, coarsening DEM decreases the mean of slope and its variation for all watersheds. Across all watersheds in the Bull and Providence sites, the largest difference in the distribution of slopes occurs between 5 m and 90 m, and between 5 m and 150 m resolutions; there is a similar distribution of slope between 5 m and 10 m, between 5 m and 30 m.

Table 1. The watershed mean values of topographic parameters with various digital elevation model (DEM) resolution (5 m, 10 m, 30 m, 90 m and 150 m) for the Providence sites and the Bull sites.

Watershed	Watershed Mean Value of Topographic Parameters [1]					
	Parameter	DEM Resolution				
		5 m	10 m	30 m	90 m	150 m
P301	Elevation (m)	1975.9	1976.6	1976.7	1975.5	1982.1
	Slope (°)	12.3	11.9 ***	10.7 ***	9.2 ***	7.8 ***
	Aspect [2] (°)	258.8	259.0	256.5	256.3	266.6
	Wetness (m)	5.9	6.2 ***	7.0 ***	7.8 ***	8.4 ***
P303	Elevation (m)	1894.8	1894.5	1895.4	1890.9	1901.7
	Slope (°)	14.0	13.7 ***	12.6 ***	11.6 ***	10.7 ***
	Aspect [2] (°)	214.4	214.9	214.4	212.4	218.2
	Wetness (m)	6.0	6.5 ***	7.2 ***	7.8 ***	8.0 ***
P304	Elevation (m)	1898.1	1896.8 *	1898.1	1894.3	1905.5
	Slope (°)	13.8	13.5 ***	12.5 ***	10.7 ***	8.5 ***
	Aspect [2] (°)	165.8	166.5	167.1	162.5	169.7
	Wetness (m)	5.9	6.2 ***	6.8 ***	7.7 ***	8.0 ***
D102	Elevation (m)	1772.0	1774.8 **	1772.8	1767.4	1785.4
	Slope (°)	19.2	18.6 ***	17.4	15.8 ***	14.8 ***
	Aspect [2] (°)	200.8	200.9	200.8	199.6	203.4
	Wetness (m)	5.7	6.1 ***	6.9 ***	7.6 ***	7.7 ***
B201	Elevation (m)	2253.8	2253.7	2254.4	2251.9	2248.4
	Slope (°)	12.5	12.3 ***	11.7 ***	10.0 ***	9.3 ***
	Aspect [2] (°)	217.2	217.2	215.8	215.4	214.8
	Wetness (m)	6.3	6.5 ***	6.9 ***	7.7 ***	7.9 ***
B203	Elevation	2371.9	2371.6	2372.4	2372.7	2369.6
	Slope	12.1	11.9 ***	11.3 ***	9.7 ***	8.3 ***
	Aspect	189.4	189.5	188.9	184.9	184.0
	Wetness	6.4	6.7 ***	7.2 ***	7.8 ***	8.0 ***
B204	Elevation	2360.3	2360.0	2360.7	2361.0	2357.3
	Slope	12.1	11.9 ***	11.1 ***	9.0 ***	8.2 ***
	Aspect	178.2	177.8	176.7	173.2*	172.6*
	Wetness	6.3	6.5 ***	7.0 ***	7.8 ***	8.4 ***
T003	Elevation	2286.5	2287.0	2285.8	2283.2	2292.8
	Slope	15.7	15.5 ***	14.5 ***	11.7 ***	9.7 ***
	Aspect	304.0	304.1 *	305.1 ***	309.3 ***	308.8 ***
	Wetness	6.0	6.2 ***	6.6 ***	7.4 ***	8.2 ***

Notes: [1] Watershed-scale parameter values; [2] Aspect is calculated with Grass GIS program (r.slope.aspect): 90° is North, 180° is West, 270° is South, and 360° is East. The aspect having zero is used to indicate undefined aspect in flat areas with slope having zero. p-value of Wilcox rank-sum test; Asterisks indicate a significant difference in mean values between the topographic parameters computed using the 5 m DEM and those using the coarser DEMs (* $p < 0.1$; ** $p < 0.05$; *** $p < 0.01$).

Coarsening the DEM increases the mean of the wetness index, but inconsistently changes the variance of the wetness index (Figure 4). The changes in the wetness index distribution with resolution are not linear and different resolutions often have different shapes of the wetness index distribution. TSWs and SDWs have a substantial change in the distribution of the wetness index at resolutions coarser than 10 m and 30 m, respectively. These results suggest that the DEM resolution may have a larger effect on local moisture estimates and lateral flow drainage patterns.

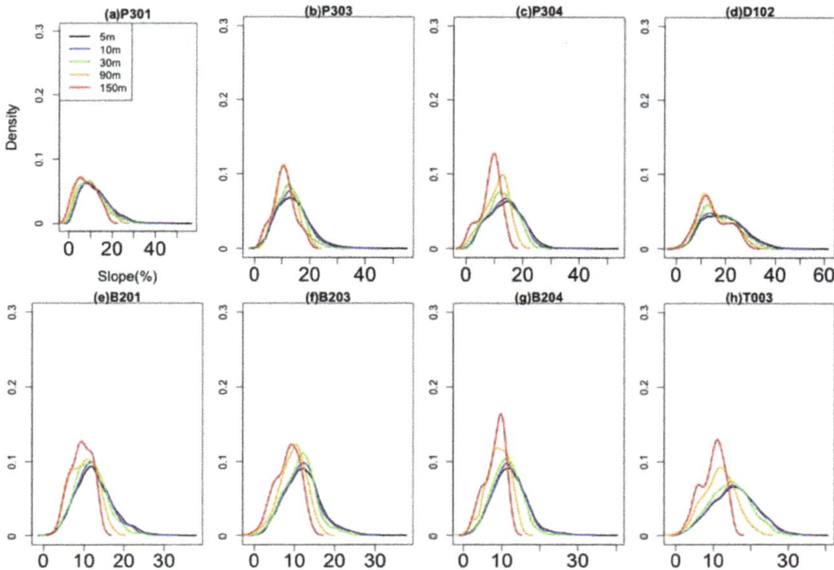

Figure 3. Distribution of slope computed using different DEM resolutions: (**a**) P301; (**b**) P303; (**c**) P304; (**d**) D102; (**e**) B201; (**f**) B203; (**g**) B204 and (**h**) T003. The first column displays the TSWs, and the second column displays the SDWs.

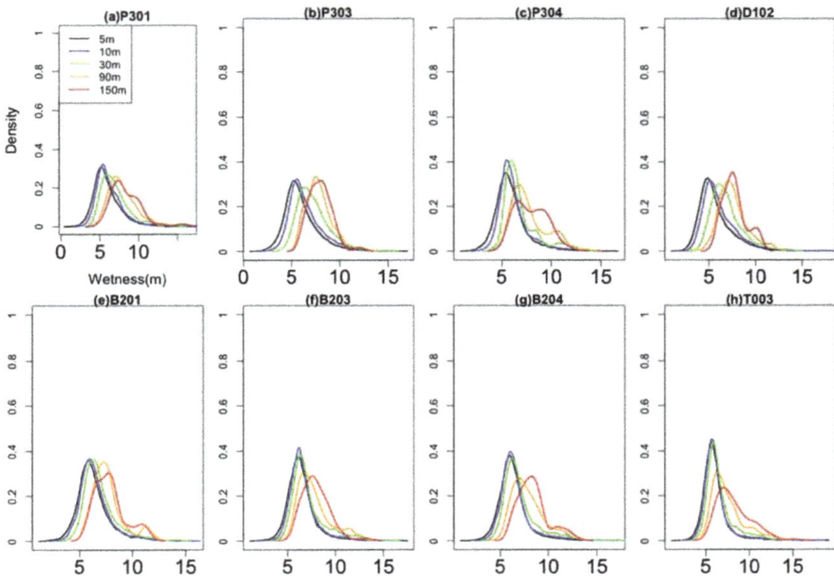

Figure 4. Distribution of wetness index using different DEM resolutions: (**a**) P301; (**b**) P303; (**c**) P304; (**d**) D102; (**e**) B201; (**f**) B203; (**g**) B204 and (**h**) T003. The first column displays the TSWs, and the second column displays the SDWs.

This study also estimated the watershed-scale mean and standard deviation of slope and wetness index computed using the different DEM resolutions (Figure 5). At 5 m, D102 and T003 have the largest mean slope of 19.2 and 15.7 for TSWs and SDWs, respectively. Among TSWs, P304 has the largest change in mean slope (5.3, 38%), and among SDWs, T003 has the largest change in mean slope (6.0, 38%) with different DEM resolutions. TSWs tend to have larger spatial variance of slope at 5 m than SDWs, but the change in the spatial variance of slope with coarsening DEM is similar between TSWs and SDWs. SDWs tend to have a larger mean wetness index at 5 m than TSWs, but TSWs tend to have larger spatial variance of wetness index at 5 m than SDWs. Among TSWs, the spatial variances of the wetness index for P303 and D102 tend to decrease with coarsening DEM. Their variances for P301 and P304 increase with coarsening resolution up to 30 m, and their variances decrease at 90 m and 150 m, respectively. Among SDWs, the spatial variances of wetness index for B203 and B204 increase for resolutions up to at 30 m, and then decrease. Spatial variance of the wetness index for B201 slightly decreases with coarsening DEM. However, its variance for T003 increases with coarsening DEM. In summary, increasing the DEM resolution generally decreases the mean and standard deviation of the slope, and increases the mean wetness index for all watersheds. The changes in the standard deviation of the wetness index with coarsening the DEM have more complex patterns than other parameters.

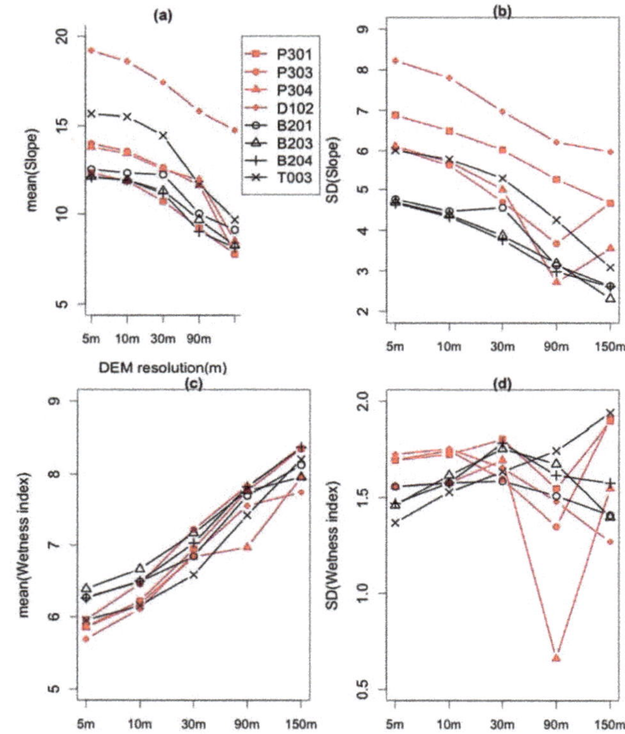

Figure 5. Watershed-scale statistical properties (mean and standard deviation) for slope and wetness index in the eight Sierra Nevada watersheds computed with increasingly coarse DEM resolution. (**a**) watershed mean values of slope; (**b**) watershed standard deviation of slope; (**c**) watershed mean values of wetness index; and (**d**) watershed standard deviation values of wetness index. Red color line refers to TSWs; black color line refers to SDWs.

3.2. Model Description

The RHESSys [16] was used to investigate the effect of DEM resolution on the model estimates of ecohydrologic fluxes of the eight small Sierra Nevada watersheds in California. The RHESSys is a physically-based, distributed ecohydrologic model. The RHESSys is under continuous development. In this study, version 5.15.r326 was used. A detailed description of the model is provided in [16].

The RHESSys has a hierarchic spatial structure to partition the landscape: basins, hillslope, zones, patches and canopy strata. Patch and canopy strata are the smallest spatial units, and these can be derived using various layers: landcover, soil and elevation maps. In this study, a patch was defined as a grid cell with uniform elevation. Hydrological variables including snow, soil moisture and evapotranspiration are computed at the patch level. Within a patch, multiple horizontal and vertical canopy layers can be potentially used, but in this study we used a simple, single, vertical-layer canopy strata with the same horizontal (spatial) resolution as the patch. In addition to a canopy layer, patches also contain a litter layer. Zones were used to characterize the spatial and temporal distribution of climate inputs including precipitation, air temperature and solar radiation. To account for the fine-scale variability of the climate within a watershed, we defined zones using the same spatial data (elevation grids) that are used to define the patches. In the RHESSys, basins are the largest spatial unit, and it is generally defined as a hydrologically closed drainage area. In this study, the basins were created based on the stream gauging station, using r.water.outlet (GRASS GIS, (http://grass.osgeo.org/)). Hillslope maps were created using r.watershed, and the multiple flow direction algorithm [19] was used to create the flow direction maps, and a hillslope drains a single stream reach. Lateral flow is organized within a hillslope, and is computed at the patch level.

The minimum climate data required for model simulation include daily precipitation, and daily maximum and minimum air temperature data. Other climate data (including solar radiation, saturation vapor pressure, relative humidity, etc.) are computed using a climate interpolation model (MT-CLIM, Running et al. [20]). Mountain watersheds have frequently experienced a lack of available climate data usable for hydrologic modeling, largely due to high climate variability along steep topographic gradients. MT-CLIM has been tested and improved using field data, and has successfully reproduced the field measured climate data [21–23].

Energy, wind and water are attenuated through the aboveground canopy, using standard approaches such as Beer's Law for radiation extinction as a function of vegetation leaf area index. Snowmelt is estimated using a combination of an energy budget approach for radiation-driven melt and advective-driven melt (rain on snow) with a temperature index-based approach for sensible and latent heat exchange. The partitioning of total precipitation into snow versus rain is calculated based on linear temperature threshold values. Transpiration from the canopy and evaporation of intercepted water and soil evaporation are computed using the Penman-Monteith [24] approach, where stomatal conductance for vegetation is computed using a Jarvis multiplicative model of radiation, vapor pressure deficit, rooting zone soil moisture and temperature controls [25]. Net primary productivity is estimated as the difference between gross photosynthesis and respiration. Gross photosynthesis is estimated using the Farquhar model [26]. Respiration is computed separately for different plant components (leaves, live/dead wood and roots) as a function of biomass, nitrogen content and air temperature [27]. Infiltration and vertical drainage between unsaturated and saturated stores is a function of soil hydraulic parameters. A lateral shallow groundwater flux is calculated based on hydraulic gradients (determined by surface topography) and soil hydraulic conductivity, and is explicitly routed between patches. The explicit routing scheme is based on topographic slope and soil transmissivity. The model also includes a bypass flow mechanism to simulate direct drainage through macropores from surface to deep groundwater storage. The flow from deep groundwater storage is calculated based on a linear storage equation.

This study compared RHESSys estimates from model implementations using five different resolutions (5 m, 10 m, 30 m, 90 m and 150 m). For example, for RHESSys, the DEM is used to derive topographic parameters (elevation, aspect, slope and flow drainage parameter (e.g., wetness index)) that determine the distribution of the microclimate (radiation, temperature, etc.), and local and lateral moisture distribution. Changing the DEM resolution is therefore expected to affect the RHESSy model estimates, including SWE, ET, NPP and streamflow. Each resolution model had the same vegetation definition map. The vegetation type for all watersheds was assigned as mixed conifer. Associated vegetation type parameters were taken from RHESSys parameter libraries (http://fiesta.bren.ucsb.edu/~rhessys/index.html). Leaf area index (LAI) was derived from the LIDAR point cloud using a deterministic approach [28] and was used to initialize vegetation carbon and nutrient stores. To minimize the effect of LAI resolution on model estimates, 30 m LAI was used for all resolution models. We recalibrated the model for each DEM resolution—thus, each resolution has a unique set of parameters. We chose this option, rather than running with the same parameterization, because this better reflects how models are typically implemented—and the focus of the paper is exploring how model implementations (which include calibration) influence estimates. While in this paper we focus on differences in model estimates based on an optimal parameter set, we note, however, that the different resolutions may also be associated with differences in parameter uncertainty.

3.3. Model Calibration

We used a 5 m resolution model as a baseline model. Snow-related parameters were estimated by comparing model SWE predictions of the 5 m resolution model with measured snow depths (or measured SWE) at the climate stations. Table 2 shows estimated air temperature and snow-related parameters for the Providence and the Bull sites. The two climate stations are located at the top and bottom of the watersheds, respectively. Thus, air temperature lapse rates for the Providence and Bull sites were estimated using the difference of the elevations and the difference of the measured air temperatures at the two climate stations, respectively. Positive lapse rates for maximum daily temperature and negative lapse rates for minimum daily temperature are obtained for both sites. Since the climate stations (Lower Providence and Lower Bull) at lower elevations are located in a valley or a potential cold pool drainage area, the measured air temperature data at the station may reflect the nighttime temperature inversion. The two snow-related parameters, the temperature melt coefficient and the temperature threshold value for the partitioning of total precipitation into rain and snow, were estimated by adjusting the parameter values until the model prediction was similar to the measured snow depth or SWE. The Providence and Bull sites have similar temperature lapse rates and the same temperature melt coefficients and temperature threshold values. To compare the model estimate of the SWE and measured snow depths, the day of complete snowmelt in the four climate stations was calculated (Table 2). Model estimates were compared to measurements taken by acoustic snow-depth sensors. The model reproduces the timing of observed snowmelt at the four climate stations. The comparison of the model estimates with measured values results in R^2 values of 0.92, 0.86 in the Providence stations and the Bull stations. The comparison of the modeled SWE with measured SWE at the Upper Providence and the Upper Bull stations results in R^2 values of 0.91 and 0.83, respectively. In general, the model accuracy for SWE estimation is slightly better at the Providence station (transient snow watersheds) than at the Bull station (snow-dominated watersheds).

After snow-related parameters were estimated, soil parameters were calibrated by comparing daily estimates of model streamflow to measured streamflow. The calibrated soil parameters are anisotropic horizontal and vertical saturated hydraulic conductivity (Ksat_h, Ksat_v), the decay coefficient of saturated hydraulic conductivity with depth (m), the proportion coefficient of macro-pore drainage into deep groundwater storage (gw1), air entry pressure (ae) and pore size index (psi). The linear coefficient of deep groundwater storage (gw2) is fixed as zero to reflect deep groundwater losses that are not captured by the stream gauge. In addition, to account for the observed difference of the rooting depth across watersheds, Providence watersheds are assigned a 2 m rooting depth, and Bull watersheds are assigned a 1 m rooting depth [17].

Table 2. Calibrated snow-related parameters and the model accuracy of snow predictions for the Providence and Bull watersheds.

Watershed	Snow-Related Parameters			Model Accuracy of Snow Predictions	
	Temperature Lapse Rates [1] (tmax/tmin) (°C/m)	Temperature Threshold for Rain vs. Snow [2] (°C)	Temperature Melt Coefficient [3] (m/°C)	Day of Snow Melt [4]	SWE [5]
Providence	0.0063/−0.0064	−3-3	0.005	0.92	0.91
Bull	0.0068/−0.0060	−3-3	0.005	0.83	0.83

Notes: [1] Since fine-spatial-scale air temperature is not available in the two watersheds, air temperature within a watershed is spatially interpolated with the given elevation and the calculated temperature laps rates; [2] To partition total precipitation into snow and rain, we use the air temperature as a proxy variable, and the proportion of snow and rain in the total precipitation is linearly interpolated based on the minimum and maximum temperature values; [3] Temperature melt coefficient accounts for snowmelt due to latent heat and sensible heat; [4] The day of snowmelt is estimated using observed snow depths and the modeled SWE value, and the correlation coefficient is measured; [5] The measured SWE data are available in the upper Providence and upper Bull station.

To evaluate the model streamflow accuracy, this study adopted a multi-objective approach. Many hydrologic modeling studies have found that using a single accuracy measure can bias the evaluation of the model performance [29]. This study's accuracy measures are listed in Equations (1)–(4). Each measure focuses on a particular aspect of flow variation. Nash-Sutcliffe efficiency (NSE) [30] focuses on peak streamflow. Log Nash-Sutcliffe efficiency (LNSE) is the log value of the Nash-Sutcliff efficiency and focuses on recession and low flow. Percent Error (PerErr) is the percent volume error and focuses on flow bias. The three accuracy measures are combined to evaluate the model streamflow accuracy robustly (Equation (4)). This accuracy measure ranges from 0 to 1 with 1.

$$NSE = 1 - \frac{\sum_i \left(Q_{obs,i} - Q_{sim,i}\right)^2}{\sum_i \left(\overline{Q_{obs}} - Q_{sim,i}\right)^2} \tag{1}$$

$$LNSE = 1 - \frac{\sum_i \left(\log(Q_{obs,i}) - \log(Q_{sim,i})\right)^2}{\sum_i \left(\log(\overline{Q_{obs}}) - \log(Q_{sim,i})\right)^2} \tag{2}$$

$$PerErr = \frac{\left(\overline{Q_{obs}} - \overline{Q_{sim}}\right)}{\overline{Q_{obs}}} \tag{3}$$

$$Total\ Accuracy = NSE \times LNSE \times PerErr \tag{4}$$

where $Q_{obs,i}$ is the observed streamflow and $Q_{sim,i}$ is the simulated flow at daily time step (i), and $\overline{Q_{obs}}$, $\overline{Q_{sim}}$ are the long-term average of observed daily streamflow and simulated streamflow, respectively.

Mean annual values for each flux (flow, ET and NPP) were computed to quantify the long-term average ecohydrologic responses to climate. The coefficient of variation (COV) was also calculated to quantify the inter-annual variation of each flux for climate sensitivity. The sensitivity of the mean and COV for each flux to DEM resolution was calculated. A long-term historical climate period (>50 years) is required to investigate the sensitivity of model estimates to inter-annual climate variability. However, at the time of this study the KREW streamflow and basic climate data were relatively short, just five years. Therefore, this study used the long-term climate data of the Grant Grove station located 28 km south of the Bull sites. The Grant Grove station has similar temporal precipitation and temperature patterns to the Providence and the Bull climate stations. The long-term climate data for the Providence and the Bull climate stations were estimated by fitting the local climate station data to the Grove Grant station data. The mean annual precipitation of the Bull and Providence stations for the period of 2003 to 2007 and 2002 to 2006, respectively, was divided by the mean annual precipitation of the Grant Grove station in order to estimate precipitation scaling factors for each station. To generate long-term daily precipitation data for the Providence and the Bull stations, the respective precipitation scaling factor

(1.22 and 1.21) was applied to the Grant Grove station's daily precipitation data. To generate long-term daily maximum and minimum air temperature data for the two watersheds, linear regression models were estimated by fitting the local temperature data from the Providence and Bull climate stations to the Grant Grove station's climate ($0.73 < R^2 < 0.89$).

3.4. Effect of DEM Resolution on Model Accuracy and Long-Term Ecohydrologic Responses to Climate

This study tested the effect of DEM resolution on the accuracy of modeled streamflow (Equations (1)–(4)), and on the sensitivity of estimated ecohydrologic response to inter-annual climate variability. Three hypotheses were developed with respect to differences in the sensitivity of the model estimates to DEM resolution for transient snow watersheds (TSWs) and snow-dominated watersheds (SDWs). The first hypothesis is that the flow estimates for TSWs are more sensitive to DEM resolution than the flow estimates for SDWs. The underlying assumption for this hypothesis is that TSWs may have a larger change in precipitation phase (snow vs. rain) with the changing spatial resolution of DEM because TSWs more frequently experience air temperature close to 0 °C (threshold temperature, Table 2) than SDWs. The change in precipitation phase affects snow accumulation, which influences melt rates and streamflow estimates. Flow estimates in TSWs therefore will be more sensitive to DEM resolution than flow estimates in SDWs. The second hypothesis is that the change in the spatial variance of the wetness index is the dominant topographic parameter that determines the flow sensitivity to the DEM resolution. When the DEM resolution is coarser, a large change in the spatial variance of the wetness index for the eight watersheds was observed (Figure 5d). Even though RHESSys does not use the wetness index directly for calculating the flow, change in the spatial variance of the wetness index with a coarsening DEM may reflect the influence of DEM resolution on the flow network in RHESSys and ultimately the accuracy of flow estimates [31].

This study also compared the sensitivity of four key ecohydrologic estimates to DEM resolution—annual streamflow, summer streamflow (August flow), annual ET and annual NPP. The third hypothesis is that the sensitivity of the annual and summer streamflow estimates is more sensitive to DEM resolution than annual ET and NPP estimates. Here it is assumed that flows are controlled by topographic variation; the dominant controls of ET and NPP are climate and vegetation properties.

4. Results

4.1. Effect of DEM Resolution on Snow Predictions

To evaluate the effect of DEM resolution on SWE estimates, we calculated the watershed-scale peak SWE at the five different resolutions and the mean absolute difference in the watershed-averaged daily SWE estimates between the 5 m resolution and other coarser resolutions (Figure 6). Peak SWE across the watersheds ranges from 303 mm to 607 mm for the 5 m resolution model. The change in peak SWE estimates between the different resolutions is always less than 3% (Figure 6a). The mean absolute difference in the watershed-averaged SWE between 5 m and other coarser resolutions varies between watersheds (Figure 6b). The differences range from 0.3 to 4.5 mm. Their difference is relatively indistinguishable compared with the peak SWE that ranges from 303 mm to 608 mm. Therefore, the difference in SWE change with coarsening DEM is minor for the eight watersheds.

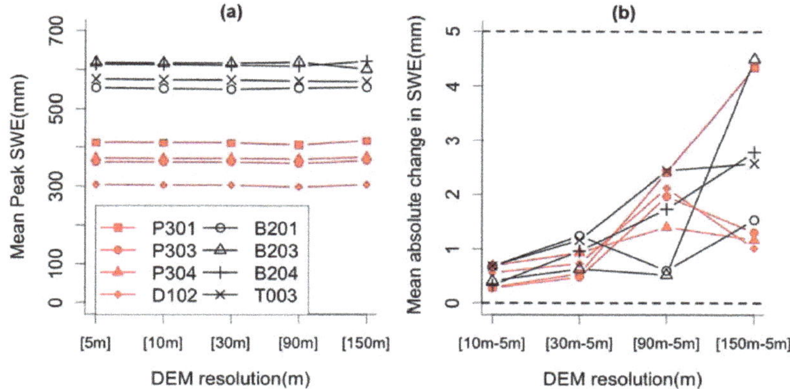

Figure 6. Watershed peak SWE estimates among various resolution and absolute difference in watershed mean SWE estimates between the 5 m resolution and the coarser resolutions (10 m, 30 m, 90 m and 150 m) in the eight Sierra Nevada watersheds. (**a**) watershed peak SWE; and (**b**) the difference in mean absolute watershed SWE estimates between 5 m and coarser resolutions.

4.2. Effect of DEM Resolution on Streamflow Prediction Accuracy

To quantify the effect of DEM resolution on streamflow predictions, we calculated the change in the four different streamflow accuracy measures with coarsening the DEM and compared the change in the streamflow accuracies between the eight watersheds. Figure 7 shows that the model accuracies declined with coarsening DEM resolutions for all watersheds. In general, there is a threshold resolution (10 m) above which coarser resolutions have a larger effect on streamflow prediction accuracy. Among TSWs, P304 and D102 equally have the largest reduction in streamflow total accuracy (Equation (4)) from 0.30 to 0.09 (71%) between 10 m and 30 m, and from 0.30 to 0.09 (71%) between 5 m and 90 m, respectively. P301 has the smallest reduction in streamflow accuracy (Equation (4)) from 0.48 to 0.36 (25%) between 5 m and 90 m. Among SDWs, B203 has the largest reduction in streamflow accuracy from 0.64 to 0.29 (55%) between 5 m and 150 m. T003 has the smallest reduction in streamflow accuracy (Equation (4)) from 0.55 to 0.47 (15%) between 10 m and 90 m. In general, streamflow estimates for the SDWs are more accurate and less sensitive to coarsening DEM resolution than the TSWs are.

Among the accuracy measures, PerErr changes most with coarsening DEM compared to the other individual measures (Figure 7b–d). The high sensitivity of PerErr may reflect the model error due to the impact of resolution on evapotranspiration estimates. LNSE values for most watersheds are least sensitive to DEM resolution (Figure 7c). NSE values for the TSWs are more sensitive to changes in DEM resolution than NSE values for the SDWs (Figure 7b). Among TSWs, NSE and LNSE values for P301 are the least sensitive to DEM resolution. The streamflow estimates for P304 are also the least accurate by all measures compared to the other TSWs (Figure 7d). Most watersheds have the highest streamflow accuracy at 5 m; however, in P304, the 5 m resolution model failed to reproduce observed streamflow at an acceptable level (Equation (4) > 0). In RHESSys, surface topography is assumed to reflect the subsurface topography, but this assumption may not be valid in this watershed. Among TSWs, P304 with poor model performance has the highest summer flow rate [17], and shows different sediment load variability, sources and erosion rates than other TSWs [32].

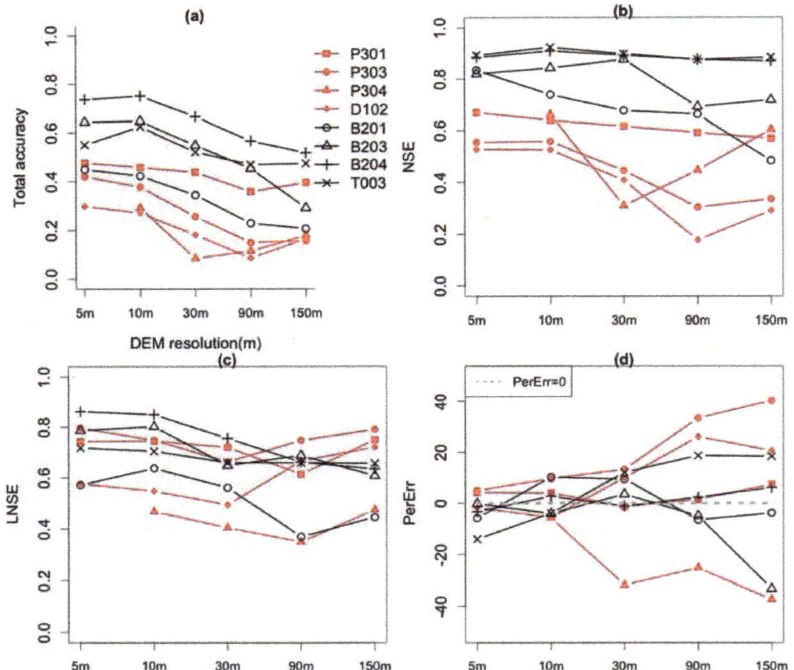

Figure 7. The model performance of streamflow prediction with different DEM resolutions. (**a**) total accuracy measure combining NSE, LNSE and PerErr; (**b**) NSE; (**c**) LNSE and (**d**) PerErr (percent error).

The 150 m resolution models for transient snow watersheds P301, P304 and D102 have higher accuracy than the 90 m resolution models. Similar results are observed for some SDWs where the 5 m resolution model has lower streamflow accuracy than the 10 m resolution model. Among SDWs, the streamflow accuracy for B201 is similar to those for TSWs, especially regarding the LNSE measure. NSE and LNSE for B201 are more sensitive to DEM resolution compared to the other watersheds. Other research [33] suggested that B201 has a smaller subsurface flow component than other Bull sites, and the authors hypothesized that the bedrock geomorphology at B201 may be different from the other watersheds. These results suggest that the model accuracy of streamflow depends on how well surface topography at different resolutions emulates the subsurface topography for individual watersheds. In addition to issues related to subsurface controls on streamflow, other unobserved differences in vegetation and drainage properties may have contributed to these differences in accuracy.

Since RHESSy discretizes the watersheds based on DEM, and explicitly routes the flow and nutrient fluxes per grid or patch, increasing the DEM resolution will lead to increasing the computation running time. In our experiment, running times of the daily time step per year for T003, the largest watershed (2.3 km^2), using a MacBook Pro (2.7 GHz Intel Core i5, 8 GB memory) were 1087 s (5 m), 621 s (10 m), 67 s (30 m), 7 s (90 m), 1 s (150 m). Therefore, the improved model results at 10 m require a longer time period to run the model than a scale of 30 m, but it still runs in 10 min which is very reasonable given the improvement.

4.3. Sensitivity of Estimated Ecohydrologic Variables to DEM Resolution

This study investigated how DEM resolution affects the long-term average ecohydrologic responses (annual flow, summer flow, annual ET, and annual NPP) to climate (Figure 8). Mean annual streamflow estimates generally increase with coarser DEMs, especially for grid sizes that are coarser than 30 m. P304 and B203 are exceptions. T003 has the largest increase (44%) in mean annual flow, and B204 has the smallest increase (8%) in mean annual flow. Of the eco-hydrologic variables that we examined, mean summer flow is the most sensitive to DEM resolution, especially for SDWs, and generally increases with coarser DEMs. Among SDWs, T003 has the largest increase (150%) in mean summer flow with coarsening DEM. Among TSWs, P303 has the smallest decrease (21%) in mean summer flow. Mean annual ET values for SDWs are more sensitive to DEM resolution than TSWs. Coarsening the DEM reduces the mean annual ET. B201 has the largest change (33% decreases) in mean annual ET, and P301 has the smallest change (16% decreases) in mean annual ET. For most watersheds, coarsening DEM decreases the mean annual NPP. Compared with TSWs, mean annual NPP for SDWs is more sensitive to DEM resolution. T003 has the largest (50%) decrease in mean annual NPP, and P303 has smallest (14%) decrease in mean annual NPP.

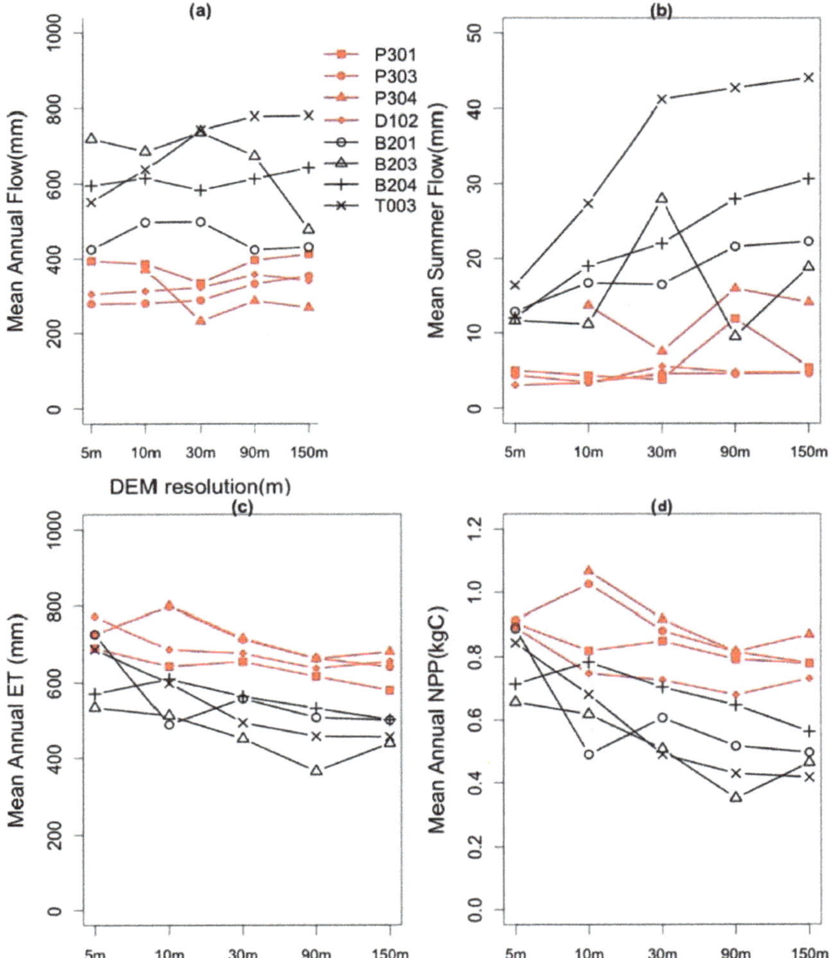

Figure 8. Annual mean of estimated ecohydrologic responses to climate with different DEM resolution. (**a**) mean annual flow; (**b**) mean summer (August) flow; (**c**) mean annual ET and (**d**) mean annual NPP.

We investigated the effect of DEM resolution on the sensitivity of each flux to inter-annual climate variation by calculating the COV at each resolution (Figure 9). TSWs tended to have a higher COV than SDWs for estimated annual streamflow, summer flow, annual ET, and annual NPP at 5 m. Coarsening the DEM has varied effects on the COVs among watersheds, and on the variables of interest. SDWs have higher changes in COV with the coarsening DEM resolution than TSWs. SDWs have larger increases in COV for annual NPP and larger decreases in COV for summer flow. There is not a large difference in the change of COV values for annual flow and annual ET between TSWs and SDWs. The COV values of annual NPP have the largest change with coarsening DEM, and the COV values of annual flow have the smallest change.

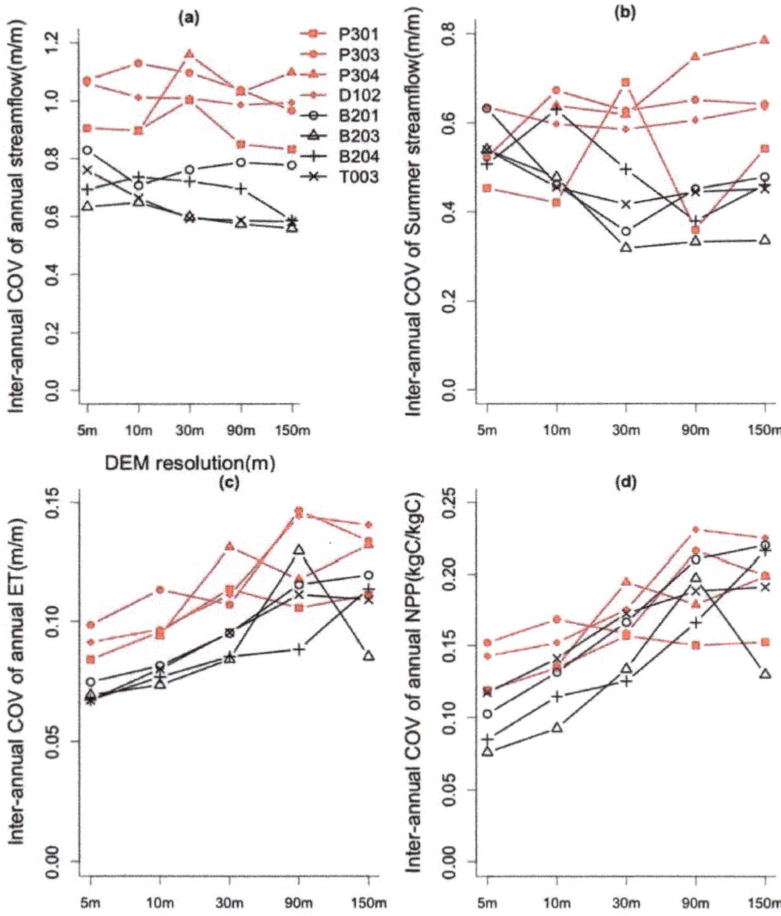

Figure 9. Inter-annual variability (coefficient of variance (COV)) of estimated ecohydrologic responses to climate with different DEM resolution (the variability was measured by coefficient of variance (COV)). (**a**) inter-annual streamflow variability; (**b**) inter-annual August streamflow variability; (**c**) inter-annual ET variability and (**d**) inter-annual NPP variability.

5. Discussion and Summary

This study was performed to improve our understanding of how DEM resolution affects ecohydrologic estimates in the context of using a model to evaluate climate change effects in small mountain watersheds. Three hypotheses were posed to test the DEM sensitivity within the TSW and SDW groups of watersheds and among the variables of interest: (1) model estimates for transient snow watersheds (TSWs) will have a higher sensitivity to DEM resolution than the model estimates for snow-dominated watersheds (SDWs); (2) changes in the spatial variation of the wetness index will explain the watershed sensitivity to DEM resolution; and (3) flow estimates will be more sensitive to DEM resolution than ET and NPP estimates.

This study showed that there is a clear threshold resolution (10 m) above which coarser resolutions have large effects on streamflow prediction accuracy (Figure 7). Among the eight watersheds, TSWs tend to have both a lower streamflow accuracy and a larger reduction of streamflow accuracy with coarsening DEM resolution (Table 3). Among TSWs, streamflow accuracy for P304 and D102 is the most sensitive to DEM resolution, but P301 is the second least-sensitive watershed to DEM resolution between the eight watersheds. The first hypothesis, that sensitivity to DEM resolution is closely linked to snow accumulation and melt characteristics, is not supported. The change in peak SWE with coarsening DEM is very minor for all eight watersheds. P301 with the lowest sensitivity to DEM resolution has the largest change in watershed absolute difference in SWE between 5 m and 150 m (Figure 5). Thus, the difference in the dominant precipitation phase between TSWs and SDWs does not lead to consistent differences in the sensitivity of flow estimates to changes in the model resolution.

Table 3. Watershed sensitivity to DEM resolution.

Watershed Group	Watershed	Change in Spatial Variance of Wetness Index (%)	Change in Streamflow Accuracy (Equation (4)) (%)	Model-Based Rank [5]
TSWs	P301	−9 [1] (3) [2]	−25 [3] (−14) [4]	7
	P303	−31 (−14)	−64 (−44)	3
	P304	−1 (5)	−71 (−42)	1
	D102	−26 (−11)	−71 (−41)	2
SDWs	B201	−5 (−1)	−54 (−33)	5
	B203	−4 (11)	−55 (−25)	4
	B204	7 (12)	−30 (−15)	6
	T003	11 (25)	−15 (−5)	8

Notes: [1] The largest change in spatial variance of wetness index between the five resolution models; [2] The mean change in spatial variance of wetness index between the five resolution models; [3] The largest change in streamflow accuracy between the five resolution models (Equation (4)); [4] The mean change in streamflow accuracy between the five resolution models (Equation (4)); [5] Ranked from highest to lowest sensitivity to DEM resolution, based on change in modeled streamflow accuracy.

Among topographic parameters, we hypothesized that the change in the spatial variation of the wetness index can explain the watershed sensitivity to DEM resolution. Changing the spatial variance of the wetness index has a complex relationship with coarsening DEM, and varies between watersheds. However, the lowering in the spatial variance of the wetness index with coarsening DEM corresponds with a reduction of the streamflow accuracy (Table 3). For example, when the 5 m resolution model was compared with coarser resolution models, P301 and D102 had the smallest reduction (−9%) and the largest reduction (−26%) of the spatial variance of the wetness index, respectively, which corresponds to the smallest and largest reductions of streamflow accuracy for the watersheds (−25% and −71%, respectively). Among the eight watersheds, T003 has the smallest reduction (−15%) of the streamflow accuracy, and that watershed shows an increase (11%) of the spatial variance of the wetness index. RHESSys does not use the wetness index directly to calculate lateral flow. However, the wetness index includes the component of topographic slope and flow-accumulating area. RHESSys actually uses these components to determine the lateral flow paths. Previous studies using TOPMODEL [10] also showed that decreasing resolution reduces the spatial variance of the wetness index [6,34]. Pradhan et al. [34]

showed that when a coarser DEM resolution (1000 m) reproduced the cumulative distribution of the wetness index at the fine resolution (50 m), the streamflow estimates using the coarser 1000 m DEM resolution matched the simulated streamflow in the 50 m DEM resolution TOPMODEL without recalibration. Results in this study suggest that the change in the wetness index distribution will also be a good indicator of whether coarsening the DEM will lead to reduced accuracy for an explicit routing model. Kenward et al. [31] tested the impact of DEM resolution on the streamflow accuracy and spatial pattern of a predicted saturated area using DHSVM [35] which has a similar routing scheme to RHESSys. Their study also showed that the spatial distribution of the wetness index corresponded to the depth to saturation and runoff production for a rain-dominated system in the WF-38 experimental watershed at the Mathantango Creak, PA. Our study confirms that the impact of DEM resolution on flow paths is also likely to be important for snow and rain-snow transition watersheds and that the impact of model resolution on the lateral redistribution of water may be more important than its impact on snow accumulation and melt for models of low-order, headwater watersheds.

Among the model accuracy measures, PerErr has the highest sensitivity to DEM resolution. Changes in PerErr are directly related to changes in annual ET. We note that annual ET estimates and their COV are strongly sensitive to DEM resolution (Figures 8 and 9). Changes in the wetness index distribution may also be important in ET estimates, particularly in water-limited environments. The impact of DEM resolution on ET is discussed in more detail below.

Among the model estimates, we hypothesized that the flow estimate to DEM resolution will be more sensitive than ET and NPP estimates. Our modeling results found that among the four ecohydrologic estimates of interest, DEM resolution has the largest effects on the mean summer flow and COV of the annual ET and NPP (Figures 7 and 8). One of the eight watersheds, T003 had the smallest reduction in streamflow accuracy with coarsening DEM, but large changes in the mean summer flow (150%), the COV of the annual ET (65%), and the COV of the annual NPP (60%) are observed. These results emphasize that accurate streamflow prediction does not guarantee a model's ability to capture long-term ecohydrologic responses to climate change. Our study also suggests that using a fine-resolution DEM in ecohydrologic modeling is essential in order to capture the long-term observed summer flow. Since summer flow is an important water resource and has substantial implications for aquatic organisms in California, fine-scale hydrologic modeling for assessing the effect of climate change in Sierra Nevada is necessary [36].

Our modeling study showed that a coarsening DEM resolution results in an increase in the COV of both ET and NPP. This result implies that coarser-resolution models overestimate the sensitivity of these processes to climate variation. This result is important for interpreting and predicting ecosystem responses to climate change. The reduced sensitivity of ET and NPP for the finer-resolution models may be related to the substantial variation in topographic properties in mountain environments. The high variation in topographic properties may lead to spatial variation in the sizes of water storage and flow path convergence. As discussed above, coarsening the DEM tends to reduce spatial variation in the wetness index. The vegetation response to changing climate may be lower for the finer-resolution model because this spatial variation in water storage and flow path convergence provides additional opportunities for plants to access water. A higher-resolution DEM, for example, may lead to greater areas of local flow path convergence typified by riparian areas and local depressions with greater soil moisture. ET in these areas may be less sensitive to inter-annual climate variation. The higher COV of ET and NPP with coarsening DEM resolution may also illustrate the role of micro-refuge created by substantial variation in other topographic properties in mountain environments [37]. Dobrowski et al. [37] provide case studies where terrain allows for local climate conditions to be decoupled from the regional climate; when sites decouple from the regional climate, micro-refuges can occur for species. The finer-resolution model may create microclimate conditions, as well as areas of increased moisture storage, that are less sensitive to the forcing climate variability.

In summary, this study demonstrates that using fine-scale DEM in ecohydrologic modeling influences the accuracy of streamflow estimation in headwater mountain catchments and substantially alters estimates of climate-driven inter-annual variation in ET and NPP in these systems. Results emphasize that these effects may be largely due to the role of the DEM in the model estimation of hydrologic flowpaths rather than the model estimation of snow accumulation and melt. This study found that coarser-resolution models tend to have lower streamflow accuracy and overestimate climate sensitivity for ET and NPP. These results have important implications for model-based studies used to assess ecosystem responses to climate change, and, in particular, caution that coarser-resolution models may overestimate climate sensitivity. The analysis, however, demonstrates a non-linear relationship between model accuracy/sensitivity and DEM resolution and suggests that increasing resolution from 30 m to 10 m makes substantial improvements. Further increasing the resolution to 5 m results in smaller gains in performance, relative to the increase in computation cost.

Acknowledgments: We gratefully acknowledge the data collected by Kings River Experimental Watershed Project (KREW) and this study was financially supported by Southern Sierra Critical Zone Observatory Project from National Science Foundation (EAR-0725097).

Author Contributions: Kyongho Son performed the data processing and modeling and wrote the manuscript, and Christina Tague helped the RHESSys model setup, and revised the manuscript and analyzed and interpreted the modeling results. Carlyon Hunsaker provided the basic hydrologic data used in this study, and helped improve the structure and language of the manuscript.

Conflicts of Interest: The authors declare no conflict of interest.

References

1. Knowles, N.; Cayan, D.R. Potential effects of global warming on the Sacramento/San Joaquin watershed and the San Francisco estuary. *Geophys. Res. Lett.* **2002**, *29*, 1891. [CrossRef]
2. Maurer, E.; Duffy, P.P.B. Uncertainty in projections of streamflow changes due to climate change in California. *Geophys. Res. Lett.* **2005**, *32*, L03704. [CrossRef]
3. Goulden, M.L.; Bales, R.C. Mountain runoff vulnerability to increased evapotranspiration with vegetation expansion. *Proc. Natl. Acad. Sci. USA* **2014**, *111*, 14071–14075. [CrossRef] [PubMed]
4. Miller, N.L.; Bashford, K.E.; Strem, E. Potential impacts of climate change on california hydrology. *J. Am. Water Resour. Assoc.* **2003**, *39*, 771–784. [CrossRef]
5. Null, S.E.; Viers, J.H.; Mount, J.F. Hydrologic response and watershed sensitivity to climate warming in California's Sierra Nevada. *PLoS ONE* **2010**, *5*, e9932. [CrossRef] [PubMed]
6. Zhang, W.; Montgomery, D.R. Digital elevation model grid size, landscape representation, and hydrologic simulations. *Water Resour. Res.* **1994**, *30*, 1019–1028. [CrossRef]
7. Cline, D.; Elder, K.; Bales, R. Scale effects in a distributed snow water equivalence and snowmelt model for mountain basins. *Hydrol. Process.* **1998**, *1536*, 1527–1536. [CrossRef]
8. Lassueur, T.; Joost, S.; Randin, C.F. Very high resolution digital elevation models: Do they improve models of plant species distribution? *Ecol. Model.* **2006**, *198*, 139–153. [CrossRef]
9. Beven, K.J.; Kirkby, M.J. A physically based, variable contributing area model of basin hydrology/Un modèle à base physique de zone d'appel variable de l'hydrologie du bassin versant. *Hydrol. Sci. Bull.* **1979**, *24*, 43–69. [CrossRef]
10. Kuo, W.; Steenhuis, T.S.; Mcculloch, C.E.; Mohler, C.L.; Weinstein, D.A.; DeGloria, S.D.; Swaney, D.P. Effect of grid size on runoff and soil moisture for a variable-source-area hydrology model. *Water Resour. Res.* **1999**, *35*, 3419–3428. [CrossRef]
11. Musselman, K.N.; Molotch, N.P.; Brooks, P.D. Effects of vegetation on snow accumulation and ablation in a mid-latitude sub-alpine forest. *Hydrol. Process.* **2008**, *22*, 2767–2776. [CrossRef]
12. Jost, G.; Weiler, M.; Gluns, D.R.; Alila, Y. The influence of forest and topography on snow accumulation and melt at the watershed-scale. *J. Hydrol.* **2007**, *347*, 101–115. [CrossRef]
13. Chaubey, I.; Cotter, A.S.; Costello, T.A.; Soerens, T.S. Effect of DEM data resolution on SWAT output uncertainty. *Hydrol. Process.* **2005**, *19*, 621–628. [CrossRef]
14. Mahmood, T.H.; Vivoni, E.R. Breakdown of hydrologic patterns upon model coarsening at hillslope scales and implications for experimental design. *J. Hydrol.* **2011**, *411*, 309–321. [CrossRef]

15. Mo, X.; Liu, S.; Chen, D.; Lin, Z.; Guo, R.; Wang, K. Grid-size effects on estimation of evapotranspiration and gross primary production over a large Loess Plateau basin, China. *Hydrol. Sci. J.* **2009**, *54*, 160–173. [CrossRef]
16. Tague, C.L.; Band, L.E. RHESSys: Regional hydro-ecologic simulation system—An object-oriented approach to spatially distributed modeling of carbon, water, and nutrient cycling. *Earth Interact.* **2004**, *8*, 1–42. [CrossRef]
17. Hunsaker, C.T.; Whitaker, T.W.; Bales, R.C. Snowmelt runoff and water yield along elevation and temperature gradients in California's Southern Sierra Nevada. *JAWRA J. Am. Water Resour. Assoc.* **2012**, *48*, 667–678. [CrossRef]
18. Jefferson, A.J. Seasonal versus transient snow and the elevation dependence of climate sensitivity in maritime mountainous regions. *Geophys. Res. Lett.* **2011**, *38*. [CrossRef]
19. Holmgren, P. Multiple flow direction algorithms for runoff modelling in grid based elevation models: An empirical evaluation. *Hydrol. Process.* **1994**, *8*, 327–334. [CrossRef]
20. Running, S.W.; Nemani, R.R.; Hungerford, R.D. Extrapolation of synoptic meteorological data in mountainous terrain and its use for simulating forest evapotranspiration and photosynthesis. *Can. J. For. Res.* **1987**, *17*, 472–483. [CrossRef]
21. Glassy, J.; Running, S. Validating diurnal climatology logic of the MT-CLIM model across a climatic gradient in Oregon. *Ecol. Appl.* **1994**, *4*, 248–257. [CrossRef]
22. Coughlan, J.; Running, S. Regional ecosystem simulation: A general model for simulating snow accumulation and melt in mountainous terrain. *Landsc. Ecol.* **1997**, *12*, 119–136. [CrossRef]
23. Thornton, P.E.; Running, S.W.; White, M.A. Generating surfaces of daily meteorological variables over large regions of complex terrain. *J. Hydrol.* **1997**, *190*, 214–251. [CrossRef]
24. Monteith, J.L. Evaporation and environment. *Symp. Soc. Exp. Biol.* **1965**, *19*, 205–234. [PubMed]
25. Jarvis, P.G. The interpretation of the variations in leaf water potential and stomatal conductance found in canopies in the field. *Philos. Trans. R. Soc. B Biol. Sci.* **1976**, *273*, 593–610. [CrossRef]
26. Farquhar, G.D.; von Caemmerer, S.; Berry, J.A. A biochemical model of photosynthetic CO_2 assimilation in leaves of C_3 species. *Planta* **1980**, *149*, 78–90. [CrossRef] [PubMed]
27. Ryan, M.G. Effects of climate change on plant respiration. *Ecol. Appl.* **1991**, *1*, 157–167. [CrossRef]
28. Richardson, J.J.; Moskal, L.M.; Kim, S.-H. Modeling approaches to estimate effective leaf area index from aerial discrete-return LIDAR. *Agric. For. Meteorol.* **2009**, *149*, 1152–1160. [CrossRef]
29. Dunn, S.M. Imposing constraints on parameter values of a conceptual hydrological model using baseflow response. *Hydrol. Earth Syst. Sci.* **1999**, *3*, 271–284. [CrossRef]
30. Nash, J.E.; Sutcliffe, J.V. River flow forecasting through conceptual models part I—A discussion of principles. *J. Hydrol.* **1970**, *10*, 282–290. [CrossRef]
31. Kenward, T.; Lettenmaier, D.P.; Wood, E.F.; Fielding, E. Effects of digital elevation model accuracy on hydrologic predictions. *Remote Sens. Environ.* **2000**, *444*, 432–444. [CrossRef]
32. Hunsaker, C.; Neary, D. Sediment loads and erosion in forest headwater streams of the Sierra Nevada, California. In Proceedings of a Workshop Held during the XXV IUGG General Assembly in Melbourne, Melbourne, Australia, 28 June–7 July 2011.
33. Liu, F.; Hunsaker, C.; Bales, R.C. Controls of streamflow generation in small catchments across the snow-rain transition in the Southern Sierra Nevada, California. *Hydrol. Process.* **2013**, *12*, 1959–1972. [CrossRef]
34. Pradhan, N.R.; Ogden, F.L.; Tachikawa, Y.; Takara, K. Scaling of slope, upslope area, and soil water deficit: Implications for transferability and regionalization in topographic index modeling. *Water Resour. Res.* **2008**, *44*. [CrossRef]
35. Wigmosta, M.S.; Vail, L.W.; Lettenmaier, D.P. A distributed hydrology-vegetation model for complex terrain. *Water Resour. Res.* **1994**, *30*, 1665–1679. [CrossRef]
36. Tague, C.; Dugger, A. Ecohydrology and climate change in the mountains of the Western USA—A review of research and opportunities. *Geogr. Compass* **2010**, *11*, 1648–1663. [CrossRef]
37. Dobrowski, S.Z. A climatic basis for microrefugia: the influence of terrain on climate. *Glob. Chang. Biol.* **2011**, *17*, 1022–1035. [CrossRef]

Article

Assessment of the Impact of Subsurface Agricultural Drainage on Soil Water Storage and Flows of a Small Watershed

Mushombe Muma [1,*], Alain N. Rousseau [1] and Silvio J. Gumiere [2]

1 Institute National de la Recherche Scientifique, Centre Eau Terre Environnement (INRS-ETE), 490 Rue de la Couronne, Ville de Québec, QC G1K 9A9, Canada; alain.rousseau@ete.inrs.ca
2 Département des Sols et de Génie Agroalimentaire, 2480 Boulevard Hochelaga, Université Laval, Ville de Québec, QC G1V 0A6, Canada; silvio-jose.gumiere@fsaa.ulaval.ca
* Correspondence: mushombe.muma@ete.inrs.ca; Tel.: +1-418-654-2621

Academic Editors: Christopher J. Duffy and Xuan Yu
Received: 15 April 2016; Accepted: 26 July 2016; Published: 3 August 2016

Abstract: 3D hydrological modeling was performed, using CATHY (acronym for CATchment HYdrology model), with the basic objective of checking whether the model could reproduce the effects of subsurface agricultural drainage on stream flows and soil water storage. The model was also used to further our understanding of the impact of soil hydrodynamic properties on watershed hydrology. Flows simulated by CATHY were consistent with traditional subsurface drainage approaches and, for wet years, flows at the outlet of the study watershed corroborated well with observed data. Temporal storage variation analyses illustrated that flows depended not only on the amount of rainfall, but also on its distribution throughout the year. Subsurface agricultural drainage increased base and total flows, and decreased peak flows. Hydrograph separation using simulated results indicated that exfiltration was the most dominant process; peak flows were largely characterized by overland flow; and subsurface drain flow variations were low.

Keywords: CATHY model; subsurface drainage; storage variation; exfiltration; infiltration; overland flow; outlet flow; peak flow; base flow; total flow

---

1. Introduction

Subsurface drainage is a common agricultural practice to improve aeration and trafficability of soils in regions characterized by seasonal high water tables [1,2]. Subsurface drainage system helps to increase crop yield of poorly drained soils by providing a better environment for plant growth, especially during wet periods [3] and improve field conditions for timely tillage, planting and harvesting [4,5].

When subsurface drains are in place, the drainable water fraction of the soil profile is converted to short-term (detention) storage over a period of few hours, days, or weeks, depending on a number of variables [6]. These include subsurface drain size, depth and spacing, soil type, outlet size/condition, and whether or not under continuous rainfall or snowmelt conditions. When drainable water is removed from the soil profile, infiltration can then occur. This is due to available soil pore space allowing water that would otherwise be stored in the surface depressions to infiltrate and have a direct pathway to downstream flow via the subsurface drains [6].

Although agricultural production has benefited from agricultural drainage in many regions and countries, there are concerns about potential environmental impacts [7]. The most dramatic hydrological changes in a landscape occur when the latter is converted from native vegetation to intensive cropping systems. When tile drains are implemented, they can substantially alter the total water yield from a field or small watershed as well as modify the timing and shape of the hydrograph [8]. Tile drainage increases the proportion of annual precipitation that is discharged to surface waters via subsurface flow relative to the amount that is stored, evaporated or transpired [9–11].

From several studies across the Midwestern United States and Canada, discharge from subsurface drains constitutes the majority of stream flow in many agricultural watersheds. It was found that tile drainage contributed 51% of annual stream flow in a headwater watershed in Ohio [12]. Meanwhile, in a watershed in Ontario, Canada, it was estimated that 42% of annual watershed discharge originated from subsurface drain flow [13]. Culley and Bolton [14] and Xue et al. [15] also estimated that 60% and 86%, respectively, of stream flow was derived from tile drain. Although total water yield from a field tends to increase following installation of subsurface drainage, surface runoff and sediment yields are often significantly decreased [2,16,17]. Subsurface drainage reduces both peak outflows and the frequency of surface runoff events at sites characterized by high water tables or prolonged surface saturation ("ponding") under undrained conditions [16,18].

The main water quality concern about subsurface drainage is the increased loss of nitrates and other soluble constituents (i.e., pesticides and dissolved phosphorus, to name a few) that can move through soils and end up in nearby surface waters. Meanwhile, it is generally agreed that, through the installation of subsurface drainage, the amount of particulate phosphorus and soil transported by surface runoff is decreased because the volume of surface runoff is reduced [19,20]. However, dissolved phosphorus loss via artificial drainage has been shown to contribute to the accelerated eutrophication of rivers, lakes, estuaries and even coastal waters, including some of the most challenging cases of agriculturally-derived eutrophication [21]. On more permeable soils, where infiltration, soil water storage capacity, and lateral conductivity/seepage is large enough to handle a given storm event, subsurface drainage may have the opposite effect for an event of equivalent magnitude, increasing peak flows by increasing the speed of subsurface discharges [18,22]. In fields where diffuse pollution dominates, installation of artificial drainage can increase peak flows, which accounts for the majority of phosphorus loss, and hence can improve the hydrological connectivity of otherwise isolated areas of the landscape [23,24]. Increased peak flows due to agricultural tile drainage have also been shown to increase channel widening and bank erosion [25].

The impact of artificial drainage strongly depends on characteristics of an individual site, including: topography (slope), drainage system design (spacing, depth and size of drains) as well as soil type (hydraulic conductivity) [26,27]. The hydrological response of subsurface drainage during a given event varies based on event characteristics, such as rainfall amount and intensity [28], and antecedent soil moisture conditions [13,17,29]. Therefore, there is a need for research to better understand the impact of subsurface drainage at both field scale and catchment scale.

The goal of this study was, using a coupled surface water groundwater hydrological model, to quantify the impact of subsurface agricultural drainage and soil properties on the hydrologic functioning of a micro-watershed under intensive farming of annual crops such as grain corn (*Zea mays*) or soybean (*Glycine max*). The specific objectives of this study were: (1) to simulate the micro-watershed outlet flow and analyse the effect of subsurface agricultural drainage on soil water storage variation; and (2) to analyse the impact of drainage networks and soil saturated hydraulic conductivity on: (2a) micro-watershed outlet flow; (2b) surface runoff, surface water and groundwater coupling; and (2c) micro-watershed outlet flow hydrograph.

To our knowledge, although there are studies dealing with testing (evaluation) and application of the coupled surface water groundwater hydrological model used in this study, CATHY (acronym for CATchment HYdrology) [30–43], there are no exhaustive studies investigating the influence of artificial subsurface drainage on flows at the watershed level. CATHY is the type of process-based model that is required to study the spatio-temporal variability of soil moisture, groundwater flow and surface runoff. Furthermore, as a virtual laboratory, CATHY provides a powerful deterministic approach to further our understanding of the impact of subsurface drainage on stream flow.

In this respect, not only does this study make a significant contribution to the understanding of CATHY, but it is also the first investigation dealing with the impact of subsurface agricultural drains on the partitioning of flow at a watershed outlet using a 3D model.

2. Materials and Methods

2.1. Study Site

The study site is a micro-watershed of 2.4 km^2 with latitude and longitude of 46°29′00″ N and 71°14′00″ W, respectively, located in the watershed of the Bras d'Henri River which covers approximately 167 km^2, a sub-watershed of the Beaurivage River. The latter is located about 30 km southeast of Quebec City on the south shore of the Saint Lawrence River, within the Chaudière-Appalaches agro-climatic region (Figure 1). Two-thirds of this drainage area is currently dedicated to agriculture (large-scale farming, pastures, etc.) while the balance remains in its natural state, which includes wooded areas and wetlands.

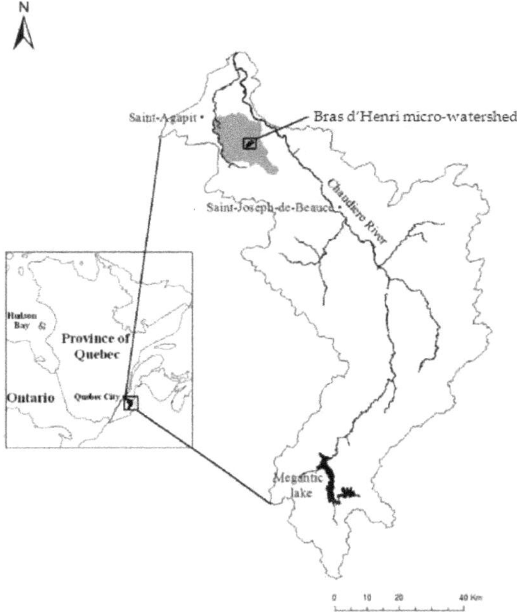

Figure 1. Location of the Bras d'Henri micro-watershed.

The region is characterized by long and cold winters, short and cool summers, and significant annual precipitation, approximately 1150 mm/year, a third of which accumulates as snow. During the summer, precipitation is generally greater than the amount of water lost to evapotranspiration. Thus, no water deficit is observed at soil level, except for a slight possibility in certain sectors of coarse sandy soil with gravel and stones which drains well to very rapidly from the plain and this exclusively during the month of July [44]. The site is covered by the "Watershed Evaluation of Beneficial Management Practices (WEBs)" program, launched nationally in 2004 and managed by Agriculture and Agri-Food Canada (AAC), which aims at measuring the environmental and economic impact of certain Beneficial Management Practices (BMPs) on the quality of water in seven hydrographic micro-watersheds in Canada [45–47].

The site is characterized by intensive agricultural production composed essentially of silage (53.1%), grain corn (*Zea mays*, 27.8%) and soybean (*Glycine max*, 8.1%). Figure 2a shows the soil codes of the different soil families from the surface to a depth of 1.25 m while the Table 1 gives their names and percentage of sand, silt and clay. The corresponding geometric means of the soil saturated hydraulic conductivities are given in Table 2. The location of the drainage systems (at a depth of 1.20 m and occupying 30% of the total area of the watershed) is presented in Figure 2b.

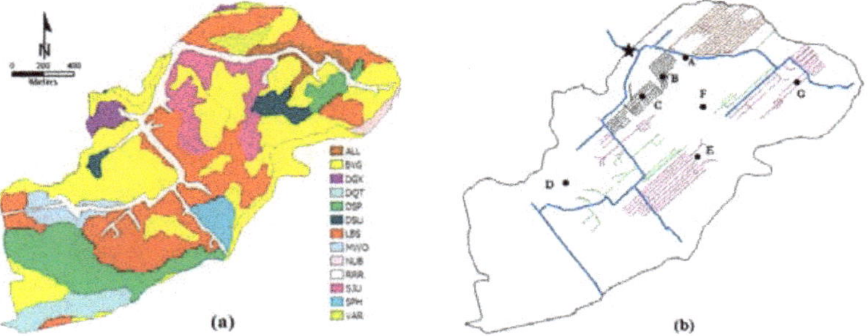

Figure 2. (**a**) Spatial distribution of different soil families; and (**b**) locations of subsurface drain systems and piezometers in the micro-watershed (blue lines are stream network, other colours are subsurface drainage systems, the star indicates the micro-watershed outlet, and dots are piezometers) (adapted from Muma et al. [48]).

Table 1. Different families of soil and their percentage of sand, silt and clay.

Soil Code	Soil Family	Sand (%)	Silt (%)	Clay (%)
BVG	Sandy	67–97	2–28	1–13
SJU	Fine sandy loam	68–94	4–31	1–4
VAR	Fine sandy loam	73–97	2–21	1–6
SPH	Silt loam	55–76	22–41	2–5
DGX	Coarse loamy	29–59	34–60	5–13
DSU	Coarse loamy	45–62	33–43	5–12
NUB	Loamy	8–60	33–64	7–32
LBS	Loamy	9–30	58–73	12–23
DQT	Coarse loamy	64–84	7–19	9–20
DSP	Coarse loamy	52–58	33–42	6–10
MWO	Coarse loamy	57–84	10–27	6–17
ALL	Loamy	45–62	33–43	5–12
RRR	Not classified	7–30	58–73	12–23

Table 2. Saturated hydraulic conductivity values (K_S), in m/s, and their geometric mean of the different soil families up to depth of 1.25 m.

Soil Code	Soil Profile (cm)				
	0–25	**25–50**	**50–75**	**75–100**	**100–125**
BVG00	5.44×10^{-5}	1.91×10^{-5}	2.94×10^{-4}	1.78×10^{-4}	1.78×10^{-4}
BVG49	2.78×10^{-5}	9.03×10^{-5}	1.39×10^{-4}	1.39×10^{-4}	1.39×10^{-4}
SJU	9.03×10^{-6}	9.03×10^{-5}	9.03×10^{-5}	9.03×10^{-6}	9.03×10^{-6}
VAR	8.42×10^{-6}	8.11×10^{-5}	9.75×10^{-5}	8.11×10^{-5}	8.11×10^{-5}
SPH	5.44×10^{-5}	1.91×10^{-5}	2.94×10^{-4}	1.78×10^{-4}	1.78×10^{-4}
DGX	2.43×10^{-5}	1.70×10^{-5}	1.39×10^{-7}	1.39×10^{-7}	1.39×10^{-7}
DSU	1.39×10^{-7}	2.22×10^{-7}	1.39×10^{-7}	2.78×10^{-8}	2.78×10^{-8}
NUB	5.56×10^{-6}	1.42×10^{-6}	1.39×10^{-7}	1.39×10^{-7}	1.39×10^{-7}
LBS	8.06×10^{-7}	1.04×10^{-5}	8.33×10^{-8}	1.39×10^{-7}	1.39×10^{-7}
DQT	1.31×10^{-4}	1.25×10^{-4}	2.73×10^{-4}	2.66×10^{-4}	2.66×10^{-4}
DSP	3.06×10^{-7}	1.21×10^{-5}	2.31×10^{-6}	4.44×10^{-7}	4.44×10^{-7}
MWO	5.86×10^{-5}	6.00×10^{-5}	1.36×10^{-4}	1.95×10^{-5}	1.95×10^{-5}
ALL	1.39×10^{-7}	2.22×10^{-7}	1.39×10^{-7}	2.78×10^{-8}	2.78×10^{-8}
RRR	2.78×10^{-6}	2.78×10^{-6}	2.78×10^{-6}	2.78×10^{-6}	2.78×10^{-6}
Geometric mean of *Ks*	6.04×10^{-6}	1.16×10^{-5}	7.07×10^{-6}	3.51×10^{-6}	3.51×10^{-6}

2.2. Data Source

The database provided by AAC includes meteorological data (average air temperature, relative humidity, saturated vapour pressure, wind speed at an altitude of 2 m, net radiation, precipitation, etc.), saturated hydraulic conductivity of soils up to a depth of 1.25 m obtained from suction tests, and water height data measured every 15 min at the outlet of the micro-watershed by a probe installed above the stream. The water flow rate data were obtained by means of calibration curves (linking flow rate to water height) determined by Ratté-Fortin [49].

The saturated hydraulic conductivities of the soil at a depth greater than 1.25 m were obtained from permeability tests or slug tests [50,51] conducted at the piezometer locations identified in Figure 2b. The saturated hydraulic conductivity values varied from 3.33×10^{-6} to 2.08×10^{-5} m/s with a geometric mean of 1.29×10^{-5} m/s.

Given the number of soil families and piezometers, the saturated hydraulic conductivity value of each horizon of the soil profile was used to calibrate the hydrological model, the geometric means of each horizon being used as initial values.

2.3. The CATHY Model

CATHY is a spatially-distributed physics-based model that integrates surface and subsurface flows [52]. The surface flow (in rills and waterway) is generally formulated by the Saint Venant equation:

$$\frac{\partial Q}{\partial t} + C_k \frac{\partial Q}{\partial s} = D_h \frac{\partial^2 Q}{\partial s^2} + C_k q_s \tag{1}$$

This equation is resolved numerically by the finite difference method. Since the CATHY model is a rill flow-based model, Equation (1) is represented in a 1D coordinate system s (L) to describe each element of the surface drainage network, where Q is the flow rate along the rivulet/waterway (L^3/T), C_k is the kinematic celerity (L/T), D_h is the hydraulic diffusivity (L^2/T), and q_s is the input flow rate (positive) in the medium or the output flow rate (negative) from the subsurface to the surface (L^3/LT).

Richards 3D equation (Equation (2)), which describes flow in a variably-saturated media (subsurface flow), is resolved numerically in space by the Galerkin finite element method using tetrahedral elements. Since the storage and conductivity terms strongly depend on pressure, Equation (2) is mostly nonlinear and, as a result, is linearized by Picard's or Newton's iterative methods [53]. Thus, the partial differential equation, which mathematically describes the flow in porous media, is:

$$S_w S_s \frac{\partial \psi}{\partial t} + \phi \frac{\partial S_w}{\partial t} = \nabla . \left[K_s K_r \left(\nabla \psi + \eta_z \right) \right] + q_{ss} \tag{2}$$

where $S_w = \theta/\theta_s$ is the saturation level, θ is the volumetric water content (-), θ_s is the saturated water content (generally equal to the porosity ϕ), S_s is the specific storage coefficient (1/L), ψ is the capillary potential (L), t is time (T), ∇ is the gradient operator, K_s is the hydraulic conductivity tensor (LT), $K_r(\psi)$ is the relative conductivity function (-), $\eta_z = (0,0,1)^T$, z is the vertical coordinate oriented towards the top (L), and q_{ss} represents the source (positive) or the well (negative) (L^3/L^3T).

CATHY is a complex hydrological model with different and varied advantages. The surface hydrology links terrain topography, hydraulic geometry and flow dynamics. Its outputs include surface pressure head (or ponding), overland fluxes, subsurface pressure head and moisture content values, and groundwater velocities. Numerous other variables can be derived from these main outputs such as aquifer recharge, catchment saturation, and stream flow. Surface and subsurface contributions to runoff can be computed at any specified surface node within the catchment, and by default also at the catchment outlet, representing the total stream flow at the outlet.

As a robust model, the use of CATHY requires many input data (parameters). In its current version, no dimension of diameter or radius is taken in account regarding the tile-drain representation in the subsurface porous medium.

To our knowledge, there has not been any study involving CATHY and dealing with the equifinality thesis (identifiability of equally-performing sets of parameter values) or estimation of uncertainty. To our knowledge, all reported papers on CATHY have shown that the model has always been calibrated using observed parameter values mainly of soil saturated hydraulic conductivity and soil porosity. This is mostly a result of the large computational time required to run CATHY as reported in our global sensitivity analysis of the model [48]. It is in this deterministic context that this study was conducted rather than an underestimation of parameter uncertainties.

The subsurface drainage system is represented in the model by assigning a pressure potential (Dirichlet boundary condition) or a flux (Neumann boundary condition) to the corresponding node. More detailed descriptions of the model can be found in the work of Camporese et al. [52].

2.4. Setting up CATHY at the Watershed Scale

2.4.1. Discretization of the Porous Medium

The application of CATHY was developed on the micro-watershed using a Digital Elevation Model (DEM) with a resolution of 20 m. The porous medium was discretized into 15 layers with thinner layers at the surface and near the nodes located closest to the drainage networks at a depth of 1.20 m (interface of the 7th and 8th layers). This was done to properly account for the interactions between surface and subsurface waters and for the influence of drains on the flow. The discretization of the watershed surface resulted in 6148 cells; each one divided into 2 triangles, producing 12,296 cells (linked by 6391 nodes). The latter cells were projected vertically on the 15 layers with each triangle creating three tetrahedrons. The porous medium was thus represented by 553,320 (12,296 $\times$ 3 $\times$ 15) tetrahedral elements linked by 102,256 nodes (6391 $\times$ (15 + 1)).

The values of the soil hydrodynamic properties, that is the saturated soil hydraulic conductivity in the horizontal (in X and Y, or $KsXY$) and vertical (KsZ) planes, the specific storage coefficient (Ss), and the porosity (ϕ), associated with each layer of the porous medium are introduced in Table 3. In the vertical direction, there are 5 groups of layers: the first group includes layers 1 and 2, the second group layer 3, the third group layer 4, the fifth group layers 5 to 8 (location of the drainage networks), and the fifth group layers 9 to 15 as a whole. The first two layers make up the superior or surface layer where the partition of available water (precipitation) into surface water and infiltration takes place, as well as the superficial transfer.

Table 3. Soil hydraulic properties of the porous medium for Scenario 1 (corrugated line between layers 7 and 8 indicating the location of drainage network nodes in the drained parcels.

Layer Number	ΔZ (m)	Z (m)	KsXY (m/s)	KsZ (m/s)	Ss (m^{-1})	ϕ (-)
01 02	0.10 0.15	0.10 0.25	6.06×10^{-5}	6.06×10^{-6}	3.75×10^{-3}	0.52
03	0.25	0.50	3.80×10^{-4}	3.80×10^{-5}	1.15×10^{-3}	0.48
04	0.25	0.75	4.04×10^{-5}	4.04×10^{-6}	3.75×10^{-3}	0.37
05 06 07 08	0.25 0.10 0.10 0.10	1.00 1.10 1.20 1.30	6.06×10^{-5}	6.06×10^{-6}	3.75×10^{-3}	0.48
09 10 11 12 13 14 15	0.30 0.45 0.50 0.60 0.64 0.70 0.85	1.60 2.05 2.65 3.26 3.90 4.60 5.45	4.20×10^{-4}	4.20×10^{-5}	1.15×10^{-3}	0.50

In this study, four scenarios of saturated hydraulic conductivity values were used. The first scenario (Sc. 1), considered as the baseline scenario, is a non-homogenous medium with anisotropic values in each group of layers as given in Table 3 and derived from measured saturated hydraulic conductivity values. They were used to calibrate (saturated hydraulic conductivity was the parameter mostly adjusted during this process) and validate the model. In order to analyse the impact of drainage networks and soils on: (a) the flow at the outlet of the micro-watershed; (b) the coupling of surface water and groundwater, and surface runoff; and (c) the hydrograph at the outlet of the micro-watershed, in addition to the baseline scenario, three additional values of saturated hydraulic conductivity were taken into account based on the fact that, when measuring the saturated hydraulic conductivities in situ or in laboratory, their values can fluctuate within a certain range of order 10 (Table 4). To simplify their visual interpretation, these are represented by circular- or elliptic-shaped diagrams where horizontal and vertical axes represent KsXY and KsZ, respectively (Figure 3). Scenario 2 (Sc. 2) has a non-homogenous medium with anisotropic layers, Scenario 3 (Sc. 3) is a non-homogenous medium with isotropic layers, and Scenario 4 (Sc. 4) has a non-homogenous medium with isotropic layers. All of these scenarios were applied to the calibration period (year 2006).

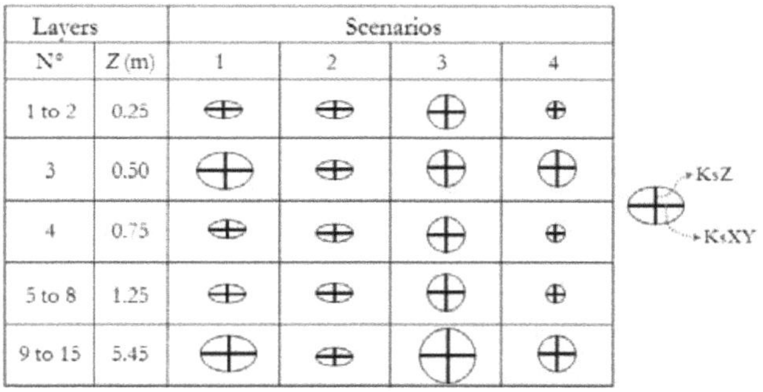

Layers		Scenarios			
N°	Z (m)	1	2	3	4
1 to 2	0.25				
3	0.50				
4	0.75				
5 to 8	1.25				
9 to 15	5.45				

Figure 3. Representation of saturated hydraulic conductivity values representation in the different layers of the porous medium in the form of diagrams.

Table 4. Isotropic and anisotropic properties of soil saturated hydraulic conductivity (*) analysed.

Layers			Scenarios			
N°	**ΔZ (m)**	**Z (m)**	**1 (Reference)**	**2**	**3**	**4**
01	0.10	0.25	$6.06 \times 10^{-5}/6.06 \times 10^{-6}$	$6.06 \times 10^{-5}/6.06 \times 10^{-6}$	$6.06 \times 10^{-5}/6.06 \times 10^{-5}$	$6.06 \times 10^{-6}/6.06 \times 10^{-6}$
02	0.15					
03	0.25	0.50	$3.80 \times 10^{-4}/3.80 \times 10^{-5}$			$3.80 \times 10^{-5}/3.80 \times 10^{-5}$
04	0.25	0.75	$4.04 \times 10^{-5}/4.04 \times 10^{-6}$			$4.04 \times 10^{-6}/4.04 \times 10^{-6}$
05	0.25	1.30	$6.06 \times 10^{-5}/6.06 \times 10^{-6}$			$6.06 \times 10^{-6}/6.06 \times 10^{-6}$
06	0.10					
07	0.10					
08	0.10					
09	0.30	5.45	$4.20 \times 10^{-4}/4.20 \times 10^{-5}$		$4.20 \times 10^{-4}/4.20 \times 10^{-4}$	$4.20 \times 10^{-5}/4.20 \times 10^{-5}$
10	0.45					
11	0.50					
12	0.60					
13	0.64					
14	0.70					
15	0.85					

Notes: * All hydraulic conductivity values are expressed in m/s. In each column, the value on the left of the slash is the horizontal hydraulic conductivity (*X* and *Y*), while the value on the right is the vertical hydraulic conductivity (*Z*). The broken line between layers 7 and 8 indicates the location of drainage network nodes in the subsurface drained plots.

In their sensitivity analysis study of CATHY to the soil hydrodynamic properties, Muma et al. [53] noticed that the saturated hydraulic conductivity of the deeper layers (fifth group of layers) had a significant impact on to drain discharge and outlet of the micro-watershed flow. Furthermore, they revealed that the vertical saturated hydraulic conductivity in the two surface layers (first group of layers) as well as the vertical and lateral saturated hydraulic conductivity in the layers where the subsurface drains are located deserved special attention due to their strong interaction with other parameters with regards to drain discharge. Based on these findings, the hydraulic conductivity in the porous medium is in decreasing order as follows: Sc. 3 > Sc. 1 > Sc. 2 > Sc. 4.

2.4.2. Boundary Conditions and Initial Humidity Conditions in the Soil

The study period stretches from 1 May (121 JD, Julian date) to 31 October (304 JD) of each one of the following years: 2006, 2007, 2008 and 2009. This period, which corresponds to the growing season [54], is characterized by surface runoff, infiltration, evapotranspiration and intense agricultural activities in the micro-watershed.

Boundary conditions at the surface of the watershed are given by the effective precipitation; that is, real precipitation minus potential evapotranspiration. The latter was calculated by the FAO Penman–Monteith reference evapotranspiration weighted by crop coefficient [55].

The values of total effective precipitation for the months of May to October are 343, 212, 337 and 382 mm for years 2006, 2007, 2008 and 2009, respectively. For 2006, precipitation was abundant near the end of the simulation period; that is, around mid-autumn (fall). Year 2007 was characterized by low precipitation and the lack of precipitation was seen over a large portion of the spring and summer. The wettest summer was in 2008, with high peaks of precipitation in spring, whereas year 2009 presents a lack of precipitation in summer.

Regional values of mean and median effective precipitation, as well as percentiles 25% and 75%, for the past forty years (1971 to 2010) for the period from 1 May to 31 October are 242, 220, 146, 329 mm, respectively. These regional values were established from the meteorological data in the database for the HYDROTEL model [56–58] applied to the Beaurivage watershed [59]. In this last application, precipitation and temperature were weighted means of the three stations nearest to the study site and the potential evapotranspiration was calculated with the equation developed at Hydro-Québec [57]. It can be observed that only the percentile 25% is lower than all effective precipitation values for all the years under study. Among the other three statistical measures of effective precipitation (mean, median and percentile 75%), only year 2007 has the lowest value. This confirms once again that not only year 2007 was the driest among the years being studied, but also it was regionally among the driest years over the past forty years.

At the beginning or first time step of each simulation, the initial soil water conditions are given by a water table set at 20 cm under the soil surface. This choice was based on the fact that the beginning of the simulations coincides with the spring season, which corresponds with the thawing period of the soil. It is known that, when the thaw ends, the water accumulated in the depressions infiltrates and reaches the water table to bring it closer to the soil surface in the Saint Lawrence lowlands [60]. The water table is generally near the surface early in spring, drops significantly during the summer months, and then rises again in the fall [61]. Furthermore, from a modelling point of view, under these spring conditions drainage systems fulfill their role when the initial water table is so high.

Subsurface drainage systems, located at 1.20 m from the soil surface (at the interface of the 7th and 8th layers), are represented by nodes to which was assigned a zero head pressure (i.e., atmospheric pressure), known as Dirichlet condition.

2.4.3. Variables Analyzed

Figure 4 illustrates the different variables analysed. These are subsurface drain flow and micro-watershed outlet flow, storage variation, surface runoff, infiltration and exfiltration.

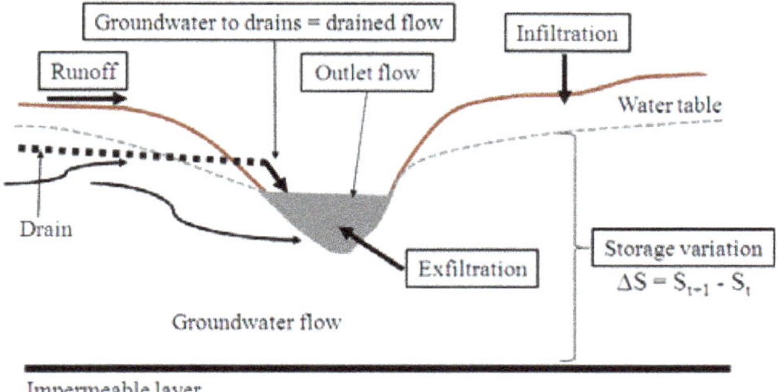

Figure 4. Illustration of the analysed variables: outlet flow, drained flow, runoff, infiltration, exfiltration and storage variation (ΔS with S_{t+1} the storage at current time step and S_t the storage at previous time step).

2.4.4. Evaluation of Model Performance

To evaluate the performance of the model, three criteria introduced in Table 5 were used: the error in flow rate peaks (*EP*), error in flow volumes (*EV*), and Nash–Sutcliffe Evaluation criterion (*NSE*).

Table 5. Evaluation of model performance.

Criteria	Statistical Approach	Best Value	Goal and Interpretation
Error in peak flow rate Error in flow volume	$EP = \left[\frac{\max(O)-\max(P)}{\max(O)}\right]_{i=1}^{n}$ $EV = \frac{\sum_{i=1}^{n} C_i - \sum_{i=1}^{n} P_i}{\sum_{i=1}^{n} O_i}$	0	- Minimize the criteria value. - Low (absolute) value: better performance, i.e., more precise simulation. - Zero (0) indicates that there is no difference between predicted and observed values.
Nash coefficient	$NSE = 1 - \frac{\sum_{i=1}^{n}(P_i-O_i)^2}{\sum_{i=1}^{n}(O_i-\overline{O})^2}$	1	- Maximize the criteria value. - One (1) means that the model is perfect: simulated values are in perfect agreement with observed ones.

Note: n is the number of days during the simulation period. O_i and P_i are observed (measured) and predicted (simulated) discharges on Julian day i, respectively. $\overline{O}$ is the arithmetic mean of observed flows, and max and min are the maximum and minimum flow rates for the entire simulation period, respectively.

To account for the errors associated with observed flows from the measurement of water heights up to the establishment of the rating curve, a modified Nash criteria (NSE_m) was used [62], allowing ±20% error on the observed values as mentioned above. The formula reads as follows:

$$NSE_m = 1 - J \tag{3}$$

where NSE_m is the modified Nash criterion and J is defined as follows:

$$\begin{cases} \text{if } P_i < O_{i30} \text{ then } J_{i30} = \frac{\sum_{i=1}^{N_{i30}}(O_{i30}-P_i)^2}{\sum_{i=1}^{N_{i30}}(O_i-\overline{O})^2} \\ \text{if } P_i > O_{i70} \text{ then } J_{i70} = \frac{\sum_{i=1}^{N_{i70}}(O_{i70}-P_i)^2}{\sum_{i=1}^{N_{i70}}(O_i-\overline{O})^2} \end{cases} \tag{4}$$

where P_i and O_i represent the simulated and observed flows on day J, respectively, and $\overline{O}$ is the arithmetic mean of observed flows during the simulation period. As mentioned above, if a simulated flow is included in the $N_{i30–70}$ values (for those included between O_{i30} and O_{i70}), it is deemed sufficient and the value of the numerator $J_{i30–70}$ is zero by definition. Thus, NSE_m is merely the sum of J_{i30} and J_{i70}.

Flow measurement from year 2006 was used for the calibration process, whereas years 2007, 2008, and 2009 for the validation process with respect to the performance of the model and the storage variation behaviour according to different effective precipitation shapes.

3. Results

3.1. Simulation of the Flow at the Micro-Watershed Outlet and Analysis of the Effect of Subsurface Drainage on the Storage Variation

3.1.1. Calibration and Validation of the Model: Micro-Watershed Outlet Flow

Table 6 introduces the values of the model performance statistics while Figure 5 presents measured and simulated flows, along with the uncertainty zone of ±20% on the observed values (in grey) depicted by the confidence interval of 30%–70% for the calibration year (2006) and the validation periods: (2007, 2008, and 2009). Since year 2006 has the smallest values of EP (0.16) and EV (−0.04), the resulting hydrographs demonstrate a good visual agreement between measured and simulated flows (Figure 5), corresponding to a good performance in terms of NSE (0.71). Over a large part of the hydrograph of years 2007 and 2009, measured flows are larger than those simulated, implying EP values (0.26 and 0.41, respectively) and EV (−0.33 and −0.74, respectively) values greater than 25%. From the fact that flow rates of year 2009 are greater than those of year 2007, the model performance was good (NSE = 0.67) for year 2009 and not acceptable for year 2007 (NSE = 0.17). Given that year 2008 has a lower gap between measured and simulated hydrographs for low flow rates compared to those of years 2007 and 2009, it has the smallest EV value (−0.05). Therefore, the model performance was good (NSE = 0.72).

Looking at the uncertainty zone (in grey) of ±20% with the 30%–70% confidence interval, years 2006 and 2008 are those where the simulated hydrographs are more or less within the uncertainty band. For low observed flow rates, the simulated hydrograph is generally not within the uncertainty zone especially for years 2007 and 2009. Thereby, the model performance increased from 0.71, 0.17, 0.72, and 0.67 in terms of NSE to 0.89, 0.69, 091, and 0.81 in terms of NSE_m for years 2006, 2007, 2008 and 2009, respectively. In this work, the model was calibrated according to the measured flow rates at the outlet. The distributed impact of subsurface drainage over catchment behaviour will be addressed in another paper.

Table 6. Model performance statistics.

Process	Year	EP	EV	NSE	NSE_m
Calibration	2006	0.16	−0.04	0.71	0.89
Validation	2007	0.26	−0.33	0.17	0.69
	2008	0.36	−0.05	0.72	0.91
	2009	0.42	−0.74	0.67	0.81

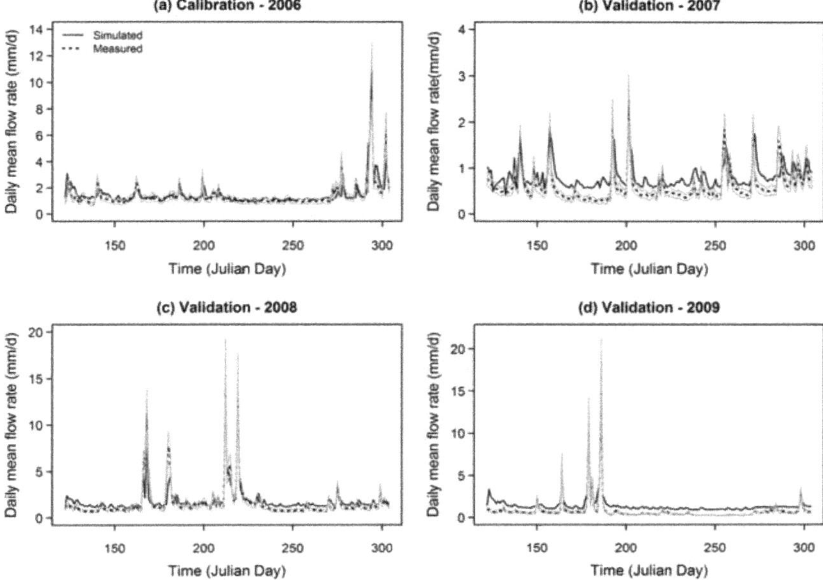

Figure 5. Measured and simulated flows along with the uncertainty zone of ±20% on the observed values (in grey) depicted by the confidence interval of 30%–70% for the (**a**) calibration year (2006) and the validation periods: (**b**) 2007; (**c**) 2008; and (**d**) 2009.

3.1.2. Effect of Subsurface Drainage on Storage Variation

The daily mean storage variation (ΔS) cumulative curves with and without drainage systems in the porous media for each year and effective precipitation are illustrated in Figure 6 where two distinct behaviours are observed. The first relates to years 2006, 2007 and 2009, during which summer is characterized by low precipitation. These years, with low precipitation during summer period, present ΔS cumulative curves with and without drainage systems that intersect at a point on the simulation period. To the left of the point of intersection, the curve without drainage systems is above the one with drainage systems. To the right of the point of intersection, the situation reverses: the curve without drainage systems passes below that with drainage systems. The absence of flow from the subsurface drain networks in summer explained the inversion of the two curves (Figure 6a,b,d). The second behaviour concerned year 2008, which is the only one having registered the largest precipitation in the summer. The two curves with and without drainage systems remained almost parallel over the entire simulation period, with the second curve always above the first one (Figure 6c). According to time interval during which the curve with drainage systems is below that without subsurface drainage, more water was drained from the porous medium during 2008 while year 2007 yielded less drained water.

Considering a given ΔS variation (Y-axis) in the time intervals where the curve with drainage systems is below that with drainage systems, it can be seen that the time corresponding to the first curve is lower than the one for the second curve. This illustrates that drainage networks evacuate precipitation faster towards the outlet. Likewise, considering any time (X axis), it was observed that the ΔS curve with subsurface drainage systems is below that without subsurface drainage systems as said earlier (Figure 6c).

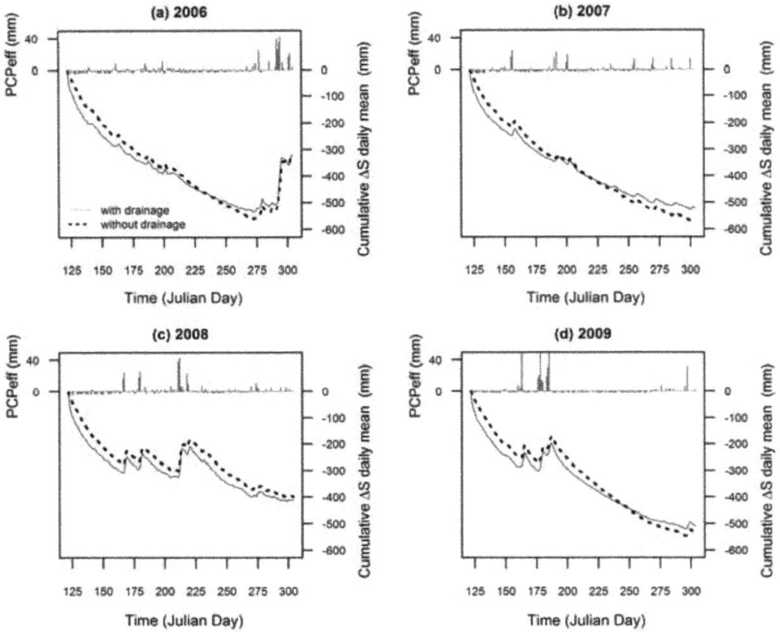

Figure 6. Effect of subsurface drain systems on the cumulative storage variation in the micro-watershed for: (**a**) the calibration period (2006) and the validation periods: (**b**) 2007; (**c**) 2008; and (**d**) 2009.

3.2. Impact of Subsurface Drainage and Soil Hydraulic Conductivity on Micro-Watershed Outlet Flow

Figure 7 presents the simulated cumulative outlet flow with and without subsurface drainage systems. It is noted that: (1) the flow in presence of subsurface drain systems is larger than in the situation without subsurface drainage for the four scenarios; (2) the lower the saturated hydraulic conductivities, the lower the flows (with and without subsurface drains); and (3) the gap or the difference between the hydrographs with and without subsurface drainage increases when saturated hydraulic conductivities are higher. Considering the cumulative flow values at the end of the simulations, for Sc. 1 to Sc. 4 those with subsurface drainage increase by 12.5%, 4.5%, 22.0% and 3.0%, respectively. Compared to those of Sc. 1, for Sc. 2 and Sc. 4 the flows decrease 60.0% and 75.0%, respectively, and increase to more than 10.0% for Sc. 3. The decreases or increase in flows with subsurface drainage were higher than those of flows without drainage.

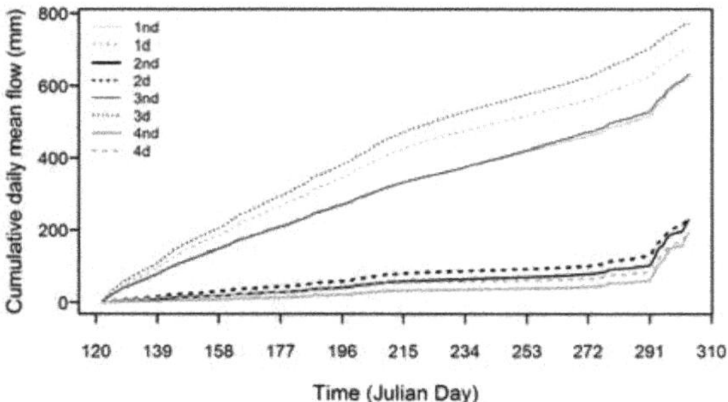

Figure 7. Cumulative daily mean flow at the micro-watershed outlet for Scenarios 1 to 4 with labels "d" and "nd" representing simulations with and without drainage systems, respectively.

Figure 8 presents the flow characteristics for Scenarios 1 to 4; namely: (a) daily mean flow expressed in form of box plot of logarithmic values; (b) maximum flow; and (c) minimum flow. For all of these scenarios, it was observed generally that the maximum flow without subsurface drainage is larger than that with subsurface drainage. Meanwhile, the reverse was observed for the minimum flow: the minimum flow without subsurface drainage is lower than that with drainage. As for cumulative flows at the end of each simulation, it was found that differences between extreme flows (maximum and minimum with and without drainage) are larger when the soil is more conductive (higher saturated hydraulic conductivity).

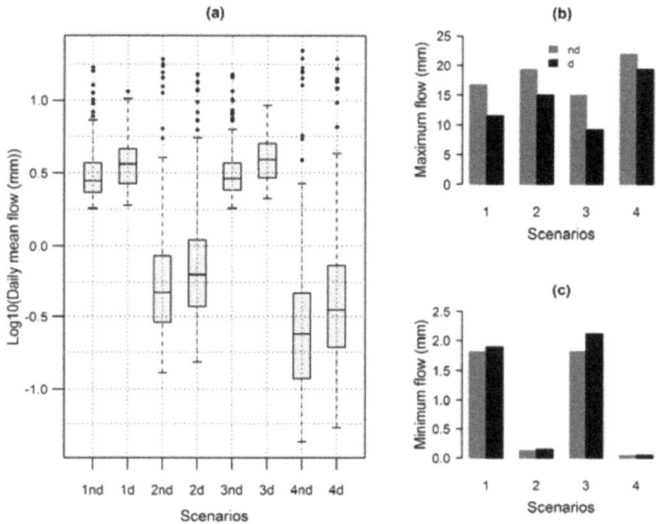

Figure 8. Micro-watershed outlet daily mean flows: (**a**) log-transformed daily mean flow; (**b**) their maximum values; and (**c**) their minimum values ("*d*" and "*nd*" expressing simulation with and without subsurface drainage, respectively) for Scenarios 1 to 4.

3.3. Impact of Subsurface Drainage and Soil Saturated Hydraulic Conductivity on Surface Runoff, Surface Water and Groundwater Coupling

Figure 9 presents the different flow values of exfiltration, infiltration and surface runoff for simulations with and without subsurface drainage systems for Sc. 1 to Sc. 4 in the form of box plots. From these results, it appears that generally exfiltration and surface runoff flows for simulations without subsurface drainage are higher than those with subsurface drainage. This situation is reversed when considering the infiltration process is concerned: flows are higher when simulations are with subsurface drain systems. Having the largest saturated hydraulic conductivity values, Sc. 3 presented highest variations in total exfiltration, infiltration and surface runoff between simulations with and without subsurface drainage systems. Since Sc. 4 had the smallest hydraulic conductivity values in the entire soil profile, it also had the smallest variations of exfiltration, infiltration and surface flows. It was also observed that scenarios having isotropic layers in the first group of layers (surface layers), namely Sc. 3 and Sc. 4, yielded infiltration and surface runoff flow variations that are almost equal.

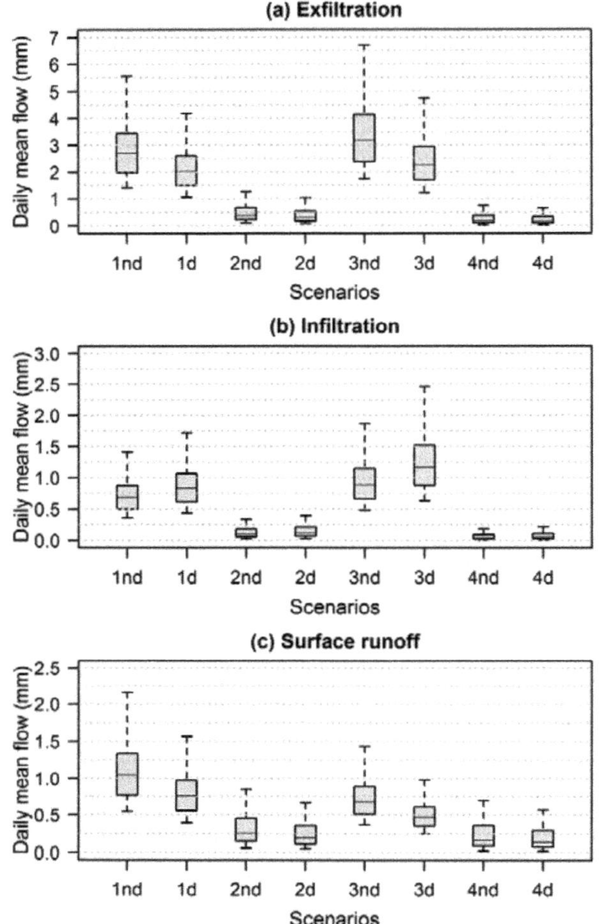

Figure 9. Impact of subsurface drain networks and soils on exfiltration (**a**); infiltration (**b**) and surface (**c**) flows in form of box plots (outlier omitted) with *"d"* and *"nd"* expressing simulation with and without subsurface drainage, respectively for Scenarios 1 to 4.

3.4. Impact of Subsurface Drainage and Soil Saturated Hydraulic Conductivity on the Outlet Flow Hydrograph

The hydrograph of the daily mean flow at the micro-watershed outlet and its different components for Sc. 1 are presented in Figure 10. It was observed that: (1) exfiltration is the most dominant flow; (2) peaks of the hydrograph are largely dominated by surface runoff; and (3) subsurface drain flow variations are low and tend to zero during the period where there precipitation was very low.

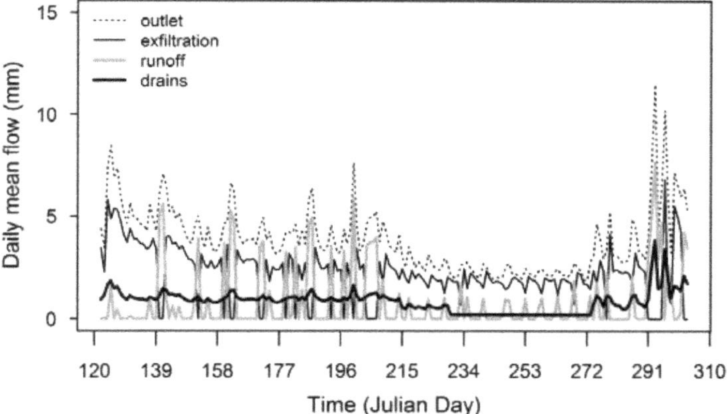

Figure 10. Hydrograph of daily mean flow at micro-watershed outlet and its different components (exfiltration, runoff, and drains) for Scenario 1.

The cumulative flow at the outlet of the micro-watershed and its various components, and its relation with them for Sc. 1 to Sc. 4 are illustrated in Figure 11. For Sc. 1, these curves show that exfiltration, subsurface drain flow and surface runoff represent 56%, 23% and 21% of the cumulative flow at the outlet of the micro-watershed, respectively. Given that the micro-watershed is artificially drained over 30% of its area, the contribution of the subsurface drain networks to the micro-watershed outlet is significant.

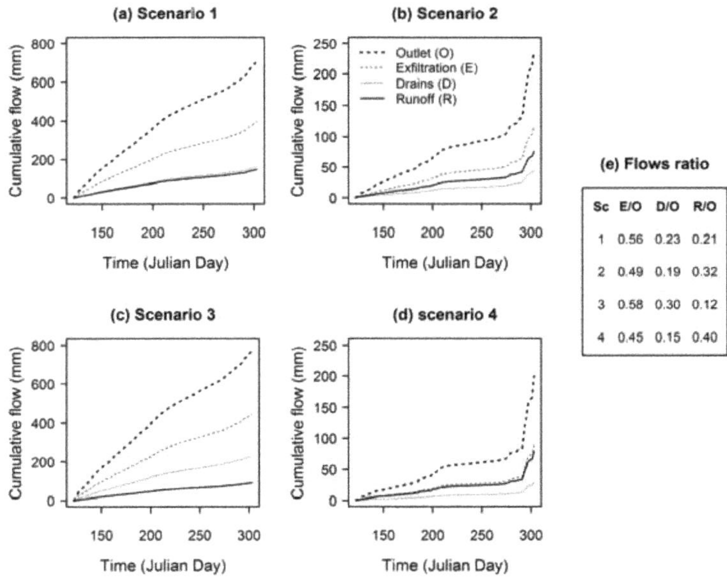

Sc	E/O	D/O	R/O
1	0.56	0.23	0.21
2	0.49	0.19	0.32
3	0.58	0.30	0.12
4	0.45	0.15	0.40

Figure 11. Curves of cumulative flow at the micro-watershed outlet (*O*) and its different components (exfiltration: *E*, drains: *D* and runoff: *R*) of: (**a**) Scenario 1; (**b**) Scenario 2; (**c**) Scenario 3; (**d**) Scenario 4; and (**e**) ratios of *E*, *D*, *R* over *O*.

Considering the values of exfiltration (E) and subsurface drain flow (D) over outlet flow (O) (i.e., E/O and D/O) in decreasing order, the scenarios can be classified as follows: Sc. 3, Sc. 1, Sc. 2 and Sc. 4. On the other hand, from the point of view of the decrease in the ratio of the surface runoff (R) over outlet flow (O) (i.e., R/O), they can be classified in the order Sc. 4, Sc. 2, Sc. 1 and Sc. 3. Thus, it was observed that when the soil is more conductive, the exfiltration and subsurface drain flows are higher while the surface runoff flow is lower.

When the last two contributing flows are further analysed (subsurface drain flow and surface runoff), it should be noted that for Sc. 1, subsurface drain flow is slightly larger than surface runoff. For Sc. 3, where the soil is more conductive than for Sc. 1, subsurface drain flow is significantly larger than surface runoff. For Sc. 2 and Sc. 4, where soils are less conductive than for Sc. 1, the surface runoff is larger than the subsurface drain flow.

4. Discussion

4.1. Simulation of the Flow at the Micro-Watershed Outlet and Analysis of the Effect of Subsurface Drainage on the Storage Variation

Given the simulated flow results obtained at the micro-watershed outlet, the performance of CATHY was deemed satisfactory. It was better when the effective precipitation was larger; corresponding to large observed and simulated flows. Subsurface drainage is a needful agricultural practice during periods with high water tables under the effect of high precipitation or irrigation [1,2]. This study demonstrated that CATHY can reliably predict subsurface drainage flow during wet periods; that is the time when it is imperative to know whether a drainage system can meet design criteria.

When considering the dynamics of the cumulative storage variation, results showed that the behaviour depended not only on the amount of precipitation but also on the associated distribution throughout the year. For a year with more summer precipitation, the simulation curve with subsurface drains remained in general below the curve without subsurface drains, indicating that there was subsurface-drain flow throughout the simulation period. In the presence of low summer precipitation, the first curve was below the second one, illustrating the absence of flow through the subsurface drainage networks. When the first curve was below the second one, the implication is that not only did subsurface agricultural drains evacuate precipitation very fast, but they increased the flow at the outlet of the micro-watershed. This means that subsurface drains increase the amount of water flowing from the soil toward the watershed outlet by diminishing the water content of the soil [63]. Konyha et al. found that the total flow exiting from a drained field could increase by 10% [64]. Since the drains route a portion of the subsurface flow faster toward surface waters (outlet), this results in a decrease of storage in the porous media-less water being actually stored in the soil profile-and an increase in the micro-watershed outlet flow. The implementation of a subsurface drainage network generally entails a reduction of the response time of agricultural plots [65,66].

4.2. Impact of Subsurface Drainage and Soil Hydraulic Conductivity on the Micro-Watershed Outlet Flow

Tile drains can substantially alter the total water yield from a field or small watershed as well as modify the timing and shape of the hydrograph [8]. Tile drainage increases the proportion of annual precipitation that is discharged to surface waters via subsurface flow relative to the amount that is stored, evaporated or transpired [9–11]. Total water yield from a field tends to increase with the installation of subsurface drainage while surface runoff and sediment yield are often significantly decreased [2,16,17].

By increasing the drainage capacity of the soils, it was observed at the micro-watershed outlet that: (a) agricultural subsurface drains increased the cumulative mean daily and base (minimum) flows, and decreased peak flows (maximum); (b) the smaller were the saturated hydraulic conductivities, the smaller were the flows (with or without subsurface drainage); (c) the gap between the cumulative flow curve with drainage and that without drainage increased when saturated hydraulic conductivities were higher; and (d) the differences between extreme flows (maximum and minimum with and without drainage) were higher when the soil was more conductive (greater values of saturated hydraulic conductivity).

4.3. Impact of Subsurface Drainage and Soil Saturated Hydraulic Conductivity on Surface Runoff, Surface Water and Groundwater Coupling

Several studies have dealt with the impact of subsurface agricultural drainage on the hydrology of fields and watersheds. The implementation of a network of drains generally entails a reduction in surface runoff by increasing infiltration of precipitation [67–70]. The changes associated with subsurface agricultural drainage according to soil properties have an impact on the hydrologic budget by altering subsurface water flow, in particular, flows to the surface (watershed outlet) by exfiltration. That being said, soil properties, in particular saturated hydraulic conductivity, may affect the extent to which subsurface drainage can influence subsurface flow [67].

When considering surface and subsurface waters, drains reduced exfiltration and surface runoff, and increased infiltration. The more conductive were the soils, the more exfiltration and subsurface drain flows were higher, whereas surface runoff was lower, and for isotropic surface layers, the variations in surface runoff and infiltration were almost equal. Most soils are heterogeneous and anisotropic; that is, hydraulic properties vary in space and are higher in a given direction [71]. When considering isotropy and anisotropy of the soil layer groups, it is evident that these properties play an important role on exfiltration, surface runoff and infiltration of water, and consequently on flows (maximum, minimum and total) at the micro-watershed outlet.

4.4. Behavior of the Outlet Hydrograph Components

Groundwater discharge to streams is commonly expressed as base flow, which is streamflow that occurs during dry times of the year [72]. Groundwater contribution is very low during the flood periods but can be very important at the annual scale and surface runoff is the most significant component in observed flows during floods [73]. When analysing the outlet hydrograph, it was noticed that: (a) exfiltration was the dominant process; (b) peaks were largely dominated by surface runoff; and (c) variations in subsurface drain flow were low.

5. Conclusions

The corner stone of this study was the application of CATHY, a 3D hydrological model, to assess the influence of artificial subsurface drainage on flows at the watershed level. Flows simulated by CATHY at the outlet of the study watershed corroborated well with observed data mostly during wet years. For these years, subsurface water flows easily towards the micro-watershed outlet via drains, allowing the model to properly simulate the flows at the outlet of micro-watershed. Furthermore, simulations were consistent with the traditional approaches of subsurface drainage effects on surface and subsurface components such as base and peak flows, surface runoff, infiltration, and exfiltration.

The development of scenarios aims to support management practices such as the effect of tillage on infiltration or percolation of rain in soil layers. Next to that, when measuring the saturated hydraulic conductivities in situ or in laboratory, their values can fluctuate within a certain range of order 10. The aim of this study was conducted in this context.

These types of studies are needed to build our capacities to manage water resources in agricultural watersheds, where water flow dynamics are often manipulated to increase productivity. A good control of the impact of these manipulations will result in a better understanding of the flows and improved design of management actions to reduce the risks of degrading the quality of surface waters.

Acknowledgments: The authors wish to thank Eric van Bochove, Georges Thériault, Catherine Bossé, Geneviève Montminy, Michel Nolin, Luc Lamontagne, and Mario Deschênes of Agriculture and Agri-food Canada for providing data. Special thanks to Brook Harker and David Kiely of AAC for their coordination of WEBs (Watershed Evaluation of Beneficial Management Practices) project. This project has received financial support from AAC (Alain N. Rousseau, principal investigator of the project "Hydrological-economic modelling of Beneficial Managements Practices impact on water quality in an agricultural watershed") and OMAFRA (Ontario Ministry of Agriculture, Food and Rural Affairs) (Wanhong Yang of the University of Guelph, principal investigator of the project "Evaluating the Cost Effectiveness of Multiple Best Management Practices in Agricultural Watersheds").

Author Contributions: All the authors conceived and designed the numerical experiments; Mushombe Muma performed the numerical experiments and the permeability tests or slug tests; Mushombe Muma analyzed the data and wrote the paper under the supervision of Alain N. Rousseau and Silvio J. Gumiere.

Conflicts of Interest: The authors declare no conflict of interest.

References

1. Eggelsmann, R. *Dränanleitung für den Landbau, Ingenieurbau und Landschaftsbau*, 2nd ed.; Verlag Paul Parey: Hamburg, Germany, 1981.
2. Skaggs, R.W.; Brevé, M.A.; Gilliam, J.W. Hydrologic and water quality impacts of agricultural drainage. *Crit. Rev. Environ. Sci. Technol.* **1994**, *24*, 1–32. [CrossRef]
3. Eidman, V. Minnesota farmland drainage: Profitability and concerns. *Minn. Agric. Econ.* **1997**. No 688. Department of Applied Economics, University of Minnesota, St. Paul, MN, USA, 1997. Available online: http://ageconsearch.umn.edu/bitstream/13165/1/mae688a.pdf (accessed on 2 June 2015).
4. Fausey, N.R. Drainage: Inadequacy and crop response. In *Encyclopedia of Water Science*; Stewart, B.A., Howell, T., Eds.; Taylor and Francis: Oxford, UK, 2003; pp. 132–134.
5. Fausey, N.R. Drainage management for humid regions. *Int. Agric. Eng. J.* **2005**, *14*, 209–214.
6. Basin Technical and Scientific Advisory Committee (BTSAC). Water Management Options for Subsurface Drainage: Briefing Paper 2. 2012. Available online: http://www.rrbdin.org/archives/4520 (accessed on 28 June 2014).
7. St-Hilaire, A.; Duchesne, A.S.; Rousseau, A.N. Floods and water quality in Canada—A review of the interactions with urbanization, agriculture and forestry. *Can. Water Resour. J.* **2015**, *41*, 273–287. [CrossRef]
8. Schilling, K.E.; Helmers, M. Effects of subsurface drainage tiles on streamflow in Iowa agricultural watersheds: Exploratory hydrograph analysis. *Hydrol. Process.* **2008**, *22*, 4497–4506. [CrossRef]
9. Serrano, S.E.; Whiteley, H.R.; Irwin, R.W. Effects of agricultural drainage on streamflow in the Middle Thames River, Ontario, 1949–1980. *Can. J. Civ. Eng.* **1985**, *12*, 875–885. [CrossRef]
10. Magner, J.A.; Payne, G.A.; Steffen, L.J. Drainage effects on stream nitrate-n and hydrology in south-central Minnesota (USA). *Environ. Monit. Assess.* **2004**, *91*, 183–198. [CrossRef] [PubMed]
11. Tomer, M.D.; Meek, D.W.; Kramer, L.A. Agricultural practices influence flow regimes of headwater streams in Western Iowa. *J. Environ. Qual.* **2005**, *34*, 1547–1558. [CrossRef] [PubMed]
12. King, K.W.; Fausey, N.R.; Williams, M.R. Effect of subsurface drainage on streamflow in an agricultural headwater watershed. *J. Hydrol.* **2014**, *519*, 438–445. [CrossRef]
13. Macrae, M.L.; English, M.C.; Schiff, S.L.; Stone, M. Intra-annual variability in the contribution of tile drains to basin discharge and phosphorus export in a first-order agricultural catchment. *Agric. Water Manag.* **2007**, *92*, 171–182. [CrossRef]
14. Culley, J.L.B.; Bolton, E.F. Suspended solids and phosphorus loads from a clay soil: II. Watershed study. *J. Environ. Qual.* **1983**, *12*, 498–503. [CrossRef]
15. Xue, Y.; David, M.B.; Gentry, L.E.; Kovacic, D.A. Kinetics and modeling of dissolved phosphorus export from a tile-drained agricultural watershed. *J. Environ. Qual.* **1998**, *27*, 917–922. [CrossRef]
16. Robinson, M.; Rycroft, D.W. The impact of drainage on stream flows. *Agron. Monogr.* **1999**, *38*, 767–800.
17. Dolezal, F.; Kulhavy, Z.; Soukup, M.; Kodesova, R. Hydrology of tile drainage runoff. *Phys. Chem. Earth B Hydrol. Ocean Atmos.* **2001**, *26*, 623–627. [CrossRef]
18. Robinson, M. *Impact of Improved Land Drainage on River Flows*; Institute of Hydrology Center for Ecology and Hydrology: Edinburgh, UK, 1990.
19. Bengtson, R.L.; Carter, C.E.; Fouss, J.L.; Southwick, L.M.; Willis, G.H. Agricutural drainage and water quality in the Mississippi delta. *J. Irrig. Drain. Eng.* **1995**, *121*, 292–295. [CrossRef]
20. Zucker, L.A., Brown, L.C.E., Eds.; *Agricultural Drainage: Water Quality Impacts and Subsurface Drainage Studies in the Midwest*; The Ohio State University: Columbus, OH, USA, 1998.

21. Kleinman, P.J.A.; Smith, D.R.; Bolster, C.H.; Easton, Z.M. Phosphorus Fate, Management, and Modeling in Artificially Drained Systems. *J. Environ. Qual.* **2015**, *44*, 460–466. [CrossRef] [PubMed]
22. Wiskow, E.; van der Ploeg, R.R. Calculation of drain spacings for optimal rainstorm flood control. *J. Hydrol.* **2003**, *272*, 163–174. [CrossRef]
23. King, K.W.; Williams, M.R.; Faussey, N.R. Contributions of systematic tile drainage to watershed-scale phosphorus transport. *J. Environ. Qual.* **2015**, *44*, 486–494. [CrossRef] [PubMed]
24. Smith, D.R.; King, K.W.; Johnson, L.; Francesconi, W.; Richards, P.; Baker, D.; Sharpley, A.N. Surface runoff and tile drainage transport of phosphorus in the midwestern United States. *J. Environ. Qual.* **2015**, *44*, 495–502. [CrossRef] [PubMed]
25. Schottler, S.P.; Ulrich, J.; Belmont, P.; Moore, R.; Lauer, W.J.; Engstrom, D.R.; Almendinger, J.E. Twentieth century agricultural drainage creates more erosive rivers. *Hydrol. Process.* **2014**, *28*, 1951–1961. [CrossRef]
26. Dunn, S.M.; Mackay, R. Modelling the hydrological impact of open ditch drainage. *J. Hydrol.* **1996**, *179*, 37–66. [CrossRef]
27. Holden, J.; Chapman, P.J.; Labadz, J.C. Artificial drainage of peatlands: Hydrological andhydrochemical process and wetland restoration. *Prog. Phys. Geogr.* **2004**, *28*, 95–123. [CrossRef]
28. Vidon, P.; Cuadra, P.E. Phosphorus dynamics in tile-drain flow during storms in the US Midwest. *Agric. Water Manag.* **2011**, *98*, 532–540. [CrossRef]
29. Kung, K.J.S.; Kladivko, E.J.; Gish, T.J.; Steenhuis, T.S.; Bubenzer, G.; Helling, C.S. Quantifying preferential flow by breakthrough of sequentially applied tracers: Silt loam soil. *Soil Sci. Soc. Am. J.* **2000**, *64*, 1296–1304. [CrossRef]
30. Bixio, A.C.; Orlandini, S.; Paniconi, C.; Putti, M. Modeling groundwater-surface water interactions including effects of morphogenetic depressions in the Chernobyl exclusion zone. *Environ. Geol.* **2002**, *42*, 162–177. [CrossRef]
31. Orlandini, S. On the spatial variation of resistance to flow in upland channel networks. *Water Resour. Res.* **2002**, *38*, 1197. [CrossRef]
32. Paniconi, C.; Troch, P.A.; van Loon, E.E.; Hilberts, A.G.J. Hillslope-storage Boussinesq model for subsurface flow and variable source areas along complex hillslopes: 2. Intercomparison with a three-dimensional Richards equation model. *Water Resour. Res.* **2003**, *39*, 1317. [CrossRef]
33. D'Haese, C.M.F.; Putti, M.; Paniconi, C.; Verhoest, N.E.C. Assessment of adaptive and heuristic time stepping for variably saturated flow. *Int. J. Numer. Methods Fluids* **2007**, *53*, 1173–1193. [CrossRef]
34. Gauthier, M.J.; Camporese, M.; Rivard, C.; Paniconi, C.; Larocque, M. A modeling study of heterogeneity and surface water-groundwater interactions in the Thomas Brook catchment, Annapolis Valley (Nova Scotia, Canada). *Hydrol. Earth Syst. Sci.* **2009**, *13*, 1583–1596. [CrossRef]
35. Sulis, M.; Meyerhoff, S.B.; Paniconi, C.; Maxwell, R.M.; Putti, M.; Kollet, S.J. A comparison of two physics-based numerical models for simulating surface water-groundwater interactions. *Adv. Water Resour.* **2010**, *33*, 456–467. [CrossRef]
36. Sulis, M.; Paniconi, C.; Camporese, M. Impact of grid resolution on the integrated and distributed response of a coupled surface–subsurface hydrological model for the des Anglais catchment, Quebec. *Hydrol. Process.* **2011**, *25*, 1853–1865. [CrossRef]
37. Sulis, M.; Paniconi, C.; Rivard, C.; Harvey, R.; Chaumont, D. Assessment of climate change impacts at the catchment scale with a detailed hydrological model of surface-subsurface interactions and comparison with a land surface model. *Water Resour. Res.* **2011**, *47*, W01513. [CrossRef]
38. Broda, S.; Paniconi, C.; Larocque, M. Numerical investigation of leakage in sloping aquifers. *J. Hydrol.* **2011**, *409*, 49–61. [CrossRef]
39. Zanello, F.; Teatini, P.; Putti, M.; Gambolati, G. Long term peatland subsidence: Experimental study and modeling scenarios in the Venice coastland. *J. Geophys. Res. Earth Surf.* **2011**, *116*, F04002. [CrossRef]
40. Dagès, C.; Paniconi, C.; Sulis, M. Analysis of coupling errors in a physically-based integrated surface water-groundwater model. *Adv. Water Resour.* **2012**, *49*, 86–96. [CrossRef]
41. Guay, C.; Nastev, M.; Paniconi, C.; Sulis, M. Comparison of two modeling approaches for groundwater–surface water interactions. *Hydrol. Process.* **2013**, *27*, 2258–2270. [CrossRef]
42. Passadore, G.; Monego, M.; Altissimo, L.; Sottani, A.; Putti, M.; Rinaldo, A. Alternative conceptual models and the robustness of groundwater management scenarios in the multi-aquifer system of the Central Veneto Basin, Italy. *Hydrogeol. J.* **2012**, *20*, 419–433. [CrossRef]

43. Niu, G.-Y.; Paniconi, C.; Troch, P.A.; Scott, R.L.; Durcik, M.; Zeng, X.; Huxman, T.; Goodrich, D.C. An integrated modelling framework of catchment-scale ecohydrological processes: 1. Model description and tests over an energy-limited watershed. *Ecohydrology* **2013**, *7*, 427–439. [CrossRef]
44. Lamontagne, L.; Martin, A.; Nolin, M.C. *Étude Pédologique du Bassin Versant du Bras d'Henri (Québec)*; Laboratoires de Pédologie et D'agriculture de Précision, Centre de Recherche et de Développement sur les sols et les Grandes Cultures, Service National D'information sur les Terres et les eaux, Direction Générale de la Recherche, Agriculture et Agroalimentaire Canada: Sainte-Foy, QC, Canada, 2010.
45. Agriculture & Agri-Food Canada. Greencover Canada. Government of Canada. 2004a. Available online: http://www.agr.gc.ca/env/greencover-verdir/index_e.phtml (accessed on 20 May 2015).
46. Agriculture & Agri-Food Canada. Watershed Evaluation of BMPs (WEBs). Government of Canada. 2004b. Available online: http://www.agr.gc.ca/env/greencover-verdir/webs_abstract_e.phtml (accessed on 20 May 2015).
47. Agriculture & Agri-Food Canada. Watershed Evaluation of BMPs (WEBs). Government of Canada. 2004c. Available online: http://www.agr.gc.ca/AAFC-AAC/displa-afficher.do?id=1228497657135&lang=eng (accessed on 20 May 2015).
48. Muma, M.; Gumiere, S.J.; Rousseau, A.N. Analyses de sensibilité globales du modèle CATHY aux propriétés hydrodynamiques du sol d'un micro-bassin agricole drainé. *Hydrol. Sci. J.* **2014**, *59*, 1606–1623. [CrossRef]
49. Ratte-Fortin, C. *Développement d'une Méthode D'évaluation de L'impact de Pratiques de Gestion Bénéfiques sur les Flux de Contaminants Agricoles: Cas du Micro-Bassin Versant D'intervention du Bras d'Henri, Québec, Canada*; Mémoire de Maîtrise, Institut National de la Recherche Scientifique—Centre Eau Terre Environnement (INRS-ETE): Ville de Québec, QC, Canada, 2014.
50. Bouwer, H.; Rice, R.C. A slug test for determining hydraulic conductivity of unconfined aquifers with completely or partially penetrating wells. *Water Resour. Res.* **1976**, *12*, 424–428. [CrossRef]
51. Pandit, N.S.; Miner, R.F. Interpretation of slug test data. *Ground Water* **1986**, *24*, 743–749. [CrossRef]
52. Camporese, M.; Paniconi, C.; Putti, M.; Orlandini, S. Surface-subsurface flow modeling with path-based runoff routing, boundary condition-based coupling, and assimilation of multisource observation data. *Water Resour. Res.* **2010**, *46*, W02512. [CrossRef]
53. Paniconi, C.; Putti, M. A comparison of Picard and Newton iteration in the numerical solution of multidimensional variably saturated flow problems. *Water Resour. Res.* **1994**, *30*, 3357–3374. [CrossRef]
54. Ministère de l'Agriculture, des Pêcheries et de l'Alimentation du Québec (MAPAQ). *Profil de la région Chaudière-Appalaches (région 12)-Climat* [en ligne]. Ministère de l'Agriculture, Pêcheries et Alimentation du Québec: Canada, 2007. Available online: http://www.mapaq.gouv.qc.ca/Fr/Regions/chaudiereappalache/vraiprofil/ (accessed on 26 June 2015).
55. Allen, R.G.; Pereira, L.S.; Raes, D.; Smith, M. *Crop Evapotranspiration: Guidelines for Computing Crop Water Requirements*; Irrigation and Drainage, FAO: Roma, Italy, 1998.
56. Fortin, J.P.; Turcotte, R.; Massicotte, S.; Moussa, R.; Fitzback, J.; Villeneuve, J.P. A distributed watershed model compatible with remote sensing and GIS data. Part 1: Description of the model. *J. Hydrol. Eng.* **2001**, *6*, 91–99. [CrossRef]
57. Fortin, J.P.; Turcotte, R.; Massicotte, S.; Moussa, R.; Fitzback, J.; Villeneuve, J.P. A distributed watershed model compatible with remote sensing and GIS data. Part 2: Application to the Chaudière Watershed. *J. Hydrol. Eng.* **2001**, *6*, 100–108. [CrossRef]
58. Fortin, J.P.; Moussa, R.; Bocquillon, C.; Villeneuve, J.P. Hydrotel, un model hydrologique distribué pouvant bénéficier des données fournies par la télédétection et les systèmes d'information géographique. *Rev. Sci. de l'Eau* **1995**, *8*, 97–124. [CrossRef]
59. Rousseau, A.N.; Savary, S.; Hallema, D.W.; Gumiere, S.J.; Foulon, E. Modeling the effects of agricultural BMPs on sediments, nutrients and water quality of the Beaurivage River watershed (Quebec, Canada). *Can. Water Resour. J.* **2013**, *38*, 99–120. [CrossRef]
60. Lagacé, R. Infiltration et Drainage, Notes de Cours: Bilan Hydrique au Québec. Département des Sols et de Génie Agroalimentaire, Université Laval: Québec City, QC, Canada, 2012. Available online: http://www.grr.ulaval.ca/gaa_7003/Documents/Notes_cours_2012/CH_10_Bilan_QC.pdf (accessed on 12 January 2015).
61. Savoie, V. *Le Drainage de Surface*; Formation OAQ, MAPAQ: Centre du Québec, QC, Canada, 2009.

62. Bessière, H. Assimilation de Données Variationnelles Pour la Modélisation Hydrologique Distribuée des Crues à Cinétique Rapide. Thèse de Doctorat, Université de Toulouse, Institut National Polytechnique de Toulouse, Toulouse, France, 2008.
63. Larson, C.L.; Moore, I.D. Hydrologic impact of draining small depressional watersheds. *J. Irrig. Drain. Div.* **1980**, *104*, 345–363.
64. Konyha, K.D.; Skaggs, R.W.; Gilliam, J.W. Effects of drainage and water management practices on hydrology. *J. Irrig. Drain. Eng.* **1992**, *118*, 807–819. [CrossRef]
65. Bailey, A.D.; Bree, T. Effect of improved land drainage on river flood flows. In *Flood Studies Report-Five Years on*; Thomas Talford: London, UK, 1981; pp. 95–106.
66. Rycrott, D.W.; Massey, W. *The Effect of Field Drainage on the River Flood Flows*; Field Drainage Experimental Unit, MAFF: London, UK, 1975.
67. Hill, A.R. The environmental impacts of agricultural land drainage. *J. Environ. Manag.* **1976**, *4*, 251–274.
68. Irwin, R.W.; Whitely, H.R. Effects of land drainage on stream flow. *Can. Water Resour. J.* **1983**, *8*, 88–103. [CrossRef]
69. Thomas, D.L.; Perry, C.D.; Evans, R.O.; Izuno, F.T.; Stone, K.C.; Gilliam, J.W. Agricultural drainage effects on water quality in Southeastern U.S. *J. Irrig. Drain. Eng.* **1995**, *121*, 277–282. [CrossRef]
70. Watelet, A.; Johnson, P.G. Hydrology and water quality of the Raisin River: Overview of impacts of recent land and channel changes in Eastern Ontario. *Can. J. Water Qual. Res.* **1999**, *34*, 361–390.
71. Kasenow, M. *Determination of Hydraulic Conductivity from Grain Size Analysis*; Water Resources Publications, L.L.C.: Highlands, CO, USA, 2010.
72. Winkler, R.D.; Moore, R.D.; Redding, T.E.; Spittlehouse, D.L.; Carlyle-Moses, D.E.; Smerdon, B.D. Hydrologic Processes and Watershed Response. *Compendium of Forest Hydrology and Geomorphology in British Columbia*; Pike, R.G., Redding, T.E., Moore, R.D., Winkler, R.D., Bladon, K.D., Eds.; B.C. Min. For. Range, For. Sci. Prog., Victoria, B.C. and FORREX Forum for Research and Extension in Natural Resources, Kamloops, B.C. Land Manag. Handb. Government Publications Services: Victoria, BC, Canada, 2000.
73. Bennis, S. *Hydraulique et Hydrologie, 2e Édition Revue et Augmentée*; Presses de l'Université du Québec: Ville de Québec, QC, Canada, 2009.

Article

Characterizing Changes in Streamflow and Sediment Supply in the Sacramento River Basin, California, Using Hydrological Simulation Program—FORTRAN (HSPF)

Michelle Stern [1,*], Lorraine Flint [1], Justin Minear [2], Alan Flint [1] and Scott Wright [1]

1 California Water Science Center, United States Geological Survey, Sacramento, CA 95819, USA; lflint@usgs.gov (L.F.); aflint@usgs.gov (A.F.); sawright@usgs.gov (S.W.)

2 Geomorphology and Sediment Transport Laboratory, United States Geological Survey, Golden, CO 80403, USA; jminear@usgs.gov

* Correspondance: mstern@usgs.gov; Tel.: +1-916-278-3093

Academic Editor: Xuan Yu
Received: 14 May 2016; Accepted: 19 September 2016; Published: 30 September 2016

Abstract: A daily watershed model of the Sacramento River Basin of northern California was developed to simulate streamflow and suspended sediment transport to the San Francisco Bay-Delta. To compensate for sparse data, a unique combination of model inputs was developed, including meteorological variables, potential evapotranspiration, and parameters defining hydraulic geometry. A slight decreasing trend of sediment loads and concentrations was statistically significant in the lowest 50% of flows, supporting the observed historical sediment decline. Historical changes in climate, including seasonality and decline of snowpack, contribute to changes in streamflow, and are a significant component describing the mechanisms responsible for the decline in sediment. Several wet and dry hypothetical climate change scenarios with temperature changes of 1.5 °C and 4.5 °C were applied to the base historical conditions to assess the model sensitivity of streamflow and sediment to changes in climate. Of the scenarios evaluated, sediment discharge for the Sacramento River Basin increased the most with increased storm magnitude and frequency and decreased the most with increases in air temperature, regardless of changes in precipitation. The model will be used to develop projections of potential hydrologic and sediment trends to the Bay-Delta in response to potential future climate scenarios, which will help assess the hydrological and ecological health of the Bay-Delta into the next century.

Keywords: HSPF; watershed hydrology; suspended sediment; hydrologic modeling; water resources; San Francisco Bay-Delta; Sacramento River; sediment transport

1. Introduction

The Sacramento River Basin of northern California (Figure 1) produces 84% of the fresh water inflow and 80% of the sediment to the San Francisco Bay-Delta (Bay-Delta) [1]. The Computational Assessments of Scenarios of Change for the Delta Ecosystem (CASCaDE II) project [2] is developing a better understanding of how potential future changes in the characteristics and climate of watersheds draining into the Bay-Delta will affect water quality, ecosystem processes, and critical species. Turbidity, geomorphic change, and wetland stability all depend on sediment supply; therefore, sediment supply has critical implications for the ecology of the Bay-Delta [3]. The total sediment load has decreased by 50% during the last 50 years for a suite of interacting reasons, including the diminishment of the sediment pulse created during 19th century hydraulic mining in the Sierra Nevada, sediment trapping behind reservoirs, deposition of sediment in flood bypasses, and armoring of river channels [4].

Turbidity is related to sediment and therefore a decline in sediment supply contributes to undesirable conditions for key fish species such as the delta smelt [5]. Discharge from the Sacramento River is a primary driver of turbidity in critical delta smelt habitat and thus has implications for the future survival of delta smelt and other species which share similar ecological preferences [5].

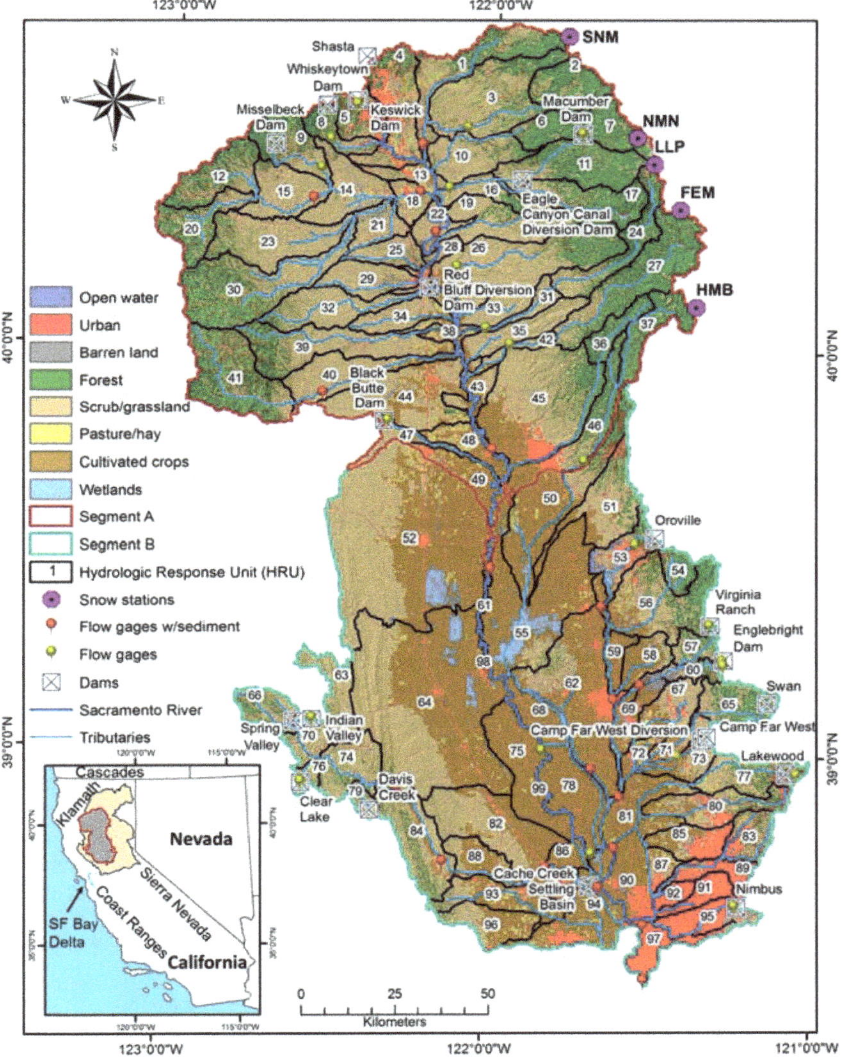

Figure 1. Location map of the Sacramento River Basin model domain, CA, indicating: land use, snow stations, major California dams/diversions, hydrologic and sediment gages, and model reaches. Total drainage area of the Sacramento River is indicated (orange of inset). California Data Exchange Center (CDEC) snow stations are abbreviated as: SNM = Snow Mountain; NMN = New Manzanita Lake; LLP = Lower Lassen Peak; FEM = Feather River Meadow; HMB = Humbug. Inset indicates major mountain ranges, and Sacramento Valley surface water drainage in orange, and model domain in gray.

The main objective of this project was to develop a spatially distributed daily streamflow and sediment transport model of the Sacramento River Basin to link with a hydrodynamic model of the Bay-Delta, and to characterize the changes in sediment transport over the historical record. The Sacramento River Basin supplies the majority of the sediment to the Bay-Delta, and representing the mechanisms of sediment transport in the watershed is important to assess the impact of potential changes in future climate regimes. Because of limited measured suspended sediment data, the model was calibrated to available data below major dams and impoundments, and will be used to simulate potential future changes in sediment delivery to the Bay-Delta under climate change. The approach of this study relied on the development of a unique combination of spatially distributed climate variables and hydraulic function tables, and the preservation of spatially distributed information during calibration. This allowed for the extrapolation of known processes to areas without calibration data, resulting in a robust calibration. This approach is generally in contrast with typical methods used in large river basins that depend on one to a few climate stations, sometimes without interpolation across the watershed, and individual calibration of hydrologic response units without consideration of distributed properties across the landscape.

The current conceptual model of sediment decline in the Sacramento River basin [6] attributes hydraulic mining and dams and reservoirs as the likely anthropogenic causes, and relies on qualitative descriptions of sediment drivers and linkages, in addition to a water and sediment budget using four years of data. The development of this model has allowed us to investigate the role of climate on sediment transport in this basin, and how changes in climate over the last 50 years have contributed to changes in sediment delivery to the Bay-Delta.

Study Area

The Sacramento River drainage is being studied because of its dominant contribution of sediment supply to the Bay-Delta. It is a 6.8×10^4 km^2 area in the northern part of the Central Valley of California (Figure 1, inset) and it comprises over half of the total drainage area of the Bay-Delta [7]. The Sacramento River is the longest river system in the state of California, and it is a complex and highly managed system. Dams located in the foothills surrounding the valley floor restrict flow in most of the tributaries to the main stem of the Sacramento River. The majority of floodwater in the lower Sacramento River is diverted using a series of passive weirs into a system of flood bypasses, which act as natural conveyance floodplains by transferring excess streamflow directly into the Bay-Delta [8]. The weir and bypass system was built before most of the large dams were constructed in 1920–1970. The dams were installed to assist in flood control, for power generation, and to provide water for downstream use, with ~38% used for agricultural irrigation, and ~4% for urban use [9]. Reservoirs behind the dams have accumulated large amounts of sediment and gravels as a byproduct of hydraulic mining in the late 19th century, in addition to land use changes within the watershed [6]. The study area within the Sacramento River Basin has an area of roughly 26,000 km^2, and begins below Keswick Dam (Figure 1). The largest contribution of flow to the Sacramento River in the model domain is the outflow from Keswick Dam, which is fed by the outflow from Shasta Dam nearly 16 km upstream from Keswick Dam (Figure 1).

2. Methods

The methods described below focus on non-standard methods and approaches used to develop this model. Standard Hydrological Simulation Program—FORTRAN (HSPF) methods used are described in the Supplementary Materials (S1–S5).

2.1. Hydrological Simulation Program—FORTRAN (HSPF)

A watershed model for the Sacramento River Basin was developed to simulate streamflow and suspended sediment for the period 1958–2008 using the Hydrological Simulation Program—FORTRAN (HSPF) [10]. HSPF was chosen for this study because it is a comprehensive process-based watershed model that can be used to simulate daily streamflow and suspended sediment for a continuous multi-year period and for a wide range of watershed sizes and conditions [11]. HSPF is useful for modeling natural and developed watersheds, and is effective for land surface and subsurface hydrology as well as water quality processes. Importantly, it enables the user to create alternative scenarios by adjusting land use or meteorological variables such as air temperature or precipitation [11].

HSPF is a spatially distributed and temporally continuous simulation model that employs lumped parameter segments. Each sub-watershed or hydrologic response unit (HRU) is considered homogeneous with the same set of parameters. Three modules in the HSPF model represent the major watershed processes. Pervious land is modeled in the PERLND module, impervious land in the IMPLND module, and the RCHRES module simulates the processes in streams or reservoirs. A comprehensive functional description of HSPF can be found in Bicknell et al., 2001 [10]. In this paper, the segments from the combined PERLND and IMPLND modules will be referred to as hydrologic response units or HRUs, and RCHRES will be referred to as reaches. Sediment sources are modeled in HSPF as wash-off detached sediment in surface storage and scour of the soil matrix of PERLNDs, and as bedload scour in RCHRESs. Sediment transport is modeled as an advective, non-reactive constituent of streamflow, with deposition and scour dependent on the simulated flow velocity and particle size of the sediment. Bedload and suspended sediment are simulated as separate components; however, longitudinal bedload transport is not simulated in HSPF.

The HSPF model was developed for the Sacramento River Basin at a daily time step to simulate streamflow and suspended sediment concentrations (SSC) in the Sacramento River and tributaries below major dams. HSPF allows for daily, hourly, and sub-hourly time intervals; however, meteorological data within the model domain were not widely available at hourly or sub-hourly time increments, and therefore a daily time step was considered appropriate for the purposes of this study.

2.2. HSPF Input

Data used to develop the HSPF model included: the 2006 land use data from the National Land Cover Database (NLCD) (mrlc.gov/nlcd2006.php) [12], elevation data from the National Elevation Dataset (NED) (ned.usgs.gov/) [13], and the hydrology and stream network from the USGS National Hydrography Dataset (NHD) (nhd.usgs.gov/) [14]. Soils data were obtained from the Soil Survey Geographic (SSURGO) [15] database, including erosion potential (k-factor), texture (percent sand silt and clay), hydrologic soil group, available water storage capacity, and surface texture. Time series data including meteorological station data, streamflow, dam releases, snow water equivalent, and suspended sediment and sediment loads were collected for the time period 1958–2008. Station data for the meteorological development (air temperature, precipitation, and solar radiation) were collected from the National Weather Service (NWS) Cooperative Observer Program (COOP, www.ncdc.noaa.gov/), Remote Automated Weather Stations (RAWS, www.raws.dri.edu/), and California Irrigation Management Information System (CIMIS, www.cimis.water.ca.gov/) stations within the study area. Solar radiation, wind speed, cloud cover and dew point were downloaded from the Fair Oaks CIMIS station. Streamflow, snow and sediment calibration and boundary condition data are described in the Supplementary Materials (S1), the development of boundary dam conditions is described in S2, and the development of hydrologic response units in S3.

2.2.1. Gridded Meteorological Inputs to HSPF

HSPF does a reasonable job of predicting hydrology, sediment, water quality, and pollutant loads; however, the inability of meteorological stations to cover the spatial variations of precipitation and the uncertainty of managed flow data (diversions and return flows) are major limiting factors of more accurate predictions [16]. A better spatial representation of watershed precipitation was found to improve model accuracy, even for a semi-lumped model [17]. HSPF typically assigns data from meteorological stations directly to each sub-basin using Thiessen polygons. Meteorological stations are sparsely located in some areas and are not all active on any given day. Typical HSPF models rely on very few climate stations, some with only one station to represent the entire watershed [18–21]. The objective of the meteorological development was to enhance the spatial distribution of daily meteorological data and therefore increase the accuracy of modeled streamflow and suspended sediment. Incorporating gridded meteorological data as an input to HSPF has been successfully demonstrated [17,22] and, as the key driver of the rainfall runoff process and sediment transport, was shown to have increased the accuracy of the hydrologic results [17].

Meteorological data from climate stations consisting of daily precipitation and air temperature were spatially interpolated over the model domain using the Gradient Inverse Distance Squared (GIDS) spatial interpolation [23]. Spatial interpolation of meteorological data can be inaccurate when there are few stations and large distances between stations. The GIDS method provides accuracies at least as good as established kriging techniques without the complexity and subjectivity of kriging and the required station density [23,24]. For every active station, GIDS develops regressions for each day including the variables northing, easting, and elevation to interpolate to each grid cell, in this case, 270 m. This approach provides a detailed and localized incorporation of topographic and regional influences on precipitation (Figure 2A).

Precipitation was summed and the temperature was averaged to transform the daily maps into monthly maps. PRISM data (Parameter-elevation Regressions on Independent Slopes Model) (prism.oregonstate.edu/) (Figure 2B) are recognized as high-quality spatial climate data sets [25] and were spatially downscaled using GIDS to the 270 m scale and used for comparison to match the measured data trend and to keep the regional monthly spatial structures intact. A ratio was developed for each grid cell using monthly PRISM values. The daily GIDS maps were multiplied by the PRISM ratio to produce daily meteorological values (Figure 2C) that sum or average to exactly match PRISM monthly values. This method was applied to precipitation and air temperature data, and the data were spatially distributed to each HRU (Figure 2D). This method is an improvement on the typical distribution of meteorological stations because the measured data spatial trend is preserved and the regional monthly structures are incorporated across the watershed to produce better interpolations when data stations are not present for any given day.

Potential evapotranspiration (PET) is an important component of the water balance equation in HSPF. PET can be estimated directly by HSPF or estimated using pre-processing methods and applied as a daily time series for each HRU and reach. For this project, estimates of daily PET were developed as a pre-processing step using the Priestley-Taylor equation [26], simulated hourly solar radiation, including topographic shading and atmospheric parameters, estimated cloudiness, and the gridded daily air temperature estimates described above. Cloudiness was estimated using the Bristow-Campbell equation [27] which was calibrated to CIMIS and RAWS daily measured solar radiation and the gridded air temperature data. This method of estimating PET is considered to be an improvement over internal HSPF calculations that use pan-evaporation measurements with an adjustment factor or other temperature-based equations to distribute PET to each HRU. The Priestley-Taylor equation is an energy balance equation that has a non-linear relationship of PET to temperature that is considered more appropriate when extrapolating PET into the future when elevated air temperatures are projected to persist [28]. Other formulas that use a linear relationship of PET to air temperature have been shown to overestimate PET under warming climates [28]. The PET estimates using the Priestley-Taylor equation were calibrated to all California Irrigation Management Information System (CIMIS) stations in the study area and then distributed as a daily time series to each HRU.

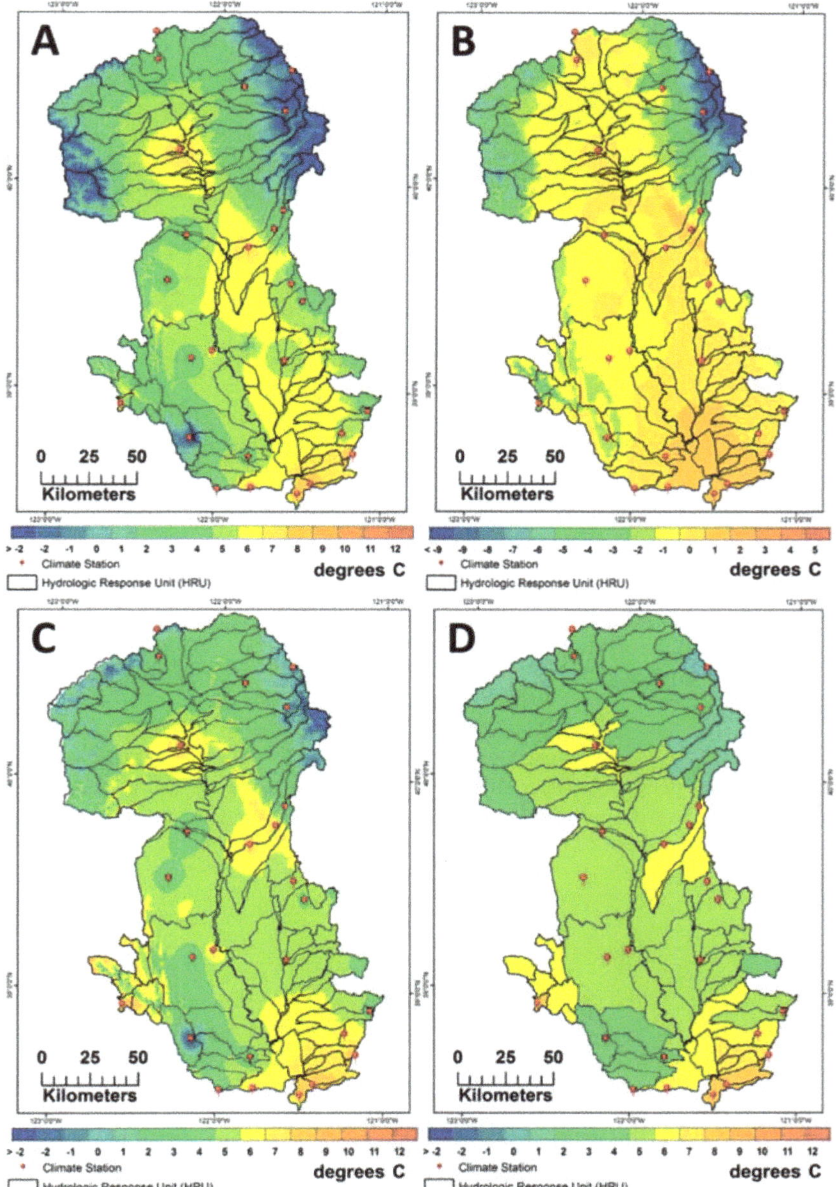

Figure 2. Example of meteorological development (minimum air temperature) for 12 January 1957 using the GIDS (Gradient Inverse Distance Squared) interpolation and PRISM (Parameter-elevation Regressions on Independent Slopes Model) data. Of 78 available stations within the region, 42 were active on this day and used in the development, and 26 are shown above within the study area boundary. The development includes (**A**) GIDS-developed regressions for each day including the variables northing, easting, and elevation to interpolate to each 270 m grid cell; (**B**) PRISM monthly map used to create a ratio for scaling maps of daily interpolated station data; (**C**) daily GIDS maps multiplied by the PRISM ratio to produce daily meteorological values that sum or average to exactly match PRISM monthly values; and (**D**) scaled daily data for each hydrologic response unit that were averaged for model input.

2.2.2. FTABLE Development

HSPF uses a hydraulic function table (FTABLE) to represent the geometric and hydraulic properties of stream reaches and reservoirs [29]. FTABLEs define the stage-area-volume-discharge relation for each RCHRES. BASINS, the user interface that accompanies the HSPF code, generates FTABLEs from a simplified reach file using preset channel geometry assumptions. Some assumptions of the channel geometry include a trapezoidal-shaped channel cross-section, channel sides with slopes of a 1:1 ratio, the flood plain width on each side of the reach being equal to the mean channel width, and the maximum depth in the FTABLE being 100 times the depth of the floodplain slope break [29].

One of the main problems with this simplified approach is that the BASINS-generated FTABLEs tend to have very few points within the range of the vast majority of actual flows. For example, in most of the HRUs there are only one to two data points in the BASINS-generated FTABLE that fall within all daily values for discharge. It is essential to develop FTABLEs that accurately reflect the major hydrological processes and water quality constituents such as sediment in order to ensure a realistic and unique simulation of the Sacramento River Basin. To enable a better simulation of suspended sediment, BASINS-generated FTABLEs were not used in the model; instead, FTABLEs were developed using hydraulic geometry. The heterogeneity of the Sacramento River watershed required two methodologies to develop FTABLEs; hydraulic geometry relationships were developed for the upland HRUs, and a one-dimensional hydraulic model was used for the lowland HRUs with very little slope.

Hydraulic Geometry Method

Regional hydraulic geometry relationships are often used to estimate channel dimensions for hydrologic models which require channel geometry data. Streamflow data from the National Water Information System (NWIS; www.waterdata.usgs.gov) contains channel dimension measurements including the width, depth, and watershed size. These measurements were used to develop FTABLEs by ranking each gage by calculated discharge frequencies over a range of discharge exceedances. Analyses were limited to 45 out of the 288 gages that had greater than 10 years of data, had not moved or changed significantly during the period of record, and had less than 5% missing data. For each of the 45 gages, field measurements of channel width, depth, and velocity over a range of streamflows were converted to at-a-station hydraulic geometry parameters using a power-law fit for width, depth, and velocity with discharge:

$$w = a \times Q^b \quad (1)$$

$$d = c \times Q^f \quad (2)$$

$$v = k \times Q^m \quad (3)$$

where w is the width (meters), d is the depth (meters), v is the velocity (meters per second), Q is the discharge (cubic meters per second) and a, b, c, f, k, and m are the power-law fit parameters [30].

The station power-law fit parameters (at-a-station) were scaled by each exceedance discharge (ranging from 0% to 100% of the maximum discharge). The power-law parameters for each exceedance discharge were then used to calculate the downstream hydraulic geometry [30]. For example, at the 50% exceedance discharge for width, the at-a-station power-law fit was used to estimate the width for all of the gages that were used to calculate a single power-law fit for the downstream hydraulic geometry for the width at the 50% exceedance flow (R^2 = 0.97). Gages with zero discharge up to relatively high exceedances were removed because they violate the assumptions of the power law. Major reach segments (greater than a 10 km^2 drainage area and 10 km in length) were identified using GIS. Discharge was related to the drainage area for each gage and exceedance flow using a linear regression (R^2 = 0.93–0.97) (Figure 3) because discharge was not known for the majority of reaches in the model domain.

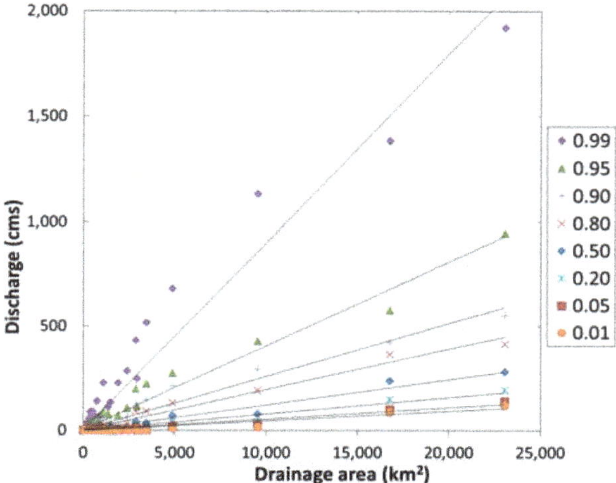

Figure 3. Discharge to drainage area relationship for stream gages within the model domain, grouped by exceedance discharge.

The hydraulic geometry was calculated for each reach and exceedance discharge using the drainage area-to-discharge regression equations and the downstream hydraulic geometry relationship. Each HRU contains multiple reaches that were represented in the model as a single RCHRES; therefore, the reach segments were summed to generate the final FTABLE assigned to the RCHRES. Several reaches on the Sacramento River include flood structures called weirs. The weirs included in the HSPF model are located on reaches 61, 98, 99, 90 and 97 (Figure 1). Adjustments were made to the FTABLEs to enable observed changes in streamflow routing during high flows. At certain high flow levels, the simulated weir will divert water and sediment to an adjacent bypass.

One-Dimensional Hydraulic Model Method

For seven of the lower-elevation HRUs in the valley floor (68, 78, 90, 94, 97, 98, 99), the hydraulic geometry method was not successful in formulating the FTABLEs. The breakdown of the hydraulic geometry method was primarily due to the low elevational gradients and large flood plains. For these lowland HRUs, the Digital Elevation Model (DEM)-derived drainage routing and watershed area frequently did not follow known river channels because out-of-levee floodplain areas are routinely near or below the river water level, and the steepest descent path did not accurately follow the river channels.

Instead of using the hydraulic geometry method for these low-elevation, low-gradient HRUs, an existing one-dimensional (1-D) hydraulic model was used to create the FTABLEs. The 1-D hydraulic model was created by the California Department of Water Resources (DWR) as part of the Central Valley Floodplain Evaluation and Delineation Program (CVFED) for the Sacramento River and its tributaries, with bathymetry and topography collected from 2001 to 2011 [31]. The CVFED hydraulic model was run with flow only within the major tributary channels (no flows overtopped the levees). The hydrologic inputs to the model were the same as those used for generating the aforementioned hydraulic geometry FTABLES. The FTABLES were created from the 1-D hydraulic model by calculating the channel surface area, depth and volume for each reach over a range of streamflows.

3. Calibration

The HSPF model calibration was done in three steps; snow was calibrated first, followed by hydrology and finally sediment. Standard calibration techniques were used to develop snow (S4), and to develop initial hydraulic parameters (S5). Novel or non-standard calibration techniques are described below.

3.1. Hydrologic and Sediment Calibration

The HSPF model was calibrated for the period 1998–2008. The calibration period was chosen to include the land use data from 2006 and a wide range of annual precipitation. The calibration period included four wet years, one above-normal year, two below-normal years, three dry years, and one critically dry year [32]. Calibration runs were the initial test of the model, and using only data from 1998 to 2008, the parameters were changed iteratively to find an acceptable fit of modeled to observed data. Validation of the model was performed by calculating the goodness-of-fit statistics using a period of record other than what was used for model calibration. A successful model validation is indicated by the goodness-of-fit statistics being comparable to the results obtained for model calibration. The validation was completed using data only within the period 1980–1995. The validation period included five wet years, two above-normal years, four dry years and five critically dry years [32].

Hydrologic parameters were developed for all HRUs using spatially distributed physical properties to maintain a more accurate spatial representation of the model parameters in locations without calibration data. For example, infiltration capacity (INFILT) is one of the parameters to which modeled streamflow is the most sensitive [33], and initial values were assigned based on a spatially distributed map of the hydrologic soil group (HSG). Hydrologic soil groups range from "A" to "D", where "A" soils are characterized by the highest infiltration rate and the lowest runoff potential and "D" soils have a very slow infiltration rate and the highest runoff potential. A range of values were scaled to the "typical" parameters listed in the BASINS Technical Note 6 (0.254–6.35) [34]. During the iterative process of calibration these values were raised or lowered in unison. Each hydrologic parameter was determined with this method using land use, soils data, and/or slope. There are "typical" and "possible" ranges for each parameter, although parameter values should reflect conditions in the watershed [34]. Every effort was made to derive parameter values that reflected the spatial characteristics of the watershed, even though they are lumped segments.

After the hydrologic calibration was complete, the sediment calibration was performed using a separate set of parameters related to sediment sources and transport capabilities in HRUs and reaches. Initial sediment parameters were based on GIS layers of soil properties, land use, and slope. The same iterative technique used to calibrate the hydrology was used to calibrate the sediment. Daily modeled sediment loads (tons/day) and suspended sediment concentrations (mg/L) were compared to observed data at NWIS gage locations with sediment data (Table S1).

3.2. Model Performance Evaluation

Goodness-of-fit statistics were calculated to quantitatively compare modeled and observed streamflow at stream gages selected for calibration and validation (Table S1). Three main statistical methods were used to assess the goodness-of-fit of the model results: the coefficient of determination (R^2), the Nash-Sutcliffe coefficient of efficiency (NSE) [35], and mean error percent (ME%). The coefficient of determination (R^2) is the square of the Pearson's product-moment correlation coefficient (R) and ranges from 0 to 1, with higher values indicating better agreement between modeled and observed data [36]. The NSE is the ratio of one minus the mean square error to the variance and is a widely used and reliable statistic for determining the goodness-of-fit for hydrologic models.

A value of 1 for NSE indicates a perfect match between observed and modeled data; values equal to zero indicate that the model is predicting no better than using the average of the observed data. Values less than zero indicate that the model does not do as good as the sample mean in predicting the observed values, and therefore indicates a poor model. NSE and R^2 are sensitive to the extreme values [36]. A mean error (ME%) value closest to zero indicates a better simulation result. Hydrologic calibration results for the 20 gages were assessed for daily and monthly flows using the four statistical methods. The "model performance" is related to guidelines outlined by Donigian [37] (Table 1).

Table 1. Performance rating guidelines for daily and monthly flow statistics [37]. R^2 = coefficient of determination; ME% = mean error; NSE = Nash Sutcliffe efficiency.

Performance Rating	NSE		R^2		ME%	
	Daily	Monthly	Daily	Monthly	Daily	Monthly
Excellent	≥0.85	≥0.90	≥0.85	≥0.90	≤±10	≤±5
Very good	0.75–0.85	0.80–0.90	0.75–0.85	0.80–0.90	±10–±15	±5–±10
Good	0.65–0.75	0.70–0.80	0.65–0.75	0.70–0.80	±15–±25	±10–±15
Fair	0.50–0.65	0.50–0.70	0.50–0.65	0.50–0.70	±25–±30	±15–±25
Poor	<0.50	<0.50	<0.50	<0.50	≥±30	≥±25

4. Results

4.1. Hydrology Calibration Results

Qualitative comparisons of modeled to observed data were performed using hydrographs and flow-duration curves. Visual inspection of flow hydrographs for the calibration (Figure 4) and validation (Figure S2) periods showed a good relationship of modeled to observed data. The modeled summer low-flows tend to be higher than the observed data, and some of the smaller peak flows are underestimated. The highest modeled peaks generally correlated well with observed data but are overestimated in some cases.

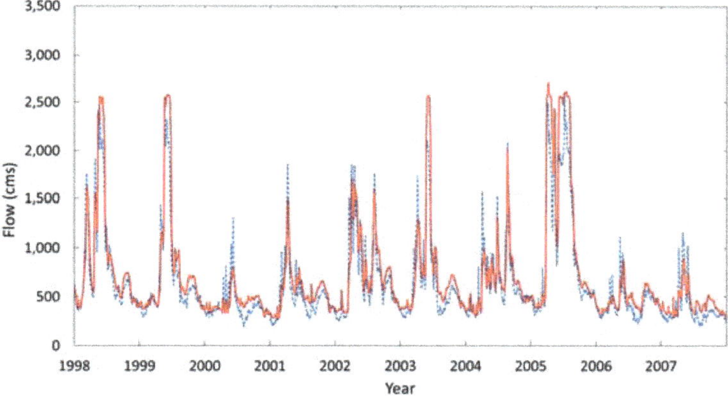

Figure 4. Modeled daily flow at reach 97 (Figure 1, solid red) compared to observed flow for Sacramento River at Freeport (11447650) (dashed blue) for the calibration period (1998–2008).

Flow-duration curves were also compared to ensure the realistic replication of hydrologic processes, with an emphasis on the higher flow regimes that typically move more sediment through the system. The flow-duration curves for the Sacramento River show that the overall relationship is similar to the observed flow, although the modeled flows tend to be higher than those observed and there is a divergence between the 2%–10% exceeded range likely due to how weirs are simulated in the model. The average R^2 values for all reaches were 0.79 for daily flows and 0.91 for monthly flows for the entire simulation period (Table 2). The average R^2 values for the Sacramento River during the entire simulation period were 0.92 and 0.96 for daily and monthly flows, respectively. The calibration process focused on achieving the best results for the Sacramento River rather than the contributing tributaries, because the objective of the study was to project potential future suspended sediment being discharged by the Sacramento River to the Bay-Delta, and therefore model accuracy was most critical for simulated flows of the Sacramento River. Calibration results indicated performance ratings of "Very good" to "Excellent" for the Sacramento River gages using three statistical metrics used to evaluate the HSPF model (Table 2).

Table 2. Daily and monthly flow statistics for the simulation period, and the calibration and validation periods. Values are averages of all reaches or Sacramento River reaches only. R^2 = coefficient of determination; ME% = mean error; NSE = Nash Sutcliffe efficiency.

Model Simulation Time Period		NSE		R^2		ME%	
		All Reaches [1]	Sacramento River	All Reaches [1]	Sacramento River	All Reaches [1]	Sacramento River
Simulation (1958–2008)	daily	0.69	0.90	0.79	0.92	−3%	−7%
	monthly	0.86	0.93	0.91	0.96	−7%	−7%
Calibration (1998–2008)	daily	0.70	0.89	0.74	0.92	20%	−10%
	monthly	0.78	0.92	0.84	0.96	11%	−10%
Validation (1980–1995)	daily	0.66	0.91	0.75	0.93	−15%	−11%
	monthly	0.83	0.94	0.88	0.97	−15%	−11%

Note: [1] Not including gages located on a bypass (Reaches 62, 78, and 94).

Full statistics for each gage show the variability of statistical results between gages (Table S4). Gages located on a bypass (62, 78, and 94) displayed high R^2 values but also extremely high mean error percent values which were not included in the average statistics (Table 2). The combination of high mean error percent values with high correlation values was likely due to the bypasses not being explicitly modeled in HSPF, where modeled flow did not accurately represent the observed low flow recession and conversely overestimated high peaks compared to the observed data. Calibration focused on the Sacramento River reaches and the average statistics ranged from "Very good" to "Excellent"; the tributaries ranged from "Fair" to "Very good", except for reaches with bypasses (Table 2 and Table S5). Monthly flow statistics were generally better than daily statistics, as shown in Table 2. Annual modeled streamflow at Sacramento River at Freeport generally matched well against observed data (Figure S3). The annual modeled peaks were slightly higher than observed data but were similar to the overall long-term trend. The regression lines for both data sets indicate a slight increasing trend but are not significant.

4.2. Sediment Calibration Results

Modeled suspended sediment and total sediment loads were compared to annual sediment loads and daily suspended sediment concentrations and sediment loads from NWIS stream gages that contained historical sediment time series. Daily statistics ranged considerably (Table S5), and the calibration with emphasis on the Sacramento River resulted in higher R^2 values and mean errors closer to zero for the Sacramento River gages. The modeled ratios divided by the observed ratios ranged considerably, but on average for all gages were 0.95 for sediment loads and 1.0 for SSC (Table 4). Sediment statistics were generally poorer than the hydrologic statistics because sediment transport is simulated using model components and assumptions associated with a higher degree of uncertainty, and this uncertainty is added to the uncertainty in the modeled flow values. Calibration focused on the Sacramento River resulted in a good calibration for daily and monthly sediment loads and SSC (Table 3).

The average daily R^2 for sediment loads (for all gages) was 0.49 and the average R^2 value for SSC was 0.38 (Table 3). Annual modeled sediment loads were compared for all gages on the Sacramento River with available sediment data (Table S1). Sediment outputs were converted to million metric tons (Mt) by water year (Figure 5). The relationship between modeled and observed annual sediment loads on the Sacramento River is nearly one to one, with an R^2 of 0.81. The ratio of modeled to observed sediment loads is on average 1.12, with a mean error of 4%.

Table 3. Daily and monthly sediment statistics. Values are averages of all reaches or Sacramento River reaches only. SSC = suspended sediment concentrations; R^2 = coefficient of determination; ME% = mean error.

SSC (mg/L) or Sediment Loads (Tons/Day)		Modeled/Observed		R^2		ME%	
		All Reaches	Sacramento River	All Reaches	Sacramento River	All Reaches	Sacramento River
Sediment loads	daily	0.95	0.89	0.49	0.52	400%	21%
	monthly	0.97	0.91	0.72	0.74	3131%	−9%
SSC	daily	1.0	0.92	0.38	0.37	486%	55%
	monthly	1.0	0.95	0.64	0.56	202%	23%

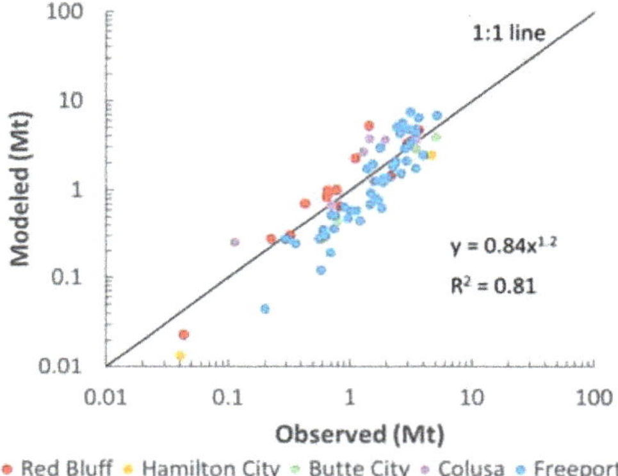

Figure 5. Comparison of modeled and observed annual sediment loads (million metric tons, Mt) for the gage locations Sacramento River at Red Bluff (1959–1970, 1977–1980), Hamilton City (1977–1979), Butte City (1978–1980), Colusa (1973, 1975, 1977–1980), and Freeport (1959–2008).

The annual modeled sediment loads for the Sacramento River at the Freeport gage reproduced the general measured trends (Figure S4) to use for the final calibration goal of demonstrating the long-term trends of sediment transport. Although the comparisons of individual water years are variable, the overall trend of modeled sediment over the past 50 years was similar to the measured data (Figure S4). There was more variability in the modeled data compared to the observed data on a year-to-year basis. The modeled annual SSC also showed a similar trend to the observed data, although there was significantly less variability in the modeled SSC data than with modeled sediment loads. Generally, modeled annual SSC slightly under-predicted observed SSC.

4.3. Historical Trends in Streamflow and Sediment

Near the confluence of the Sacramento River and the Bay-Delta, there is a notable observed decrease in suspended sediment concentrations (SSC) in the past 50 years (1957–2001) [4]. One objective of this model was to calibrate to post-dam sediment loads to determine if the observed declining trend could be replicated and if the mechanisms for the decline could be elucidated. The current conceptual model describing the mechanisms and processes contributing to the sediment supply to the Bay-Delta suggest that human activities that alter watershed sediment supply are likely to have a greater effect on the river supply to the Bay-Delta than those that modify the flow regime [6]. Thus, they propose that the decline in SSC over the past 50 years was only moderately influenced by the climate. We intend to address the influence of the climate on the historical patterns of sediment transport. Historical flow and sediment trends were assessed by dividing the annual data into the top and bottom 50% flow regimes (Figure 6).

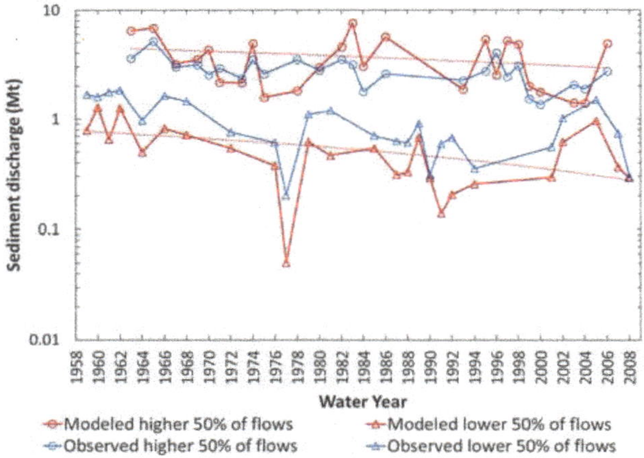

Figure 6. Comparison of modeled (**red**) and observed (**blue**) annual sediment loads (million metric tons, Mt) for the Sacramento River at Freeport gage. Flows are separated into highest and lowest 50% regimes.

Kendall's tau (τ) and Spearman's rho (ρ) statistical tests were performed to determine the statistical significance and probability of monotonic time trends. At the Sacramento River at the Freeport gage location, there were no statistically significant trends in modeled annual streamflow for the model period (1958–2008). A statistically significant decreasing trend of annual sediment discharge over the 1958–2008 period was found at the Sacramento River at the Freeport gage in the lower 50% of flows ($p = 0.01$, ρ and τ). There is also a slight decreasing trend in SSC for the lower 50% of annual flows, ($p = 0.02$, τ).

Previous studies have examined suspended sediment trends from the greater Sacramento River Basin [4,16,38], though none have examined trends between the Sacramento River below Keswick to Verona (Figure 1) due to a lack of measured data. Statistically significant decreasing trends of modeled sediment were found at the Sacramento River at the Butte City gage (reach 49, $p = 0.01$, 0.02 ρ and τ, respectively) and at the Sacramento River at the Colusa gage (reach 61, $p = 0.03$, 0.04, ρ and τ, respectively) for the lower 50% flow regime. Decreasing trends of varying magnitude were apparent at all of the gages on the Sacramento River, but were not statistically significant at the Red Bluff (reach 22) or Verona (reach 99) locations (Figure 1).

5. Historical Climate Trends

Although there was no significant mean annual change in flow or precipitation during the simulation period, simulated changes in the magnitude and frequency of precipitation, snow, and streamflow were analyzed to evaluate mechanisms that might lead to the decline of transported sediment. To determine the effects of climate change on streamflows, all inflows from dams and gages were removed to negate the influence of flow management from dams and diversions. The northern half of the model was located at the Sacramento River at the Butte City gage (BC) and the southern half of the model extended from BC to the Sacramento River at the Freeport gage (FPT). With no changes made to any spatially distributed properties over the simulation period, the 50-year simulation was divided into 1958–1983 and 1984–2008 to assess changes in flow solely related to climate. Suspended sediment transport was positively correlated to daily flow volumes with an R^2 of greater than 0.9.

As the majority of sediment is transported by higher flows, peak precipitation days (>95%) were assessed, and these decreased by 8% and 10% at BC and FPT, respectively (Table 4).

Table 4. Precipitation, snow, and streamflow changes during the model simulation period (1958–2008). Units for precipitation and streamflow are number of peak days (>95%) and mm for snow.

Location	Variable	1958–1983	1984–2008	Percent Change
Sacramento River at Butte City	precipitation	476	438	−8%
	snow	26.2	23.8	−9%
	peak streamflow	487	427	−12%
Sacramento River at Freeport	precipitation	481	433	−10%
	snow	0.18	0.04	−76%
	peak streamflow	535	379	−29%

On average, the snow water equivalent decreased 9% for BC and 76% for FPT (Table 4). Peak streamflow days considering no model inflows declined 12% and 29% at BC and FPT, respectively (Table 4). Changes in timing were also observed at both BC and FPT. When the entire Sacramento Valley Basin is added together, it is clear from Figure 7 that the peak snow water equivalent occurs a month earlier in the second time period 1984–2008, and peak flows have declined in the three months of December through February. However, peak flows increased for the four months of March through June (Figure 7).

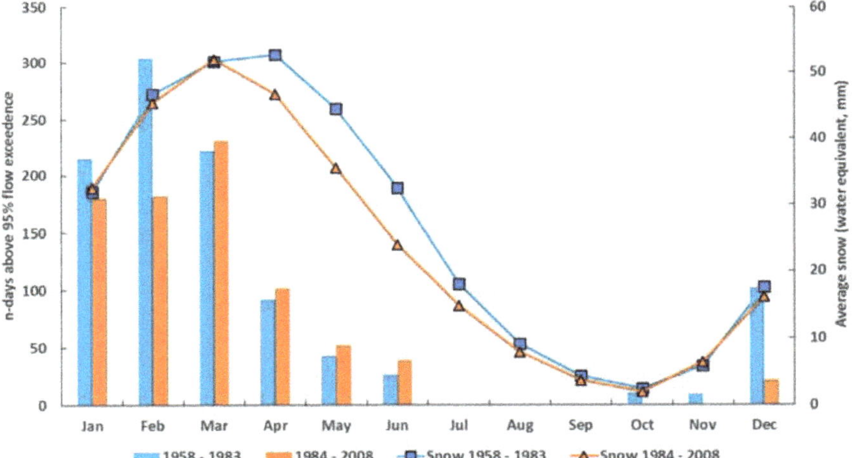

Figure 7. Modeled peak streamflow days (>95%) by month for the Sacramento River Basin with model inflows removed, for 1958–1983 (blue bars) and 1984–2008 (orange bars). Snow (water equivalent, mm) for the entire model domain is displayed on the right y-axis.

The mean temperature increased by 0.25 °C in the northern basin and 0.5 °C in the southern basin, contributing to the shift in snow accumulation and timing. The peak magnitude and frequency of precipitation and snow changed over the historical climate period, and thus peak streamflows were affected, although annual mean trends were not observed. The changes in the timing and magnitude of streamflows likely contributed to the observed decline in sediment transport for the Sacramento River Basin, thus providing more relative weight of the climate as a driver in the current conceptual model of sediment supply processes to the Bay-Delta [6].

6. Future Climate Sensitivity Analysis Using Climate Assessment Tool (CAT)

The model's sensitivity to future climate change was further explored by using the Climate Assessment Tool (CAT), a program within BASINS that provides capabilities for creating climate change scenarios using a calibrated HSPF model [39]. A wide range of "what if" questions about future weather and climate changes are possible to assess the effects on hydrology and water quality. The historical climate for 1958–2008 is considered the "base" period, and adjustments were made to precipitation, air temperature, and potential evapotranspiration to create possible "scenarios" of future climate. These scenarios were used to assess potential impacts and sensitivity to climate change on the hydrology and water quality for the Sacramento River Basin.

Several scenarios were developed to assess how sensitive hydrology and sediment transport are to climate change. While these hypothetical scenarios are useful to determine sensitivity of the hydrologic system to climate change, they are not projections or predictions of future climate variables. The scenarios chosen reflect a consensus of simulations of air temperature and precipitation change for northern California over the next century [40]. Air temperature is expected to increase by 1.5 °C under the lower emissions B1 scenario in the less conservative (in which greenhouse emissions are mitigated) climate model to an increase of 4.5 °C in the higher emissions A2 scenario with a more conservative (greenhouse emissions increase) climate model [40]. Precipitation is projected to change only slightly [40], so positive and negative changes of 10% were used to determine the sensitivity of streamflow to precipitation changes. Some future climate scenarios predicted an increase in the magnitude and frequency of large storms [41–43]; therefore, an increase of 10% for the storm magnitude and frequency was used in two of the hypothetical scenarios. A 10% increase in storm magnitude and frequency is defined in this scenario as a 10% increase in frequency and magnitude of the top 10% of storms by volume in the historical record (1958–2008).

The scenarios T1 and T2 were increases of temperature only (1.5 °C and 4.5 °C, respectively). The P1–P3 scenarios increased or decreased precipitation without changes in temperature. Scenarios W1–W4 were combinations of temperature increases (1.5 °C and 4.5 °C) and a 10% increase in precipitation or a 10% increase in storm magnitude and frequency (Table 5). The D1 and D2 scenarios were defined as a 10% decrease in precipitation with increases in air temperature of 1.5 °C or 4.5 °C. Potential evapotranspiration was recalculated for each scenario using the Hamon (1961) method, which is based on air temperature and therefore may overestimate PET. The sensitivity analysis using CAT was designed to assess how the baseline model responds to general changes in climate. For these hypothetical scenarios, the Hamon method was used because resources did not permit the recalculation of PET using the Priestley-Taylor method.

The baseline data were most sensitive to an increase of storm magnitude and frequency of 10% and no temperature change, where mean streamflow increased by 2.6%, sediment discharge increased by 9%, and SSC increased by 3.3%. Scenarios that included increases in temperature resulted in a decrease of streamflow, sediment discharge, and SSC, even scenarios with relatively large increases in precipitation. Streamflow and sediment displayed a slightly greater decline in response to a 10% increase of storm magnitude and frequency than an overall 10% increase in precipitation. Mean sediment discharge and SSC were more sensitive to changes in climate than mean streamflow. The most sensitive scenario which included a temperature change was D2, and resulted in decreases of 0.8% for streamflow, 7.9% for sediment discharge, and 8.5% for SSC (Table 5). Streamflow, sediment discharge, and SSC increased under most precipitation scenarios but began to decline at precipitation thresholds of −18%, −12%, and −19%, respectively. With increases of both temperature and precipitation, temperature had a stronger influence and therefore decreases were only seen in the T, W, and D scenarios. In the W and D scenarios, increases of the temperature from 1.5 °C to 4.5 °C only decreased streamflow and sediment by less than a percent.

Table 5. Hypothetical climate change scenarios generated using the Climate Assessment Tool (CAT). "BASE" scenario is the historical modeled flow for Reach 97 with units expressed in parentheses. Values for scenario results are expressed as a percent higher or lower than the base scenario. cms = cubic meters per second; SSC = suspended sediment concentration.

Scenario	Description	Mean Annual Streamflow (cms)	Mean Annual Sediment Discharge (Tons/Day)	Mean SSC (mg/L)
BASE	Reach 97 (1980–2008)	727	8604	41.5
		Change in Mean Streamflow (Percent)	**Change in Mean Sediment Discharge (Percent)**	**Change in Mean SSC (Percent)**
T1	Temperature +1.5 °C	−0.5	−7.7	−7.9
T2	Temperature +4.5 °C	−0.7	−7.9	−8.3
P1	Precipitation −10%	+0.5	+0.5	+0.6
P2	Precipitation +10%	+1.9	+5.1	+2.2
P3	Storm magnitude and frequency +10%	+2.6	+9.0	+3.3
W1	Temperature +1.5 °C, precipitation +10%	−0.5	−7.6	−7.9
W2	Temperature +4.5 °C, precipitation +10%	−0.6	−7.8	−8.2
W3	Temperature +1.5 °C, storm magnitude and frequency +10%	−0.6	−7.7	−8.0
W4	Temperature +4.5 °C, storm magnitude and frequency +10%	−0.7	−7.9	−8.4
D1	Temperature +1.5 °C, precipitation −10%	−0.7	−7.8	−8.1
D2	Temperature +4.5 °C, precipitation −10%	−0.8	−7.9	−8.5

7. Discussion

The main objective of this study was to characterize the hydrology and sediment transport in the Sacramento River Basin. The results provide streamflow and sediment boundary conditions to a hydrodynamic model of the Bay-Delta used to help project potential impacts to the hydrologic and ecological health of the Bay-Delta in response to climate change and variability. Spatial interpolation of meteorological data and the estimation of channel characteristics using hydraulic geometry modeling were used to develop model inputs that included many subdrainages and tributaries with sparse data. The parameterization techniques incorporated known spatial information for tributaries such as soils information, slope, and land cover. As part of the model calibration, parameters were adjusted uniformly using linear scaling to increase or decrease values in order to retain the spatial distribution of the physical characteristics of the land surface in un-gaged areas.

The hydrological calibration of the Sacramento Valley HSPF model provided statistical results for streamflow that showed a range of model accuracy for tributaries but the goodness-of-fit statistics for both calibration and validation for locations on the main branch of the Sacramento River ranged from "very good" to "excellent". Sediment calibration also resulted in a wide range of accuracy depending on location, although simulation results at gages located on the Sacramento River averaged 11% higher than observed sediment loads and 8% higher than observed SSC. Based on statistical guidelines, the calibration of flow and sediment was very successful for the Sacramento River, owing in large part to the rigorous development of spatially distributed climate, the parameterization of physical properties, and extensive development of FTABLEs.

An important objective of this model was to calibrate to post-dam sediment loads in the Sacramento River Basin to determine if the observed declining trend could be replicated which has implications for the long-term sediment yield to the San Francisco Bay-Delta. A slight but statistically significant decreasing trend was evident in the modeled suspended sediment and SSC results at the Sacramento River at the Freeport location from 1958 to 2008 in the lower 50% flow regime. No changes were made to hydrology or sediment parameters over the simulation time; therefore, changes in sediment must be attributed to managed inflows or climate. Although there was no overall change in flow over the model period, changes in timing and frequency of high flows due to warming and declining snow were observed and likely decreased the sediment yield over time.

The calibrated HSPF model was used to help evaluate the potential response of streamflow, sediment loads, and SSC to future climate change and variability using hypothetical climate scenarios. The base hydrology was most sensitive to an increase of the storm magnitude and frequency, with no temperature change. Snowpack losses are projected to increase through the next century [40], which will decrease snow pack accumulation and can cause a shift in the hydrologic timing: snow will melt earlier in spring, leaving less water during the summer months. A shift in the hydrologic timing of the peak streamflow was observed at the Butte City and Freeport locations. A significant decrease in the snow pack for the basin could explain the decrease and shift in timing of peak flows over the historical period at Butte City and Freeport. This change in hydrologic extremes and timing is likely a major driver in the decline in sediment at Freeport, since modeled sediment was more sensitive to changes in precipitation intensity than an average increase over time.

Model Uncertainty

Often the biggest measure of input uncertainty into HSPF is attributed to a scarcity of long-term climate stations as well as the quantity of stations available. The improvement of meteorological data and hydraulic function tables (FTABLEs) as well as the spatial distribution of physical properties are measures designed to reduce the amount of input-induced error into the model and rely more heavily on distributed processes to which the calibration parameters are physically related.

HSPF is a one-dimensional flow and transport model, and as such has known limitations for simulating river bank erosion, channel armoring, or longitudinal bed sediment movement. The Sacramento River is a gently sloping, meandering river with bed loads comprised predominantly of sand and gravel; therefore, the bed load contributions downstream and consequently to the Bay-Delta are minimal, and it was thus assumed that model limitations regarding more complex flow and transport processes had a minimal effect on the results in this location. In addition, many of the large tributaries below dams have experienced channel armoring, or the removal of finer sediment fractions from the bed. Therefore, bed loads in this model domain were assumed to contribute a minimal amount to the total modeled sediment loads.

HSPF is a semi-lumped model, which causes inherent error due to the assumption that each sub-watershed is physically homogeneous. In addition, the model can only process a certain number of operations (model segments and hydrologic reach segments), creating the need for larger HRUs within the model domain and, in this case, two sub-models. Considering the complexity and size of the model, uncertainty is a concern and can come from the lack of diversion data from water transfers, agricultural uses, return flows, and the pumping of groundwater, which were not explicitly modeled. Calibration results for simulated versus observed snow accumulation and melting indicated that model uncertainty might be reduced if a more detailed discretization of HRUs (smaller HRUs) was applied to the higher-elevation subdrainages to better represent orographic effects on air temperature and precipitation. The lack of long-term sediment data and accumulation or erosion information within the domain led to uncertainties of the interpretation of areas that could be either a sediment sink or source. Input data, model or user assumptions, and parameter uncertainty can also introduce errors in modeling on such a large scale.

8. Conclusions

The development of well-calibrated, spatially distributed, mechanistic models, such as this one of the Sacramento Valley below major dams, provides a valuable tool to characterize the mechanisms and climatic signals that are responsible for flow and transport in a large basin. While there are limitations to the parameterization of these models, when all parameters are held constant over time, the changes in climate and how they influence flow and transport can be scrutinized. Unique parameterization of this basin helps identify where more data is needed to further understand physical processes that affect sediment transport. The ability to isolate land use change and water management provided the opportunity to evaluate historical changes as well as potential future changes in climate. In this watershed, the changes in the historical climate, with warmer temperatures and declines in snowpack, result in fewer peak flows, which translate into a meaningful and significant mechanism responsible for the long-term declining trend in sediment transport to complement the existing conceptual model [6] that relies on anthropogenic drivers such as historical land use change and water management. This conceptual model suggests that climate has only a moderate influence on sediment transport and plays no role in the historical decline of sediment to the Bay-Delta. In contrast, our analyses of historical climate in conjunction with sensitivity analyses support an increase in the role of climate on transport, particularly the role of large storms that result in high peak flows. As a result of diminishing snowpack over the last 50 years, and the likelihood of this continuing trend with global warming, the role of climate should be emphasized alongside anthropogenic influences in sediment transport to the Bay-Delta. Simulated future trends of sediment transport in the Sacramento River basin and understanding the mechanisms behind the trends are imperative to projecting the future health of the Bay-Delta ecosystem.

For the CASCaDE II project, future climate scenarios will be used to make projections of projected hydrologic and sediment trends during the period 2010–2100. The HSPF model will provide a direct coupling of the current and future climate with the potential sediment sources to estimate sediment supply to the Bay-Delta. Outputs from this model will be used as sediment boundary conditions for a hydrodynamic model to assess the future of the San Francisco Bay-Delta estuary-watershed system under the cascading effects of Delta configurational changes and climate change.

Supplementary Materials: The following are available online at www.mdpi.com/2073-4441/8/10/432/s1.

Acknowledgments: This project was supported by the Computational Assessments of Scenarios of Change for the Delta Ecosystem (CASCaDE II) project. CASCaDE II is supported by a grant from the Delta Science Program (DSC Grant #2040). Any opinions, findings, and conclusions or recommendations expressed in this material are those of the authors and do not necessarily reflect the views of the Delta Science Program. This is CASCaDE publication #71.

Author Contributions: Lorraine Flint and Scott Wright conceived and designed the experiments; Michelle Stern performed the modeling and data collection; Michelle Stern, Lorraine Flint, Justin Minear and Scott Wright analyzed the data; Lorraine Flint, Alan Flint, Justin Minear and Scott Wright contributed inputs to the model; Michelle Stern and Lorraine Flint wrote the paper.

Conflicts of Interest: The authors declare no conflict of interest. The founding sponsors had no role in the design of the study; in the collection, analyses, or interpretation of data; in the writing of the manuscript, and in the decision to publish the results.

Abbreviations

The following abbreviations are used in this manuscript:

BASINS	Better Assessment Science Integrating Point and Nonpoint Sources program
CAT	Climate Assessment Tool
CDEC	California Data Exchange Center
CIMIS	California Irrigation Management Information System
CVFED	Central Valley Floodplain Evaluation and Delineation Program
DEM	Digital Elevation Model
DWR	California Department of Water Resources
FTABLE	Hydraulic Function Table
GIDS	Gradient Inverse Distance Squared Interpolation
GIS	Geographic Information System
HSPF	Hydrological Simulation Program: FORTRAN
HRU	hydrologic response unit
HSG	Hydrologic Soil Group
HUC	USGS Hydrologic Unit (watershed) code
IMPLND	Impervious Land Segment
INFILT	Index to the Infiltration Capacity of the Soil
km	Kilometer
m	Meter
mm	Millimeter
ME%	Mean error percent
mg/L	Milligrams per liter
Mt	Million metric tons
NED	National Elevation Dataset
NHD	National Hydrography Dataset
NID	National Inventory of Dams
NLCD	National Land Cover Dataset
NSE	Nash Sutcliffe coefficient of efficiency
NWIS	National Water Information System
PERLND	Pervious Land Segment
PET	Potential Evapotranspiration
PRISM	Parameter-elevation Regressions on Independent Slopes Model
R	Correlation Coefficient
R^2	Coefficient of Determination
RCHRES	Modeled Reach or Reservoir
SSC	suspended sediment concentration
SSURGO	Soil Survey Geographic Database
SWE	snow water equivalent
UCI	user control input file
USGS	United States Geological Survey
WY	water year

References

1. Interagency Ecological Program (IEP). Dayflow. 2005; Available online: http://iep.water.ca.gov/dayflow/index.html (accessed on 3 February 2014).
2. CASCaDE: Computational A of Scenarios of Change for the Delta Ecosystem. Available online: http://cascade.wr.usgs.gov (accessed on 4 April 2016).
3. Ganju, N.K.; Schoellhamer, D.H. Decadal-timescale estuarine geomorphic change under future scenarios of climate and sediment supply. *Estuar. Coasts* **2010**, *33*, 15–29. [CrossRef]
4. Wright, S.A.; Schoellhamer, D.H. Trends in the sediment yield of the Sacramento River, California, 1957–2001. *San Franc. Estuary Watershed Sci.* **2004**, *2*, 1–14.

5. Cloern, J.E.; Knowles, N.; Brown, L.R.; Cayan, D.R.; Dettinger, M.D.; Morgan, T.L.; Schoellhamer, D.H.; Stacey, M.T.; van der Wegen, M.; Wagner, R.W.; et al. Projected Evolution of California's San Francisco Bay-Delta River System in a Century of Climate Change. *PLoS ONE* **2011**, *6*, e24465. [CrossRef] [PubMed]
6. Schoellhamer, D.H.; Wright, S.A.; Drexler, J. A Conceptual model of sedimentation in the Sacramento-San Joaquin Delta. *San Franc. Estuary Watershed Sci.* **2012**, *10*, 1–25.
7. Singer, M.B.; Dunne, T. An empirical-stochastic, event-based program for simulating inflow from a tributary network: Framework and application to the Sacramento River basin, California. *Water Resour. Res.* **2004**, *40*. [CrossRef]
8. Singer, M.B. The influence of major dams on hydrology through the drainage network of the Sacramento River Basin, California. *River Res. Appl.* **2007**, *23*, 55–72. [CrossRef]
9. California Department of Water Resources. California Water Plan Update 2013. Available online: http://www.water.ca.gov/waterplan/docs/cwpu2013/Final/Vol2_SacramentoRiverRR.pdf (accessed on 11 April 2016).
10. Bicknell, B.R.; Imhoff, J.C.; Kittle, J.L., Jr.; Donigian, A.S., Jr.; Johanson, R.C. *Hydrologic Simulation Program—FORTRAN, User's Manual for Version 12*; U.S. EPA Environmental Research Laboratory: Athens, GA, USA, 2001; pp. 1–755.
11. Donigian, A.S.; Bicknell, B.R.; Imhoff, J.C. Hydrological Simulation Program—FORTRAN (HSPF). In *Computer Models of Watershed Hydrology*; Singh, V.P., Ed.; Water Resources Publications: Highlands Ranch, CO, USA, 1995; pp. 395–442.
12. Fry, J.A.; Xian, G.; Jin, S.; Dewitz, J.A.; Homer, C.G.; Limin, Y.; Barnes, C.A.; Herold, N.D.; Wickham, J.D. Completion of the 2006 National Land Cover Database for the Conterminous United States. *Photogramm. Eng. Remote Sens.* **2011**, *77*, 858–864.
13. Gesch, D.; Evans, G.; Mauck, J.; Hutchinson, J.; Carswell, W.J. The national map—Elevation. *U.S. Geological Survey Fact Sheet 2009–3053*; U.S. Geological Survey: Reston, VA, USA, 2009; p. 4. Available online: http://pubs.usgs.gov/fs/2009/3053/ (accessed on 5 September 2013).
14. Simley, J.; Carswell, W.J. *The National Map—Hydrography*; U.S. Geological Survey: Reston, VA USA, 2009; Volume 3054, p. 4.
15. Soil Survey Staff. Web Soil Survey. Available online: http://websoilsurvey.nrcs.usda.gov/ (accessed on 5 September 2013).
16. Chen, C.W.; Herr, J.W. *Comparison of BASINS and WARMF Models: Mica Creek Watershed*; Electrcic Power Research Institute: Palo Alto, CA, USA, 2002.
17. Nigro, J.; Toll, D.; Partington, E.; Ni-Meister, W.; Shihyan, L.; Gutierrez-Magness, A.; Engman, T.; Arsenault, K. NASA-modified precipitation products to improve USEPA nonpoint source water quality modeling for the Chesapeake Bay. *J. Environ. Qual.* **2010**, *39*, 1388–1401. [PubMed]
18. Laroche, A.M.; Gallichand, J.; Lagace, R.; Pesant, A. Simulating atrazine transport with HSPF in an agriculutral watershed. *J. Environ. Eng.* **1996**, *122*, 622–630. [CrossRef]
19. Albek, M.; Ogutveren, U.B.; Albek, E. Hydrological modeling of Seydi Suyu watershed (Turkey) with HSPF. *J. Hydrol.* **2004**, *285*, 260–271. [CrossRef]
20. Kim, S.M.; Benham, B.L.; Brannan, K.M.; Zeckoski, R.W.; Doherty, J. Comparison of hydrologic calibration of HSPF using automatic and manual methods. *Water Resour. Res.* **2007**, *43*. [CrossRef]
21. Tong, S.T.Y.; Sun, Y.; Ranatunga, T.; He, J.; Yang, Y.J. Predicting plausible impacts of sets of climate and land use change scenarios on water resources. *Appl. Geogr.* **2012**, *32*, 477–489. [CrossRef]
22. Hayashi, S.; Murakami, S.; Watanabe, M.; Bao-Hua, X. HSPF simulation of runoff and sediment loads in the Upper Changjiang River Basin, China. *J. Environ. Eng.* **2004**, *130*, 801–815. [CrossRef]
23. Nalder, I.A.; Wein, R.W. Spatial interpolation of climatic normals: Test of a new method in the Canadian boreal forest. *Agric. For. Meteorol.* **1998**, *92*, 211–225. [CrossRef]
24. Flint, L.E.; Flint, A.L. Downscaling future climate scenarios to fine scales for hydrologic and ecological modeling and analysis. *Ecol. Process.* **2012**, *1*, 1–15. [CrossRef]
25. Daly, C.; Halbleib, M.; Smith, J.I.; Gibson, W.P.; Doggett, M.K.; Taylor, G.H.; Curtis, J.; Pasteris, P.P. Physiographically-sensitive mapping of temperature and precipitation across the conterminous United States. *Int. J. Climatol.* **2008**, *28*, 2031–2064. [CrossRef]
26. Priestley, C.H.B.; Taylor, R.J. On the assessment of surface heat flux and evaporation using large-scale parameters. *Mon. Weather Rev.* **1972**, *100*, 81–92. [CrossRef]

27. Bristow, K.L.; Campbell, G.S. On the relationship between incoming solar radiation and daily maximum and minimum temperature. *Agric. For. Meteorol.* **1984**, *31*, 159–166. [CrossRef]
28. Milly, P.C.D.; Dunne, K.A. On the hydrologic adjustment of climate-model projections: The potential pitfall of potential evapotranspiration. *Earth Interact.* **2011**, *15*, 1–14. [CrossRef]
29. U.S. Environmental Protection Agency. *BASINS Technical Note 1: Creating Hydraulic Function Tables (FTABLES) for Reservoirs in BASINS*; Office of Water: Washington, DC, USA, 2007.
30. Leopold, L.B.; Maddock, T., Jr. *The Hydraulic Geometry of Stream Channels and Some Physiographic Implications*; U.S. Geological Survey Professional Paper 252; U.S. Geological Survey: Reston, VA, USA, 1953; pp. 1–57.
31. Gainey, J.; California Department of Water Resources. CVFED Program, Central Valley Floodplain Delineation. Personal communication, 25 February 2014.
32. CA DWR Water Supply Information. Available online: http://cdec.water.ca.gov/water_supply.html (accessed on 25 April 2014).
33. Fonesca, A.; Ames, D.P.; Yang, P.; Botelho, C.; Boaventura, R.; Vilar, V. Watershed model parameter estimation and uncertainty in data-limited environments. *Environ. Model. Softw.* **2014**, *51*, 84–93. [CrossRef]
34. U.S. Environmental Protection Agency. *Estimating Hydrology and Hydraulic Parameters for HSPF: EPA BASINS Technical*; Note 6; Office of Water: Washington, DC, USA, 2000; Volume 4305, p. 32.
35. Nash, J.E.; Sutcliffe, J.V. River flow forcasting through conceptual models, I, A discussion of principles. *J. Hydrol.* **1970**, *10*, 282–290. [CrossRef]
36. Legates, D.R.; McCabe, G.J. Evaluating the use of "goodness-of-fit" measures in hydrologic and hydroclimatic model validation. *Water Resour. Res.* **1999**, *35*, 233–241. [CrossRef]
37. Donigian, A.J. Watershed model calibration and validation: The HSPF experience. *Proc. Water Environ. Fed.* **2002**, *8*, 44–73. [CrossRef]
38. McKee, L.J.; Ganju, N.K.; Schoellhamer, D.H. Estimates of suspended sediment entering San Francisco Bay from the Sacramento and San Joaquin Delta, San Francisco Bay, California. *J. Hydrol.* **2006**, *323*, 335–352. [CrossRef]
39. U.S. Environmental Protection Agency. *BASINS 4.0 Climate Assessment Tool (CAT): Supporting Documentation and User's Manual*; EPA/600/R-08/088F; EPA: Washington, DC, USA, 2009.
40. Cayan, D.R.; Maurer, E.P.; Dettinger, M.D.; Tyree, M.; Hayhoe, K. Climate change scenarios for the California region. *Clim. Chang.* **2008**, *87*, 21–42. [CrossRef]
41. Cayan, D.R.; Tyree, M.; Dettinger, M.D.; Hidalgo, H.; Das, T.; Maurer, E.; Bromirski, P.; Graham, N.; Flick, R. *Climate Change Scenarios and Sea Level Rise Estimates for California 2008 Climate Change Scenarios Assessment*; PIER Research Report, CEC-500-2009-014-D; California Climate Change Center: Sacramento, CA, USA, 2009.
42. Dettinger, M.D. Climate change, atmospheric rivers, and floods in California—A multimodel analysis of storm frequency and magnitude changes. *JAWRA* **2011**, *47*, 514–523. [CrossRef]
43. Trenberth, K.E. Conceptual framework for changes of extremes of the hydrological cycle with climate change. *Clim. Chang.* **1999**, *42*, 327–339. [CrossRef]

Article

Hydrologic Alteration Associated with Dam Construction in a Medium-Sized Coastal Watershed of Southeast China

Zhenyu Zhang [1,2], Yaling Huang [1,2] and Jinliang Huang [1,2,*]

[1] Coastal and Ocean Management Institute, Xiamen University, Xiamen 361005, China; tczzy2007@gmail.com (Z.Z.); hyl@stu.xmu.edu.cn (Y.H.)

[2] Fujian Provincial Key Laboratory of Coastal Ecology and Environmental Studies, Xiamen University, Xiamen 361005, China

* Correspondence: jlhuang@xmu.edu.cn; Tel.: +86-218-3833

Academic Editor: Christopher J. Duffy
Received: 9 April 2016; Accepted: 8 July 2016; Published: 26 July 2016

Abstract: Sustainable water resource management requires dams operations that provide environmental flow to support the downstream riverine ecosystem. However, relatively little is known about the hydrologic impact of small and medium dams in the smaller basin in China. Flow duration curve, indicators of hydrologic alteration andrange of variability approach were coupled in this study to evaluate the pre- and post-impact hydrologic regimes associated with dam construction using 44 years (1967–2010) of hydrologic data in the Jiulong River Watershed (JRW), a medium-sized coastal watershed of Southeast China, which suffered from intensive cascade damming. Results showed that the daily streamflow decreased in higher flow while daily streamflow increased in lower flow in both two reaches of the JRW. The dams in the North River tended to store more water while the dams in the West River tended to release more water. The mean daily streamflow increased during July to January while decreased during February to May after dam construction in both two reaches of the JRW. After dam construction, the monthly streamflow changed more significantly and higher variability of monthly streamflow exhibited in the West River than in the North River. The homeogenizing variability of monthly streamflow was observed in both two reaches of the JRW. The earlier occurrence time of extreme low streamflow event and later occurrence time of extreme high streamflow event exhibited after dams construction. The extreme low and high streamfow both decreased in the North River while both increased in the West River of the JRW. All of the indicators especially for the low pulse count (101.8%) and the low pulse duration (−62.1%) changed significantly in the North River. The high pulse count decreased by 37.1% in the West River and the count of low pulse increased abnormally in the North River. The high pulse duration in the post-impact period increased in the two reaches of JRW. The rise rate decreased by 26.9% and 61.0%,and number of reversals increased by 40.7% and 46.4% in the North River and West River, respectively. Suitable ranges of streamflow regime in terms of magnitude, rate, and frequency were further identified for environmental flow management in the North River and West River. This research advances our understanding of hydrologic impact of small and medium dams in the medium-sized basin in China.

Keywords: damming; hydrologic alteration; eco-hydrology; environmental flow management; coastal watershed

1. Introduction

Rivers play an important role in the development of human society by providing goods and services for human beings, by which the streamflow regime in turn has been altered for thousands of years due to various human activities [1–3]. By constructing large numbers of dams, human can utilize and control rivers by changing natural streamflow variability to suit human needs [4]. As a result, the past decades have witnessed the great alteration of streamflow regime in the watersheds throughout the world for their extensive dam construction [5–8]. Identifying the environmental impacts caused by hydraulic engineering facilities (e.g., dams) has therefore become an essential component in water resources planning and management [9–12].

The construction of large modern dams produced a dramatic change in the magnitude of hydrologic, geomorphologic and ecologic impacts on rivers [12,13]. Water development, mostly related to dams and diversions, contributed to the declines of more threatened and endangered species than any other resources-related activity [14]. Previous studies show that dam regulation generally had stronger effects on hydrologic regime than other disturbances by reducing the hydrologic variability of river systems [5,15,16]. Obviously, hydrologic regime alteration is responsible for the ecological system change in the rivers [1,17,18]. Dam construction has great impacts on hydrology, therefore, it is of scientific importance to evaluate the hydrologic alteration induced by dam construction.

Many attempts have been made to explore the hydrologic consequences associated with dam construction in recent decades [13,16,19–21]. More than 170 hydrologic metrics (e.g., average flow, flood frequency, peak discharge) have been developed to elaborate the different components of streamflow regime and their contribution to ecological consequence in the river ecological system in the past decade [22]. However, studies on streamflow-related disturbances are mainly on high-steamflow and low-streamflow events [1], which just partially characterize streamflow change. The full range of natural streamflow needs to be identified for its necessarity in evaluating ecosystem health of rivers [12,23–26]. The method of indicators of hydrologic alteration (IHA) was developed by Richter et al. [27] because of their close relationship to ecological functioning as well as for their ability to reflect human induced changes to streamflow regimes for a wide range of disturbances [28]. The IHA method was employed widely to assess the hydrologic change of the dam construction in many rivers worldwide such as the Great Plains, Illinois River, Southeastern US, Yellow River, Yangtze River, and Huaihe River [8,13,16,24,26,29]. The results prove that it is possible to identify the hydrologic change by dam construction over a full range. Moreover, flow duration curve (FDC) was developed to evaluate the overall impact of the streamflow regulation [30] and further applied to effectively determine whether human activities including dams construction can modify the pattern of the ecodeficit and ecosuplus of streamflow [31,32].

Implementation of environmental flows is one of the measures taken to restore or to maintain good ecological statusof rivers [33]. Estimates suggest that by 2050 many countries will face water scarcity, placing increasing pressures on face the water-dependent ecosystems of rivers and estuaries [34]. Obviously, maintaining natural streamflow variability has become an essential principle for environment flow management [35,36]. So far, most of the studies only focus on the minimum release rule so as to maximize human benefits such as water supply or hydropower generation [37,38]. However, provision of a single minimum streamflow cannot protect the biodiversity of a river, which requires the full range of natural flows [24,39–41]. Therefore, to satisfy such strategy of environmental flow management, reservoir planners and operators should seek to minimize the degree of natural flow regime alteration along the regulated river [42]. Based on the 33 indicators of IHA, Richter et al. [27] introduced a useful approach referred to as Range of Variability Approach (RVA) to quantitatively evaluate the degree of hydrologic alteration induced by human disturbance. This method has been demonstrated as a practical and effective way to identify the reasonable range of streamflow regime for environmental flow management [24,41,42].

The hydrologic regime is an indispensable dynamic of the ecosystem change in the watershed, which requires reasonable strategies to protect the magnitude, frequency, duration, timing and rate of change in the streamflow [26,43,44]. The impacts of the dams construction on hydrologic regime show regional difference, since the effects of dams on magnitude, frequency and timing of streamflow change with the types, operations, storage capacity of dams [16,27,45,46]. For example, the groundwater levels rise significantly below ground surface after dam construction in the Tarim River [47], while the water level in the river decreased appreciably in time after dams construction in Yangtze River, Yellow River, Huaihe River [24,26,48]. Moreover, it has been reported that the pattern of the monthly streamflow change due to damming is location-dependent [7,13,49,50]. In China, the accelerating development of economy increases the demand for energy and water resource, thus raising the need for the hydraulic engineering facilities, such as dams and reservoirs [51]. So far, most studies have focused on the impacts of large dams in large basins of China including Yangtze River and Yellow River [12,26,48,52,53]. However, relatively little is known about the hydrologic impact of small and medium dams in the smaller basins. More attentions should be paid to the small and medium dams because of their abundance (with more than 800,000 throughout the world) and their vital roles in maintaining local aquatic ecosystem health and water security [54–56].

The Jiulong River Watershed (JRW), a medium-sized coastal watershed in Southeast China, suffered from intensive human activities with over 13,500 hydraulic engineering facilities including over 120 small or medium dams along the mainstream and major tributaries. Our previous study partly characterized the hydrologic impact of cascade dam in JRW [57]. However, we need more attempts to fully delineate the hydrologic alteration associated with for watershed management. The objectives of this study are: (1) to evaluate the full range of streamflow regime change induced by dam construction; (2) to identify the suitable range of streamflow regime for environmental flow management in the JRW.

2. Materials and Methods

2.1. Study Area

The Jiulong River Watershed (JRW), covering approximately 14,700 km^2 in the eastern coastal area of China (116°46′55″–118°02′17″ E, 24°23′53″–25°53′38″ N) (Figure 1). Two main tributaries, namely, the North River and West River reaches, meet in Zhangzhou, which produces an annual flow of 12 billion m^3 into the Jiulong River estuary and Xiamen-Kinmen coast. The JRW plays an extremely important role in the region economic and ecological health. Water resources in the JRW have been highly developed and supply great demand to many stakeholders, like water supply, irrigation, hydropower and industry. More than ten million residents from Xiamen, Zhangzhou and Longyan use the Jiulong River as their source of water for residential, industrial and agricultural activities. The construction of large dams along the mainstream and major tributaries of JRW greatly altered the natural streamflow regime of the river over the last several decades.

Our previous study showed that the earliest changes in streamflow regime associated with dam construction in the JRW were detected in 1992 [3,57]. As shown in Figure 2, there is distinct difference in terms of flashiness index (the ratio of absolute day-to-day fluctuations of streamflow relative to total flow in a year) between pre-impact period (namely, 1967–1991) and post-impact period (namely, 1992–2010).

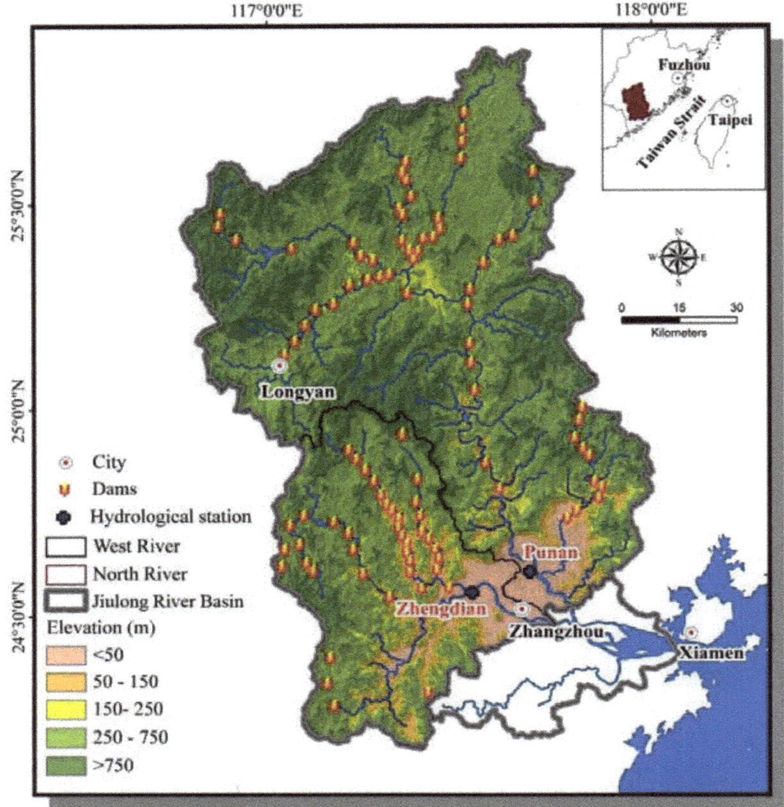

Figure 1. Study area.

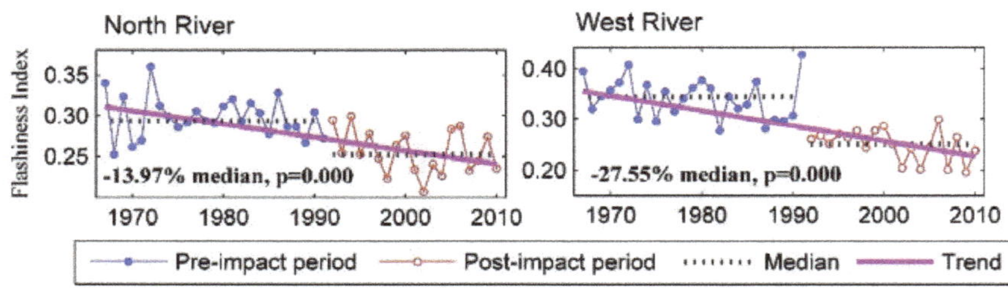

Figure 2. Temporal trend of flashiness index during 1968–2010 (modified from Huang et al., 2013).

2.2. Data Source

In this study, daily streamflow data during 1967–2010 for two downstream hydrologic stations (Punan and Zhengdian) in two reaches, namely, North River and West River were used to evaluate the effect of dam regulation on downstream streamflow in the JRW. A basic description of the two hydrologic stations is shown in Table 1.

Table 1. The streamflow for two gauging station in the JRW.

Station	Longitude	Latitude	Discharge Area (km^2)	Length (km)	Number of Dams Upstream from Stations	Pre-Impact	Post-Impact
Punan (North River)	117.67° E	24.61° N	9640	274	87	1967–1999	2000–2010
Zhengdian (West River)	117.53° E	24.56° N	3940	172	37	1967–1994	1995–2010

2.3. Method

2.3.1. Flow Duration Curve Analysis (FDC)

FDC is constructed from streamflow data over a time interval of interest and to provide a measure of the percentage of time duration that streamflow equals to or exceeds a given value. An annual FDC reflects the variability of daily streamflow during a typical period in a year. The FDC plots can be calculated by the following formula:

$$\mathrm{Pi} = i/(n + 1) \quad (1)$$

where n is the number of the days of streamflow and i is the rank.

2.3.2. Indicators of Hydrologic Alteration (IHA)

IHA was used in this study to assess the hydrologic shifts associated with dam construction in a full range. The IHA computes 32 indices that describe the hydrologic regime in the watershed through the frequency, magnitude, duration, timing and rate of the streamflow (Figure 3). The IHA indicators can be divided into five groups: monthly streamflow indices, extreme flow indices, timing indices, high-flow and low-flow indices, and rising and falling indices.

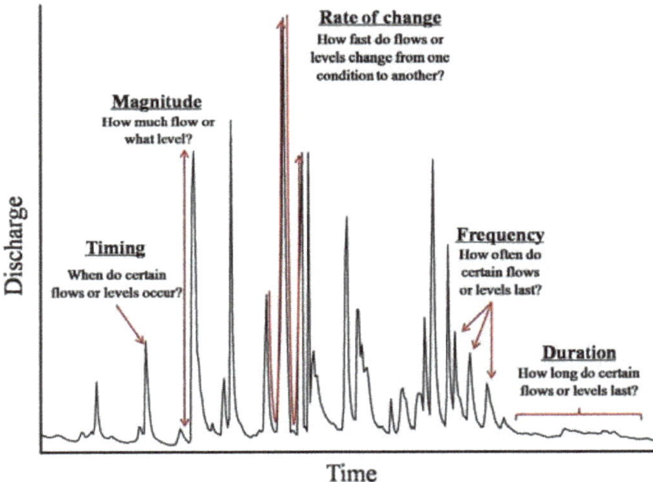

Figure 3. Quantified hydrologic regime using IHA method.

2.3.3. Range of Variability Approach (RVA)

RVA was used in this study to quantitatively evaluate the degree of hydrologic alteration induced by dam construction. In an RVA analysis, the full range of pre-impact data for each parameter is divided into three different categories. The low-level category contains all values less than or equal to the 33th percentile; the middle-level category contains all values falling in the range of the 34th to 67th percentile; and the high-level category contains all values greater than the 67th percentile. A Hydrologic Alteration (HA) factor is calculated for each of the three categories as following formula:

$$\text{HA} = (\text{observed frequency} - \text{expected frequency})/\text{expected frequency} \quad (2)$$

In this study, we divided the absolute ranges of HA factor into three classes: little or no alteration (⩽33%), moderate alteration (33%–67%) and high alteration (⩾67%) according to Richter et al. (1998) [58].

The coefficient of dispersion was a commonly used indicator to evaluate the variability of daily streamflow. It is calculated as the following formula:

$$\text{The coefficient of dispersion} = (75\text{th percentile} - 25\text{th percentile})/50\text{th percentile} \quad (3)$$

2.3.4. Method for Measuring Environmental Flow

RVA was used in this study to identify the reasonable range of streamflow regime for environmental flow management. The basic consideration using RVA method for environmental flow identification here is that the preferred environmental flow regime in terms of magnitude, rate, and frequency should maintain streamflow variability as natural as possible in order to sustain the majority of riverine ecological functions. The variability of streamflow magnitude (i.e., daily streamflow) and the three indicators on the rate and frequency of the daily streamflow (i.e., rise rate, fall rate and number of reversals) involved in the IHA indicators can effectively represent the environment change induced by hydropower operations in the river system. In practice, the suitable range of the magnitude, rate and frequency should be identified for environmental flow associated with dams regulation.

The Tennant method (also known as Montana method), by which 20% of the daily average flow was used as the minimum ecological and environmental flow [59,60], was performed in this study to calculate the environmental flow in JRW in order to make comparison with the values from RVA method.

2.3.5. Statistical Analysis

The Kolmogorov–Smirnov (K–S) goodness of fit test was first used to test for normality ofthe distribution of the indicators representing the regime of streamflow. The *t*-test was then used to determine if means for each of the indicators during pre-impact period were statistically different from one another during post-impact period where significance was defined as $p < 0.05$.

3. Result

3.1. Overall Hydrologic Impact Assessment

The effect of regulation by dams on downstream flow regime was assessed using FDC in the North River and West River. As shown in Figure 4, the variability of daily streamflow exhibited anoverall decreasing trend for both two reaches of JRW. In the North River, the daily streamflow slightly decreased in lower percentiles (i.e., higher flows), while daily streamflow slightly increased in the higher percentiles (i.e., lower flows). Comparatively, daily streamflow slightly increased in the higher percentiles (i.e., the lower flows) in the West River. The crossing point of the two curves was calculated to be 136 days and 116 days in the North River and West River, respectively. The dams in the North River stored approximately 0.42 billion m^3 water during a 136-day higher flows period while release only 0.06 billion m^3 water in the subsequent 229 days in lower flow. In the West River, the dams stored only 0.06 billion m^3 water during a 116-day higher flow period while release 0.13 billion m^3 water in the subsequent 249 days in lower flow. This phenomenon indicated that the dams in the North River tend to store more water while the dams in the West River tend to release more water.

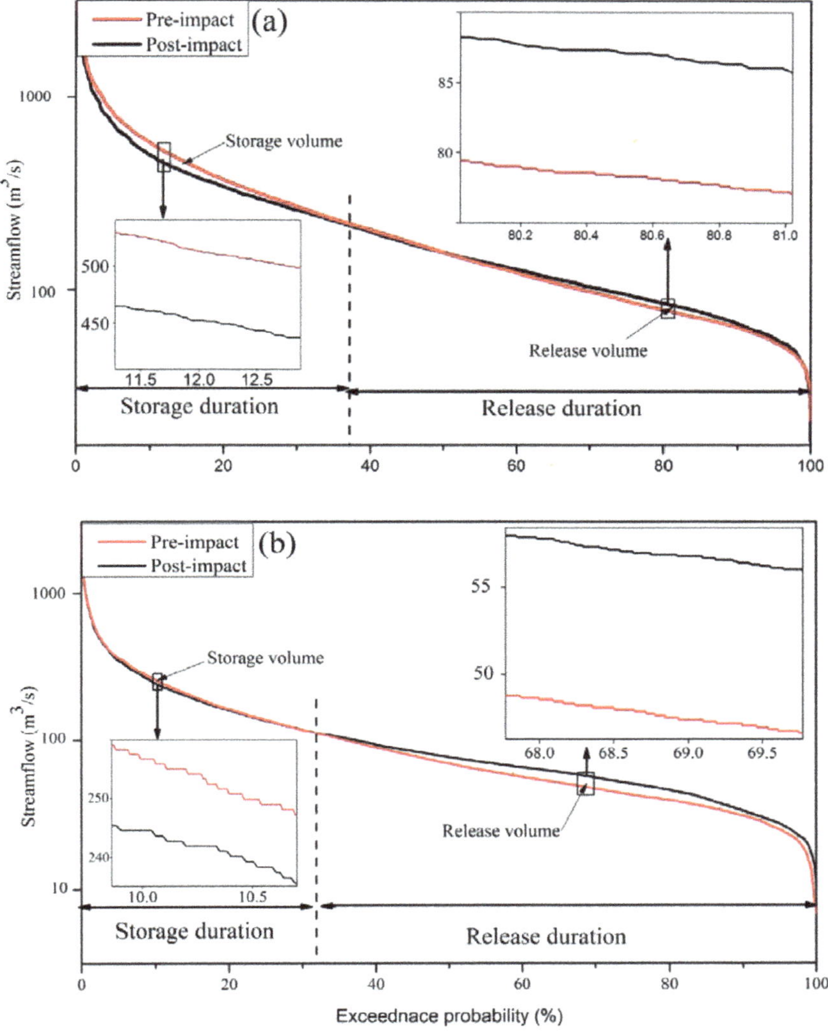

Figure 4. The flow duration curve change in JRW (**a**) North River reach; (**b**) West River reach.

3.2. Hydrologic Regime Changein a Full Range

3.2.1. Magnitude of Monthly Streamflow Regime

Themagnitude of monthly streamflow alteration was identified using IHA in the North River and West River. Similar variability patterns were observed in both two reaches of the JRW, namely, the monthly streamflow during July to January increased while decreased during February to May (Table 2). In addition, the monthly streamflow changed more significantly in the West River than in the North River. The monthly streamflow in January, August and December increased significantly in the West River while the monthly streamflow decreased significantly in May in the North River (Table 2).

Table 2. Monthly streamflow change due to dam construction in the JRW (%).

Two Reaches	Jan.	Feb.	Mar.	April	May	Jun.	Jul.	Aug.	Sep.	Oct.	Nov.	Dec.
North River	10.5	−12.1	−37.7	−10.1	−33.4 *	6.4	21.6	17.4	10.7	6.2	4.8	10.7
West River	18.7 *	−13.3	−14.1	−19.1	−23.2	−8	30.3	54.2 *	25.3	28	11.1	20.9 *

Note: * $p < 0.05$.

Figure 5 shows the homogenizing variability of monthly streamflow in the JRW. The coefficient of dispersion increased in the period with high variability of streamflow (e.g., March in the North River, Figure 5a) while decreased in the period with low variability of streamflow in the JRW (e.g., Jan in the North River, Figure 5b). Particularly, the variability of monthly streamflow tend to decrease with the coefficient of dispersion more than 0.7 whereas it increased with the coefficient of dispersion less than 0.7 in both two reaches of the JRW. Interestingly, the variability of monthly streamflow in the flood season (e.g., in July and August) in the West River increased after dam construction, which might be related to the increasing frequency of extreme weather events.

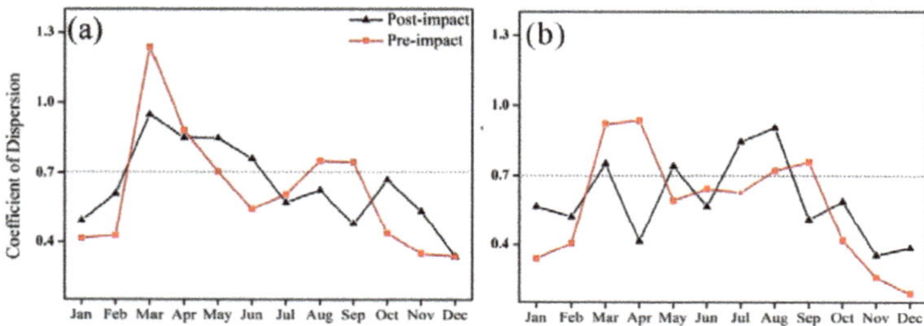

Figure 5. The variability of monthly streamflow in the JRW (**a**) North River reach; (**b**) West River reach.

Variability patterns of streamflow magnitude were delineated using RVA method and are presented in Figure 6. The similar variability patterns of monthly streamflow exhibited in the North River and West River, namely, the frequency of the monthly streamflow with high-level category increased during August to January (Figure 6) in both two reaches of JRW. Obviously, after dam construction, the monthly streamflow changed more intensively in the West River than in the North River. Particularly, the frequency of the monthly streamflow with high-level category all increased during August to January in the West River.

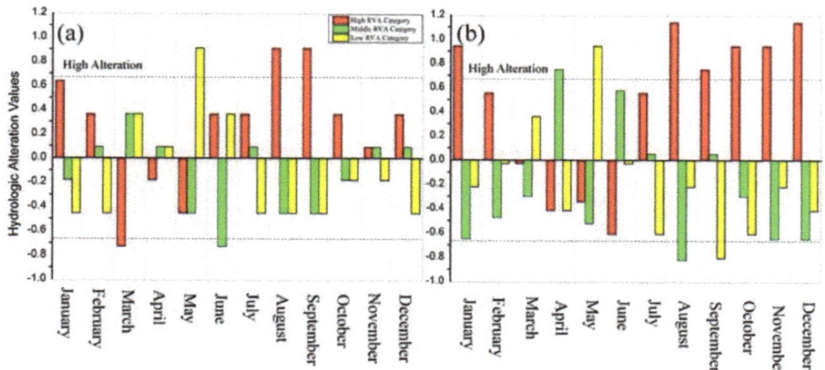

Figure 6. Hydrologic alteration of monthly streamflow in the JRW based on RVA method. (**a**): North River reach; (**b**) West River reach.

3.2.2. Magnitude, Duration and Timing of Extreme Streamflow Regime

The occurrence time of the annual extreme streamflow in the North River was approximately one month earlier than that in the West River (Table 3). The minimum and maximum streamflow occurred in early February and early July in the North River, while the minimum and maximum streamflow were in late February and early August in the West River. The Julian dates of minimum streamflow in the post-impact period for both two reaches of JRW were earlier than those in the pre-impact period, while the Julian date of maximum streamflow showed the opposite results (Table 3). In particular, the extreme low streamflow event occurred 14 days earlier than before in the West River and the extreme high streamfow event occurred 18 day later than before in the North River.

Table 3. The median of Julian date of the extreme streamflow.

Two Reaches	Pre-Impact Min Q	Post-Impact Min Q	Pre-Impact Max Q	Post-Impact Max Q
NorthRiver	34.0	33.0	171.0	189.0
West River	64.5	50.5	211.5	215.0

The baseflow index (BFI) increased by 18.8% and 17.7% in the West River and North River reaches, respectively. Given that BFI is the ratio of the 7-day minimum streamflow to the annual streamflow, the BFI increased more than the 7-day minimum streamflow, which might indicate a potential discharge decreasing trend in both two reaches of JRW (Table 4). Most of indicators of extreme streamflow decreased in the post-impact period in the North River. In contrast, most of the indicators of the extreme streamflow increased in the West River.

Table 4. The change of magnitude and duration of annual extreme streamflow condition in the JRW (%).

Two Reaches	1-Day Min	3-Day Min	7-Day Min	30-Day Min	90-Day Min	1-Day Max	3-Day Max	7-Day Max	30-Day Max	90-Day Max	BFI
North River	−20.3	−5.9	7.2	10.4	−0.4	−9.6	−7.4	−6.6	−7.5	−8.2	18.8 *
West River	8	11.7	13.7	12.7	7.1	6	−2.1	−0.2	8	9.8	17.7

Note: * $p < 0.05$.

Most of the indicators change little (HA ⩽ 33%) or moderately (HA = 33%–67%) in the JRW (Figure 7). The frequency of 30-day minimum streamflow which falled into high-level RVA category (HA ⩾ 67%) increased intensively in the North River and West River, indicating the increased frequency of minimum monthly streamflow with high-level category in both two reaches of JRW. The frequency of 30-day maximum streamflow which falled into high RVA category (HA ⩾ 67%) increased intensively in the West River, suggesting the increased frequency of maximum monthly streamflow with high-level category in the West River.

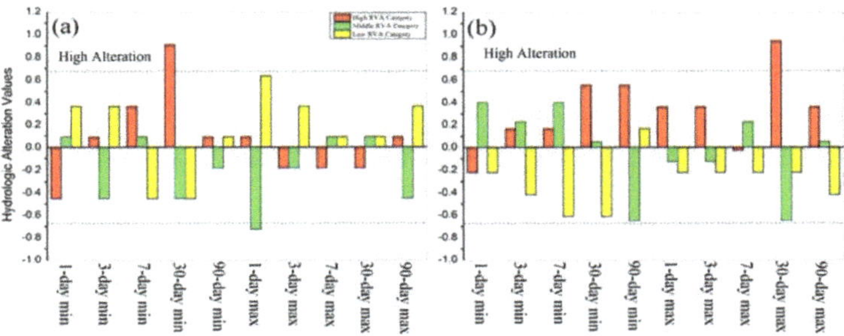

Figure 7. Hydrologic alteration of annual extreme streamflow in the JRW based on RVA method. (**a**) North River reach; (**b**) West River reach.

3.2.3. Frequency and Duration of High and Low Pulses

The frequency and duration of high and low pulses are quantified by the count and duration of the low pulse and high pulse, respectively. Except for the low pulse count, the other three indicators in Table 5 displayed the same trend after dams construction in the JRW. The frequency and duration of high and low pulses changed significantly in the last 44 years especially during the post-impact period in both two reaches of JRW. All of the indicators, especially for the low pulse count (101.8%) and the low pulse duration (−62.1%), changed significantly in the North River. The high pulse count decreased by 37.1%in the West River (Table 5). Most of the indicators increased significantly in low or middle level categories whereas decreased intensively in high-level category in the North river and West River, implying that the dams might effectively attenuate the high flow pulse event in both two reaches of JRW (Figure 8).

Table 5. The change of the frequency and duration of high and low pulses in the JRW (%).

Two Reaches	Low Pulse Count	Low Pulse Duration	High Pulse Count	High Pulse Duration
North River	101.8 **	−62.1 **	−21.4 *	19.2 *
West River	−15.5	−29.0	−37.1 **	6.7

Notes: ** $p < 0.005$; * $p < 0.05$.

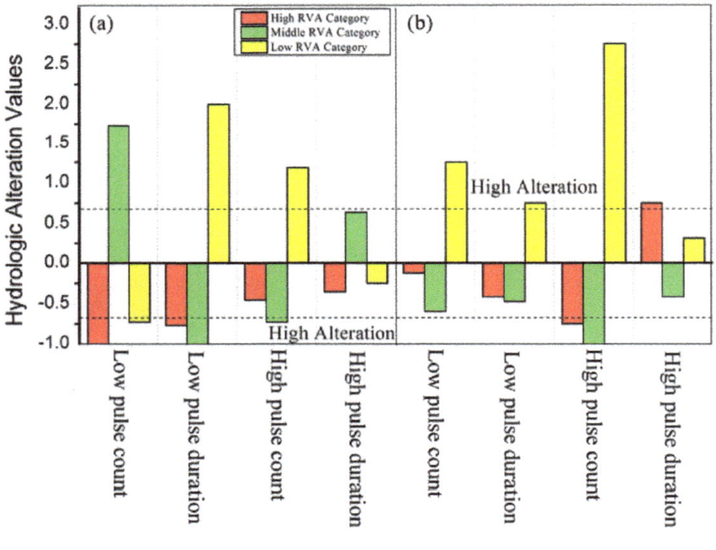

Figure 8. Hydrologic alteration frequency and duration of high and low pulses in the JRW based on RVA method. (**a**) North River reach; (**b**) West River reach.

3.2.4. Rate and Frequency of Streamflow Regime Change

The rate and frequency of the daily streamflow regime were quantified by the rise rate, fall rate and number of reversals (Table 6, Figure 9). The rate and frequency of streamflow regime changed intensively in both two reaches of JRW. The three indicators on the rate and frequency of the daily streamflow (i.e., rise rate, fall rate and number of reversals) revealed the similar trend in the post-impact period based on the observed data in the two reaches of JRW. The rise rate decreased by 26.9% and 61.0% in the North River and West River, respectively. The number of reversals increased by 40.7% and 46.4% in the North River and West River, respectively. The fall rate increased by 28.3% in the North River while increased by 0.8% in the West River (Table 6).

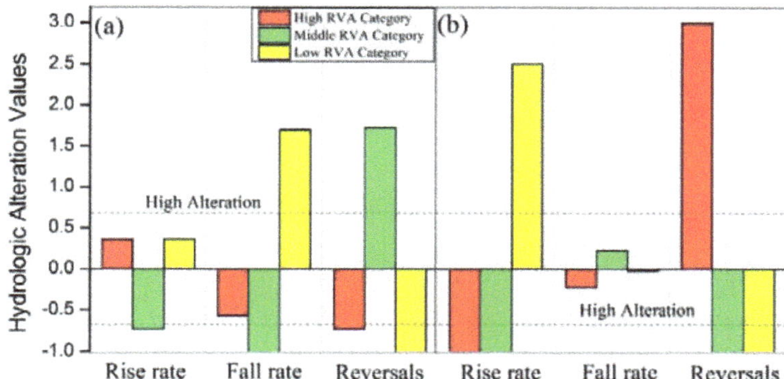

Figure 9. Hydrologic alteration of rate and frequency of the daily streamflow in the JRW based on RVA method. (**a**) North River reach; (**b**) West River reach.

Table 6. The change of the rate and frequency of the daily streamflow in the JRW (%).

Two Reaches	Rise Rate	Fall Rate	Number of Reversals
North River	−26.9 *	28.3 *	40.7 **
West River	−61.0 **	0.8	46.4 **

Notes: ** $p < 0.005$; * $p < 0.05$.

3.3. The Possible Suitable Range of Streamflow Regime

The suitable ranges of the daily streamflow are meaningful to be developed for environmental flow management in the JRW. Using Tennant method, the minimum environmental flow were 51.3 $m^3/s \cdot d$ and 24.5 $m^3/s \cdot d$ in the North River and West River. On the other hand, the suitable ranges of the daily streamflow were identified for each month using the RVA method (Figure 10). As shown in Figure 10, the monthly streamflow in August was higher than the natural streamflow boundary (i.e., RVA boundary), and the monthly streamflow in May was lower than the natural streamflow boundary in the North River. The similar result was found in the West River where the monthly streamflow in August, September, November, and December were higher than the natural streamflow boundary.

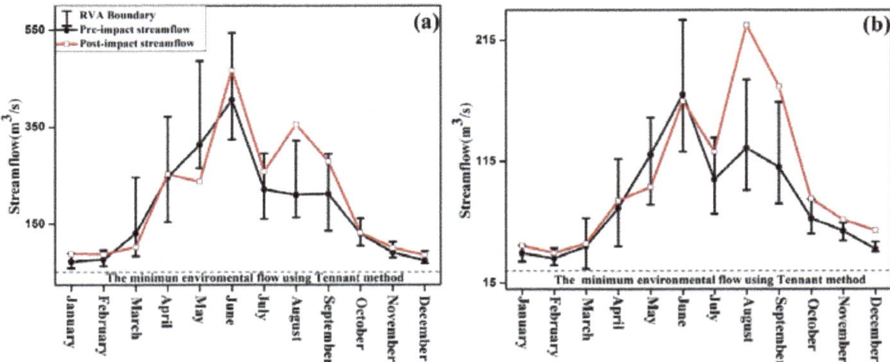

Figure 10. The suitable ranges of the monthly streamflow in the JRW. (**a**) North River reach; (**b**) West River reach.

The rate and frequency of streamflow changed largely due to dams construction in the JRW (Figure 9). From the perspective of the practical regulation, the suitable ranges for rise rate, fall rate and number of reversal in the JRW were identified based on the RVA method and the results are presented in Table 7. The fall rate in the North River and the rise rate in the West River were higher than the RVA boundary. Moreover, the number of reversal was higher than the RVA boundary in both reaches of the JRW.

Table 7. The suitable ranges of the rate and frequency in the JRW.

Hydrologic Parameters	North River			West River		
	RVA Range	**Pre-Impact**	**Post-Impact**	**RVA Range**	**Pre-Impact**	**Post-Impact**
Rise rate	16.0~33.5	24.0	17.6	10.1~19.8	15.2	5.9
Fall rate	−17.5~−12.2	−15.0	−19.2	−8.4~−5.5	−7.1	−7.2
Number of Reversals	103.0~137.0	119.9	168.7	98.3~108.8	106.7	156.2

Note: Lower and upper targets of RVA range are the 25th and 75th percentile value of the pre-impact hydrologic parameters including rise rate, fall rate, and number of reversals.

4. Discussion

4.1. Overall Impact of Streamflow Regulation on Flow Regimes

The overall streamflow regime change can be delineated through FDCs, which has been widely examined [30–32]. Our results revealed that the annual variability of daily streamflow decreased in the post-impact period in the North River and West River of JRW. The daily streamflow decreased in higher flow resulting from water storage regulated by dams while daily streamflow increased in lower flow resulting from water release regulated by dams in both two reaches of the JRW. The similar results also were found in the Han River and Yangtze River [8,31]. The dams tend to store more streamflow in the North River. For one thing, there are more dams and more length of river in the North River, which might slow down the downstream streamflow. For another, the downstream reservoirs or dams in the North River are designed to supply local residents with water for drinking, agricultural and industrial uses, which would make the downstream streamflow regulated.

4.2. Magnitude of Monthly Streamflow Regime

The effect of the dams on the monthly streamflow has been widely reported in the literature [13,16,29,61]. In this study, the mean daily streamflow increased during July to January while decreased during February to May after dam construction in the two reaches of JRW. This is similar with the previous findings in Huaihe River, Yangtze River and Langcang River [8,12,24]. Compared to the North River, the daily streamflow displayed a higher variability of monthly streamflow in the West River. This may be attributable to the smaller areas of watershed and less dams in the West River (Table 1). The effects of dam construction on water storage and release are greater for smallwatersheds than for large watersheds [62]. Generally, the impact of damconstruction in the JRW on monthly streamflow regime alteration is relatively complicated when combined with magnitude and inter-annual variability of monthly streamflow.

4.3. Magnitude, Duration and Timing of Extreme Streamflow Regime

Generally, baseflow is the most sensitive indicator to the streamflow regime change associated with dam construction [1,63–66]. The baseflow index (BFI) tends to increase after dam construction [3,8]. The increased BFI is the combined results of the increase in the 7-day minimum streamflow and the decrease in the annual streamflow due to dam construction [8]. Our study revealed that BFI value increased by 18.8 and 17.7%, respectively, in the post-impact period in the North River and West River. This suggests the general hydrologic regime alteration in the JRW due to dam construction. Our study also shows the potential discharge decreasing after construction in the JRW. The reduced discharged may be attributed to the result of increased evaporation or infiltration and loss to groundwater [13]. Our prior study shows that the 10-year average streamflowduring 2001–2010 in the North River and West Riverreaches decreased by 9.2% and 6.7%, respectively, compared to the average annual streamflow during 1967–2000, and the most important driving force for streamflow regime change in the JRW is related to population growth and economic development [3]. The potential discharge decrease after dam construction in the JRW might be related to annual streamflow decrease in recent years and the increasing water demanding in the watershed. Climatic variability in term of precipitation changes had limited effects on this situation, since that there is slightly increasing tendency of the annual precipitation in the West River while the precipitation in the North Riverseemed not to have changed in the past 50 years, as observed in our prior study [3].

The occurrence time of annual extreme condition can effectively reflect the seasonal variation of the hydrologic conditions [29]. The variation of the extreme events changed with geographic areas and climate patterns. Our study indicates the earlier occurrence time of extreme low streamflow event and later occurrence time of extreme high streamflow event after dams construction in the JRW, which is consistent with these asonal variation of the hydrologic conditions in other watersheds in South China [8,24,67]. This means the mode of dams regulation on streamflow regime in the JRW was similar to those in South China.

Previous studies in other watersheds show that minimum streamflow increased and maximum streamflow decreased after dam construction [5,46,68]. However, our study suggests the extreme low and high streamfow both decreased in the North River while both increased in the West River of the JRW. This is the results caused by the seasonal variation of streamflow regime changed in the JRW. The variability of the streamflow increased in February while decreased in July in the North River (Figure 4a), thus the minimum streamflow and maximum streamflow decreased in such two months in the North River. The situation was different in the West River, the variability of the streamflow decreased in March while increased in August (Figure 4b), therefore, the minimum streamflow and maximum streamflow increased in such two months respectively.

4.4. Pulse, Rate and Frequency Change

In response to theoperation of dams and reservoirs, the number of low pulse and high pulse will decrease [13,45,69]. Our study showed the count of low pulse in the North River increased abnormally (Table 5). This might be related to the increasing water demanding in the North River in recent years. As the water source for approximate 10 million resident in Xiamen, Zhangzhou and Longyan, large number of irrigation facilities was established, thereby inducing the low pulse count increased in the North River [3]. Similar observations were obtained in Huaihe River Basin and Tarim River Basinof China [24,47].

Due to the limited capacity of reservoirs, the duration of high pulse will change [29,70]. Our study also showed the increasing high pulse duration in the post-impact period in the JRW. This might be the result of the water storage and release regulated by the dams in JRW. The count of the high extreme high pulse decreased when water storage was operated by dams. This means the dams can store and attenuate all high flow pulse event. However, the dams tend to release the water when the water level exceeds the limited capacity of reservoir in the flood season.

The rate and frequency of streamflow can provide a measure of the rate and frequency of intra-annual environmental change and the decreased rise rate of hydrographs and increased in reversals after dams construction was widely observed [45,46]. Our study revealed that the rise rate decreased by 26.9% and 61.0% and number of reversals increased by 40.7% and 46.4% in the North River and West River, respectively. This is a byproduct of hydropower generation, wherein water is stored in the reservoir until sufficient head is attained to generate power efficiently, at which time the flow is rapidly released through the dam tubbiness [4]. Furthermore,the decreased rise rate and increased fall rate in the JRW also suggested the rate changing from high flow to low flow slowed down and the rate change from low flow to high flow speeded up, implying that the streamflow peak might be delayed (Table 3) and the variability of streamflow changed (Figure 4). Similar results were also found in Huaihe River, Yellow River, Taiwan and Great Plains [13,24,67,70].

4.5. Feasible Streamflow Regime in JRW

Streamflow regime is a primary determinant of structure and function of an aquatic and water quality in streams [3,24,26]. The magnitude of streamflow will influence the available habitat for organisms and the water quality in downstream. Too low streamflow will induce degrading water quality while too high streamflow will increase water level causing lost habitats. In order to protect native biodiversity and evolutionary potential of aquatic, riparian and wetland systems, the natural flow paradigm emphasizes the need to maintain or restore the range natural intra-annual and inter-annual variation of hydrologic regimes [24,39–41]. Our study revealed that the minimum flow requirement using Tennant method were close to the lower target identified using RVA method (Figure 10). Obviously, Tennant method only considers the lowest streamflow (i.e., the lowest streamflow in dry season) in the river but rarely considered the effective habitat quality at varying flow. Our study quantified reasonable range of the streamflow regime using RVA method in order to maintain the natural streamflow regime in the JRW.

Dams may not homogenize all river systems, but may move them outside the bounds of normal river function [16]. Our results suggest a suitable streamflow framework to generalize seasonal patterns inhydrologic alterations due to dam regulation. One the one hand, the suitable ranges in the flood season (e.g., in August) for both reaches of the JRW should be guaranteed by the reasonable regulation from downstream dams. On the other hand, more attention should be paid to the streamflow release or storage in the average season (e.g., May) and dry season (e.g., December) in the North River and West River, respectively.

Our study shows that the three indicators on the rate and frequency of the daily streamflow (i.e., rise rate, fall rate and number of reversals) were informative to delineate the critical role of dam construction on streamflow change. In this study, the fall rate in the North River and the rise rate in the West River were higher than the RVA boundary, which suggests that the dams in the JRW should release more water to get the natural targets of the rise rate and fall rate, so as to maintain the natural streamflow regime in the JRW. Moreover, the mean number of the reversal increased significantly and was higher than the RVA boundary, we therefore suggest that the dams in the JRW should decrease the frequency of store-release streamflow.

5. Conclusions

Flow duration curve analysis, indicators of hydrologic alteration, and range of variability approach were coupled in this study to evaluate the streamflow regime change induced by dam constructionin a full range in Jiulong River Watershed (JRW). The dailystreamflow decreased in higherflow resulting fromwater storage regulated by dams whiledaily streamflow increased in lower flow resulting fromwater release regulated by dams in both two reaches of the JRW. The dams in the North River tend to store more water while the dams in the West River tend to release more water. The mean dailystreamflow increased during July to January while decreased during February to May after dam construction in the two reaches of JRW. After dam construction, the monthly streamflow changed more significantly and higher variability of monthly streamflow was observed in the West River than in the North River. The homeogenizing variability of monthly streamflow exhibited in both two reaches of JRW. The earlier occurrence time of extreme low streamflow event and later occurrence time of extreme high streamflow event after dams construction. The extreme low and high streamfow both decreased in the North River while both increased in the West River of the JRW. All of the indicators especially for the low pulse count (101.8%) and the low pulse duration (−62.1%) changed significantly in the North River. The high pulse count decreased by 37.1% in the West River and the count of low pulse increased abnormally in the North River. The high pulse duration in the post-impact period increased in both two reaches of the JRW. The rise rate decreased by 26.9% and 61.0%,and number of reversals increased by 40.7% and 46.4% in the North River and West River, respectively. The fall rate increased by 28.3% in the North River.

Reasonable range of streamflow regime in terms of magnitude, rate, and frequency was identified using RVA method to sustain environmental flow management. More attention should be paid to the streamflow release or storage regulated by dams in May, August and December in the JRW. The dams in the JRW should release more water and decrease the frequency of store-releases treamflow. This research advances our understanding of hydrologic impact of small and medium dams in the medium-sized basin in China.

Acknowledgments: This study was supported by the Natural National Science Foundation of China (Grant No. 41471154), the Fundamental Research Funds for the Xiamen Universities (Grant No. 20720150129), and the National Science and Technology Support Program (Grant No. 2013BAC06B01). Anonymous reviewers supplied constructive feedback that helped to improve this paper.

Author Contributions: Jinliang Huang and Zhenyu Zhang conceived and designed the experiments; Zhenyu Zhang performed the experiments; Zhenyu Zhang and Yaling Huang analyzed the data; Zhenyu Zhang and Yaling Huang contributed reagents/materials/analysis tools; Zhenyu Zhang and Jinliang Huang wrote the paper.

Conflicts of Interest: The authors declare no conflict of interest.

References

1. Poff, N.L.; Allan, J.D.; Bain, M.B.; Karr, J.R.; Prestegaard, K.L.; Richter, B.D.; Spark, R.E.; Stromberg, J.C. The natural flow regime: A paradigm for river conservation and restoration. *BioScience* **1997**, *47*, 769–784. [CrossRef]
2. Ripl, W. Water: The bloodstream of the biosphere. *Philos. Trans. R. Soc. B* **2003**, *358*, 1921–1934. [CrossRef] [PubMed]
3. Huang, J.L.; Zhang, Z.Y.; Feng, Y.; Hong, H.S. Hydrologic response to climate change and human activities in a subtropical coastal watershed of southeast China. *Reg. Environ. Chang.* **2013**, *13*, 1195–1210. [CrossRef]
4. Richter, B.D.; Andrew, T.; Warner, J.L.; Meyer, K. A collaborative and adaptive process for developing environmental flow recommendations. *River Res. Appl.* **2006**, *22*, 297–318. [CrossRef]
5. Poff, N.L.; Olden, J.D.; Merritt, D.M.; Pepin, D.M. Homogenization of regional river dynamics by dams and global biodiversity implications. *PNAS* **2007**, *104*, 5732–5737. [CrossRef] [PubMed]
6. Li, Z.W.; Zhang, Y.K. Multi-scale entropy analysis of Mississippi River flow. *Stoch. Environ. Res. Risk Assess.* **2008**, *22*, 507–512. [CrossRef]
7. Chen, Y.Q.; Yang, T.; Xu, C.Y.; Zhang, Q.; Chen, X.; Hao, Z.C. Hydrologic alteration along the middle and upper East River (Dongjiang) basin, South China: A visually enhanced mining on the results of RVA method. *Stoch. Environ. Res. Risk Assess.* **2010**, *24*, 9–18. [CrossRef]
8. Gao, B.; Yang, D.; Zhao, T.T.G.; Yang, H.B. Changes in the eco-flow metrics of the upper Yangtze River from 1961 to 2008. *J. Hydrol.* **2012**, *448–449*, 30–38. [CrossRef]
9. Flug, M.; Seiz, H.L.H.; Scott, J.F. Multicriteria decision analysis applied to Glen Canyon dam. *J. Water Resour. Plan. Manag.* **2000**, *126*, 270–276. [CrossRef]
10. Cowell, C.M.; Stoudt, R.T. Dam-induced modifications to upper Allegheny River streamflow patterns and their biodiversity implications. *J. Am. Water Resour. Assoc.* **2002**, *38*, 187–196. [CrossRef]
11. Shiau, J.T.; Wu, F.C. Assessment of hydrologic alterations caused by Chi-Chi diversion weir in Chou-Shui Creek, Taiwan: Opportunities for restoring natural flow conditions. *River Res. Appl.* **2004**, *20*, 401–412. [CrossRef]
12. Zhao, Q.H.; Liu, S.L.; Deng, L.; Dong, S.K.; Wang, C.; Yang, Z.F.; Yang, J.J. Landscape change and hydrologic alteration associated with dam construction. *Int. J. Appl. Earth Obs. Geoinf.* **2012**, *16*, 17–26. [CrossRef]
13. Costigan, K.H.; Daniels, M.D. Damming the prairie: Human alteration of Great Plains river regimes. *J. Hydrol.* **2012**, *444–445*, 90–99. [CrossRef]
14. Losos, E.; Hayes, J.; Phillips, A.; Wilcove, D.; Alkire, C. Taxpayer-subsidized resource extraction harm species: Double jeopardy. *Bioscience* **1995**, *45*, 446–455. [CrossRef]

15. Trush, W.J.; McBain, S.M.; Leopold, L.B. Attributes of an alluvial river and their relation to water policy and management. *PNAS* **2000**, *97*, 11858–11863. [CrossRef] [PubMed]
16. McManamay, R.Y.; Orth, D.J.; Dolloff, C.A. Revisiting the homogenization of dammed rivers in the southeastern US. *J. Hydrol.* **2012**, *424–425*, 217–237. [CrossRef]
17. Bunn, S.E.; Arthington, A.H. Basic principles and ecological consequences of altered flow regimes for aquatic biodiversity. *Environ. Manag.* **2002**, *30*, 492–507. [CrossRef]
18. Moyle, P.B.; Mount, J.F. Homogenous rivers, homogeneous fauna. *PNAS* **2007**, *104*, 5711–5712. [CrossRef] [PubMed]
19. Jansson, R.; Nilsson, C.; Renofalt, B. Fragmentation of riparian floras in rivers with multiple dams. *Ecology* **2000**, *81*, 899–903. [CrossRef]
20. Chovanec, A.; Schiemer, F.; Waibacher, H.; Spolwind, R. Rehabilitation of a heavily modified river section of the Danube in Vienna (Austria): Biological assessment of landscape linkages on different scales. *Int. Rev. Hydrobiol.* **2002**, *87*, 183–195. [CrossRef]
21. Tockner, K.; Stanford, J.A. Riverine flood plains: Present state and future trends. *Environ. Conserv.* **2002**, *29*, 308–330. [CrossRef]
22. Olden, J.D.; Poff, N.L. Redundancy and the choice of hydrologic indices for characterizing stremflow regimes. *River Res. Appl.* **2003**, *19*, 101–121. [CrossRef]
23. Dudgeon, D. River rehabilitation for conservation of fish biodiversity in monsoonal Asia. *Ecol. Soc.* **2005**, *10*, 15.
24. Hu, W.W.; Wang, G.X.; Deng, W.; Li, S.N. The influence of dams on ecohydrological conditions in the Huaihe River basin, China. *Ecol. Eng.* **2008**, *33*, 233–241. [CrossRef]
25. Zeilhofer, P.; de Moura, R.M. Hydrological changes in the northern Pantanal caused by the Manso dam: Impact analysis and suggestions for mitigation. *Ecol. Eng.* **2009**, *35*, 105–117. [CrossRef]
26. Yang, Z.; Yan, Y.; Liu, Q. Assessment of the flow regime alterations in the lower Yellow River, China. *Ecol. Inform.* **2012**, *10*, 56–64. [CrossRef]
27. Richter, B.D.; Baumgarter, J.V.; Powell, J.; Braun, D.P. A method for assessing hydrologic alteration within ecosystems. *Conserv. Biol.* **1996**, *10*, 1163–1174. [CrossRef]
28. Mathews, R.; Richter, B.D. Application of the indicators of hydrologic alteration software in environmental flow setting. *J. Am. Water Resour. Assoc.* **2007**, *43*, 1400–1413. [CrossRef]
29. Lian, Y.Q.; You, J.Y.; Sparks, R.; Demissie, M. Impact of human activities to hydrologic alterations on the Illiois River. *J. Hydrol. Eng.* **2012**, *17*, 537–546. [CrossRef]
30. Vogel, R.M.; Sieber, J.; Archfield, S.A.; Smith, M.P.; Apse, C.D.; Huber-Lee, A. Relations among storage, yield and instream flow. *Water Resour. Res.* **2007**, *43*, W05403. [CrossRef]
31. Kim, N.; Lee, J.; Kim, J. Assessment of flow regulation effects by dams in the Han River, Korea, on the downstream flow regimes using SWAT. *J. Water Resour. Plan. Manag.* **2012**, *138*, 24–35. [CrossRef]
32. Brown, A.E.; Western, A.W.; McMahon, T.A.; Zhang, L. Impact of forest cover changes on annual stramflow and flow duration curves. *J. Hydrol.* **2013**, *483*, 39–50. [CrossRef]
33. Acreman, M.C.; Ferguson, A.J.D. Environmental flows and the European Water Framework Directive. *Freshw. Biol.* **2010**, *55*, 32–48. [CrossRef]
34. Petts, G.E. Instream flow science for sustainable reiver management. *J. Am. Water Resour. Assoc.* **2009**, *45*, 1071–1086. [CrossRef]
35. Jowett, I.G.; Biggs, B.J.F. Application of the 'natural flow paradigm' in a New Zealand context. *River Res. Appl.* **2009**, *25*, 1126–1135. [CrossRef]
36. Poff, N.L.; Richter, B.; Arthington, A.; Bunn, S.E.; Naiman, R.J.; Kendy, E.; Acreman, M.; Apse, C.; Bledsoe, B.P.; Freeman, M.; et al. The ecological limits of hydrologic alteration (ELOHA): A new framework for developing regional environmental flow standards. *Freshw. Biol.* **2010**, *55*, 147–170. [CrossRef]
37. Halleraker, J.H.; Sundt, H.; Alfredsen, K.T.; Dangelmaier, G. Application of multiscale environmental flow methodologies as tools for optimized management of a Norwegian regulated nation salmon water course. *River Res. Appl.* **2007**, *23*, 493–510. [CrossRef]
38. Jager, H.I.; Smith, B.T. Sustainable reservoir operation: Can we generate hydropower and preserve ecosystem values? *River Res. Appl.* **2008**, *24*, 340–352. [CrossRef]
39. Richter, B.D.; Thomas, G.A. Restoring environmental flows by modifyingdam operations. *Ecol. Soc.* **2007**, *12*, 12.

40. Naiman, R.J.; Latterell, J.J.; Pettit, N.E.; Olden, J.D. Flow variability and thebiophysical vitality of river systems. *C. R. Geosci.* **2008**, *340*, 629–643. [CrossRef]
41. Yin, X.A.; Yang, Z.F.; Petts, G.E. Reservoir operating rules to sustain environmental flows in regulated rivers. *Water Resour. Res.* **2011**, *47*, W08509. [CrossRef]
42. Yin, X.A.; Yang, Z.F.; Petts, G.E. Optimizing environmental flows below dams. *River Res. Appl.* **2012**, *28*, 703–716. [CrossRef]
43. Arthington, A.H.; Bunn, S.E.; Poff, N.L.; Naiman, R.J. The challenge of providing environmental flow rules to sustain river systems. *Ecol. Appl.* **2006**, *16*, 1311–1318. [CrossRef]
44. Richter, B.D. Re-thinking environmental flows: From allocations and reserves to sustainability boundaries. *River Res. Appl.* **2010**, *26*, 1052–1063. [CrossRef]
45. Magilligan, F.J.; Nislow, K.H. Long-term changes in regional hydrologic regime following impoundment in a humid-climate watershed. *J. Am. Water Resour. Assoc.* **2001**, *37*, 1551–1569. [CrossRef]
46. Pyron, M.; Neumann, K. Hydrologic alterations in theWabash river watershed, USA. *River Res. Appl.* **2008**, *24*, 1175–1184. [CrossRef]
47. Zhang, X.Q.; Chen, Y.N.; Li, W.H.; Yu, Y.; Sun, Z.H. Restoration of the lower reaches of the Tarim River in China. *Reg. Environ. Chang.* **2013**, *13*, 1021–1029. [CrossRef]
48. Dai, Z.J.; Liu, J.T. Impact of large dams on downstream fluvial sedimentation: An example of the Three Gorges Dam (TGD) on the Changjiang (Yangtze River). *J. Hydrol.* **2013**, *480*, 10–18. [CrossRef]
49. Räsänen, T.A.; Koponen, J.; Lauri, H.; Kummu, M. Downstream hydrological impacts of hydropower development in the Upper Mekong Basin. *Water Resour. Manag.* **2012**, *26*, 3495–3513. [CrossRef]
50. Sun, Z.; Huang, Q.; Opp, C.; Hennig, T.; Marold, U. Impacts and Implications of major changes caused by the Three Gorges Dam in the middle reaches of the Yangtze River, China. *Water Resour. Manag.* **2012**, *26*, 3367–3378. [CrossRef]
51. Cai, B.M.; Zhang, B.; Bi, J.; Zhang, W. Energy's thirst for water in China. *Environ. Sci. Technol.* **2014**, *48*, 11760–11768. [CrossRef] [PubMed]
52. Zhang, Q.; Zhou, Y.; Singh, V.P.; Chen, X.H. The Influence of dam and lakes on the Yangtze River streamflow: Long-range correlation and complexity analyses. *Hydrol. Process.* **2012**, *26*, 436–444. [CrossRef]
53. Li, S.; Xiong, L.H.; Dong, L.H.; Zhang, J. Effects of Three Gorges reservoir on the hydrologic droughts at the downstream Yichang station during 2003–2011. *Hydrol. Process.* **2013**, *27*, 3891–3993. [CrossRef]
54. World Commissionon Dams (WCD). *Dams and Development: A New Framework for Decision-Making*; Report of the World Commissionon Dams; Earthscan Publishing: London, UK, 2000.
55. Deitch, M.J.; Merenlender, A.M.; Feirer, S. Cumulative effects of small reservoirs and streamflow in Northern coastal California catchments. *Water Resour. Manag.* **2013**, *27*, 5101–5118. [CrossRef]
56. World Wide Fund for Nature (WWF). WWF's Dams Initiative: Rivers at Risk. Available online: http://wwf.panda.org/what_we_do/footprint/water/dams_initiative/ (accessed on 1 July 2012).
57. Zhang, Z.; Huang, J.; Huang, Y.; Hong, H. Streamflow variability response to climate change and cascade dams development in a coastal China watershed. *Estuar. Coast. Shelf Sci.* **2015**, *166*, 209–217. [CrossRef]
58. Richter, B.D.; Baumgartner, J.V.; Braun, D.P.; Powell, J. A spatial assessment of hydrologic alteration within a river network. *Regul. Rivers Res. Manag.* **1998**, *14*, 329–340. [CrossRef]
59. Baxter, G. River utilization and the preservation of migratory fish life. *Proc. Inst. Civ. Eng.* **1961**, *18*, 225–244.
60. Tennant, D.L. Instream flow regimens for fish, wildlife, recreation and related environmental resources. *Fisheries* **1976**, *1*, 6–10. [CrossRef]
61. Majhi, I.; Yang, D.Q. Streamflow characteristics and changes in Kolyma basin in Siberia. *J. Hydrometeorol.* **2008**, *9*, 267–279. [CrossRef]
62. Matteau, M.; Assani, A.A.; Mesfioui, M. Application of multivariate statistical analysis methods to the dams hydrologic impact studies. *J. Hydrol.* **2009**, *371*, 120–128. [CrossRef]
63. Hirsch, R.M.; Walker, J.F.; Day, J.C.; Kallio, R. *The Influence of Man on Hydrological Systems*; Wolman, M.G., Riggs, H.C., Eds.; Surface Water Hydrology, Geological Society of America, Geological Society of America: Boulder, CO, USA, 1990.
64. Stanford, J.A.; Ward, J.V.; Liss, W.J.; Frissell, C.A.; Williams, R.N.; Lichatowich, J.A.; Coutant, C.C. A general protocol for restoration of regulated rivers. *Regul. Rivers Res. Manag.* **1996**, *12*, 391–413. [CrossRef]
65. Graf, W.L. Damage control: Restoring the physical integrity of America's rivers. *Ann. Assoc. Am. Geogr.* **2001**, *91*, 1–27. [CrossRef]

66. Baker, D.B.; Richards, R.P.; Loftus, T.T.; Kramer, J.W. A new flashiness index: Characteristics and applications to Midwestern rivers and streams. *J. Am. Water Resour. Assoc.* **2004**, *40*, 503–522. [CrossRef]
67. Shiau, J.T.; Wu, F.C. Feasible diversion and instream flow release using range of variability approach. *J. Water Res. Plan. Manag.* **2004**, *130*, 395–403. [CrossRef]
68. Magilligan, F.J.; Nislow, K.H. Changes in hydrologic regime by dams. *Geomorphology* **2005**, *71*, 61–78. [CrossRef]
69. Wang, H.; Yang, Z.; Satio, Y.; Liu, J.P.; Sun, X. Interannual and seasonal variation of the Huanghe (Yellow River) water discharge over the past 50 years: Connections to impacts from ENSO events and dams. *Glob. Planet. Chang.* **2006**, *50*, 212–225. [CrossRef]
70. Yang, T.; Zhang, Q.; Chen, Y.Q.; Tao, X.; Xu, C.Y.; Chen, X. A spatial assessment of hydrologic alteration caused by dam construction in the middle and lower Yellow River, China. *Hydrol. Process.* **2008**, *22*, 3829–3843. [CrossRef]

Article

Estimation of Active Stream Network Length in a Hilly Headwater Catchment Using Recession Flow Analysis

Wei Li [1,2,*], Ke Zhang [3,*], Yuqiao Long [1] and Li Feng [4]

1 Nanjing Hydraulic Research Institute, Hydrology and Water Resources Department, Nanjing 210029, China; yqlong@nhri.cn
2 Department of Bioproducts and Biosystems Engineering, University of Minnesota, Twin Cities, MN 55108, USA
3 State Key Laboratory of Hydrology-Water Resources and Hydraulic Engineering, and College of Hydrology and Water Resources, Hohai University, Nanjing 210098, China
4 Science and Technology Novelty-Checking Institute, Hohai University, Nanjing 210098, China; fengli_tsg@hhu.edu.cn
* Correspondence: liw@nhri.cn (W.L.); kzhang@hhu.edu.cn (K.Z.); Tel.: +86-25-8582-8520 (W.L.); +86-25-8378-7112 (K.Z.)

Academic Editor: Xuan Yu
Received: 21 March 2017; Accepted: 13 May 2017; Published: 16 May 2017

Abstract: Varying active stream network lengths (ASNL) is a common phenomenon, especially in hilly headwater catchment. However, direct observations of ASNL are difficult to perform in mountainous catchments. Regarding the correlation between active stream networks and stream recession flow characteristics, we developed a new method to estimate the ASNL, under different wetness conditions, of a catchment by using streamflow recession analysis as defined by Brutsaert and Nieber in 1977. In our study basin, the Sagehen Creek catchment, we found that aquifer depth is related to a dimensionless parameter defined by Brutsaert in 1994 to represent the characteristic slope magnitude for a catchment. The results show that the estimated ASNL ranges between 9.8 and 43.9 km which is consistent with direct observations of dynamic stream length, ranging from 12.4 to 32.5 km in this catchment. We also found that the variation of catchment parameters between different recession events determines the upper boundary characteristic of recession flow plot on a log–log scale.

Keywords: active stream network; recession flow; headwater catchment; parameter estimation

1. Introduction

Recession flow analysis has been used to estimate catchment-scale hydraulic parameters for a long time. Brutsaert and Nieber [1] first linked the solution of the Boussinesq equation to a power law relationship as follows:

$$-\,dQ/dt = \alpha Q^{\beta} \quad (1)$$

where Q is discharge during the period of streamflow recession and dQ/dt is the recession rate of discharge. Parameters α and β can be estimated from a $-dQ/dt$ vs. Q plot on a log–log scale. Specifically, α is related to aquifer hydraulic properties by comparing with several analytical solutions of the Boussinesq equation for a horizontal aquifer. Despite some limitations of this method when applied to sloping aquifers [2–4], this methodology has subsequently been expanded and applied in many mountainous catchments such as the Yosocuta Watershed in Mexico [5], the Dongjiang Basin in China [6], the arid Coquimbo Region of Chile [7], the Beca River Basin in Portugal [8], and the Massif Central in France [9].

For a given catchment, parameter α in Equation (1) has been found to vary significantly across recession events. It has been shown that aquifer antecedent storage (represented by the average depth of saturated aquifer) has an obvious influence on streamflow recession dynamic and thus on the variation of α [10–13]. Besides antecedent storage, the dynamic of an active stream network might also lead to the variation of α across recession events. A number of studies have characterized the changes of active stream networks at different spatial and temporal scales. Recently, thorough summaries of this phenomenon were given by Godsey and Kirchner [14] and Shaw [15]. Variances of ASNL have not been widely taken into account in recession flow analysis, although it is a common phenomenon, especially in mountainous catchment. Biswal and Marani [16] proposed that recession flow was a function of the flow per unit of river length and the dynamic stream network, and assumed that recession flow was predominantly controlled by the dynamic stream network in a sloping basin. As a result, the parameters in Equation (1) are related to the signatures of catchment geomorphology [11,17–19]. Vannier et al. [9] distinguished the length of permanent and temporary streams in the estimation of catchment-scale soil properties, but did not estimate variation of ASNL across different recession events.

Although whether or not a change of active stream network has a causal effect on the dynamics of streamflow recession is still under debate, there is no doubt that there is a correlation between active stream network and stream recession flow characteristics [14–16]. Thus, we propose to estimate ASNL by using an analysis of recession flow. Our study was carried out in Sagehen Creek, a hilly headwater watershed located in the Sierra Nevada mountain range of California.

2. Description of the Study Catchment and Associated Data

Sagehen Creek is located on the east side of the northern Sierra Nevada (Figure 1). The drainage area of the watershed is 27 km^2. The land surface elevation in the watershed ranges from 1935 to 2653 m, and the average slope of surface is about 15.8%. The geology of the Sagehen Creek catchment consists of granodiorite bedrock overlain by volcanics, which are overlain by till and alluvium. Although very little is known regarding the depths of different geologic formations, the volcanics are assumed to range in thickness between 50 and 300 m. Alluvium is assumed to range in thickness between 0 and 10 m [20]. Sagehen has a Mediterranean-type climate with cold, wet winters and warm, dry summers [21]. Mean annual temperature from 1980 to 2002 was 4 °C, at an altitude of 2545 m. Mean annual precipitation from 1960 to 1991 was 970 mm. Approximately 80% of precipitation falls as snow. Daily mean streamflow data were obtained at the gage near the outlet of the catchment (gauging station number: 10343500). Mean daily streamflow was approximately 1 m^3/s during snowmelt in May and June for 1953–2003. Sagehen Creek is a documented example of the application of the GSFLOW model to a watershed [20]. The GSFLOW model is composed of a surface water model—PRMS—coupled with MODFLOW, a groundwater flow model [20]. In the Markstrom et al. study [20], the GSFLOW model was calibrated to simulate the observed streamflow given the observed meteorological data collected for Sagehen Creek watershed. The calibrated hydraulic conductivity and specific yield for the alluvium have ranges of 0.026–0.39 m/d and 0.08–0.15, respectively. The geometric means of hydraulic conductivity and specific yield are 0.075 m/d and 0.1, respectively. A period of 16 water years, 1 October 1980 to 30 September 1996 was taken in this research. This time period was previously evaluated by the GSFLOW model. As such, recession flow analysis can be performed by using more reliable hydrologic fluxes (infiltration, actual evapotranspiration) simulated by GSFLOW. Digital elevation model data, at 30 m resolution, were obtained from USGS (available at: http://nationalmap.gov/viewer.html) and were used to stream network analysis in Section 3.2.

3. Method

3.1. Theory

One of the most commonly used analytical solutions for discharge recession of sloping aquifers was formulated by Brutsaert [22]. This solution gives the outflow from the hillslope as

$$Q = \frac{4BDKf}{L^3} \cdot \sum_{n=1,2,\ldots}^{\infty} \frac{z_n^2\left[\left(2e^{-aL}\cos z_n\right)-1\right]\exp\left[-K\left(z_n^2/L^2+U^2/4K^2\right)t\right]}{\left(z_n^2/L^2+U^2/4K^2+U/2KL\right)} \tag{2}$$

where Q is the discharge out of the aquifer, $K = kpDcosi/f$, $U = ksini/f$, $a = -U/2K$, p is a constant assumed to be 0.3465 [1,2,23], i is the slope angle of the aquifer, k is the hydraulic conductivity, f is the specific yield, B is the active stream network length, D is the initial saturated thickness of the aquifer, L is the breadth of the aquifer, and z_n is the nth root of Equation (3), which can be computed numerically with

$$\tan(z) = z/aL \tag{3}$$

In the case where $t \to 0$ in Equation (2), the discharge (Q) from the aquifer is simplified to

$$Q = 2B(kpfcosi/\pi)^{1/2}D^{3/2}t^{-1/2} \tag{4}$$

Then the recession rate of the discharge can be written as

$$\frac{-dQ}{dt} = \frac{1.133}{B^2kD^3fcosi}Q^3 \tag{5}$$

Thus

$$\alpha = \frac{1.133}{B^2kD^3fcosi} \tag{6}$$

$$\beta = 3 \tag{7}$$

As was shown by Rupp and Selker [2] and Pauritsch et al. [4], when dQ/dt and Q derived from Equation (2) are plotted on a log–log scale, the recession curve has an obvious transition point, which separates the curve into an early time domain and a late time domain. The curve in the early time domain is concave. A similar concave recession pattern was also found based on observed streamflow in Sagehen Creek (shown later in Section 3.3). We consider the early time domain to be the more typical case, at least in Sagehen Creek, because in many cases rainfall events prevent recessions from extending beyond the early time period. Since the short time behavior illustrated by Equations (4)–(7) should be the initial segment of early time domain, we used Equation (6) and the early time solution of Equation (2) to perform the following parameter estimation.

3.2. Estimation of Aquifer Breadth

In general, the breadth of the aquifer is calculated as

$$L = A/2B \tag{8}$$

where A is the catchment area. When we treat the stream network length as a dynamic rather than a stable variable, the area also should be taken as the contribution area related to the active stream instead of the total area. Otherwise, Equation (8) will lead to an unrealistically large value of L when ASNL shrinks significantly.

Based on the empirically observed dynamic of the stream network [24], the hillslopes including order-one streams stop contributing to streamflow first, indicating the shrinkage of the contribution area. As the catchment drainage continues, only the areas connected to high order streams contribute.

Following this general idea as shown in Figure 1, we used GIS to generate a stream network with a total length of 30.6 km, including three Strahler orders by ignoring several minor details. A total of 16 first-order tributaries with the combined length of 20.8 km were found (dashed lines in Figure 1). Each of them relates to a hillslope contribution area (the gray zone in Figure 1). With the assumption that the hillslope will stop contributing to higher order stream as soon as the related order-one tributary shrinks totally, we generated a series of ASNL and contributing areas with different combinations of the shrinkage of order-one streams. Figure 2 shows the calculated breadth based on Equation (8), with the dynamic contributing area and ASNL. As a result of our calculations, the breadth has a narrow range, from 413 m to 517 m, with a mean of 452 m and a standard deviation of 16.2 m. This illustrates that the breadth has a relatively stable value with the variation of the active stream network. This is not surprising, since the breadth of the aquifer is a transformation of drainage density, which tends to reach a state of equilibrium [25,26].

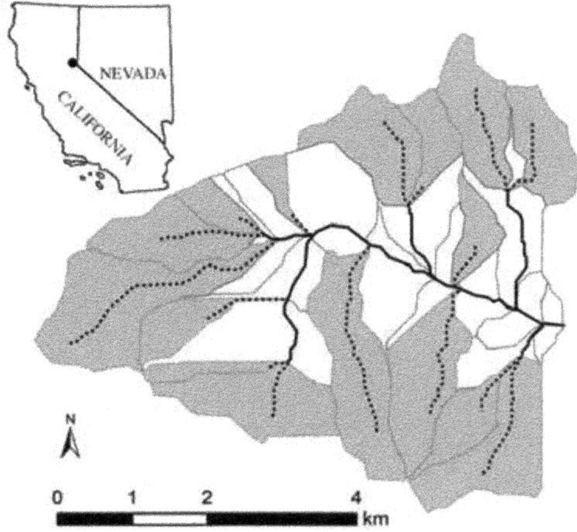

Figure 1. Map of the Sagehen Creek catchment. The stream network is extracted using the flow accumulation threshold of 300 with GIS. The dash lines illustrate the order-one streams. Each order-one stream is associated with a hillslope contribution area (gray zone). The continuous black lines illustrate the order-two and order-three streams.

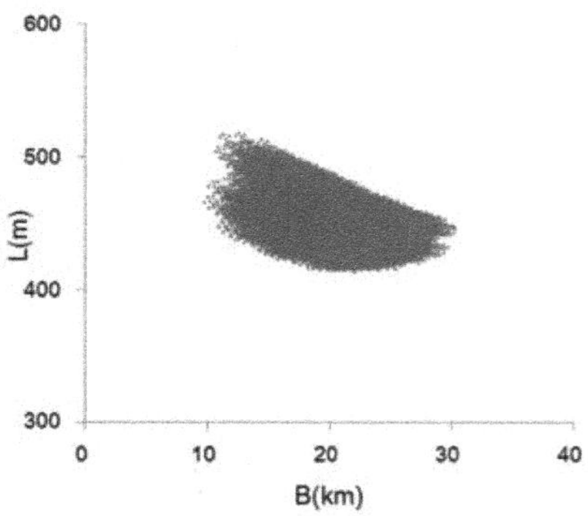

Figure 2. The calculated breadth (L) based on Equation (8) with different ASNL (B).

3.3. Recession Flow Analysis

Recession analysis was carried out based on the observed streamflow over the 16-year period. The backward difference approximation of dQ/dt (=($Q_{t-\triangle t}$ − Q_t)/△t) was calculated using consecutive daily flows and plotted against the arithmetic mean (($Q_{t-\triangle t}$ + Q_t)/2) of the corresponding flows [1]. Next, along with simulated actual daily evapotranspiration (AET) from the soil, and infiltration (AI) into the soil—both calculated by the GSFLOW model—recession stream flows were selected according to the following criteria [27–29]: (1) data points with no positive values of dQ/dt, and (2) data points when the absolute ratio of (AI-AET) to observed Q is smaller than 0.1. As a result, 51 individual recessions with at least four continuous days' length were derived. As is shown in Figure 3, the plot of these individual recessions produces a concave shape.

Two details should be considered when applying these selected recession curves. One is the influence of snowmelt. Snowmelt water can exceed the infiltration capacity of the soil, and GSFLOW does calculate the infiltration and surface runoff generated by snowmelt. Since the selection of recession data was based on the absences of significant net infiltration (AI-AET) during the recession period, we assumed there was no significant snowmelt during the selected recessions. Thus, the influence of snowmelt may be ignored, although minor snowmelt may discharge to the stream directly. The second important detail is the inverse U shape of several recession curves at the beginning of the recession. Consistent with Pauwels and Troch [30], the inverse U shape can be explained by the effect of recharge. The (AI-AET) metric calculated by the GSFLOW model was not completely consistent with the infiltration in the real catchment, although the model has been calibrated. Hence the recessions derived from the observed streamflow still can be influenced by real recharge. These initial data points were ignored during the following analysis.

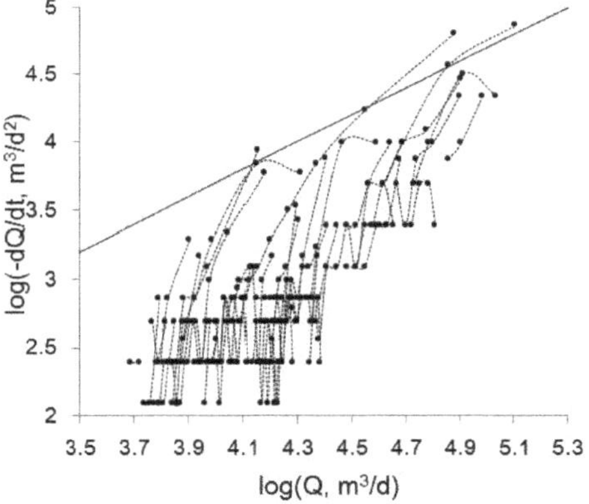

Figure 3. Plot of log(−dQ/dt) vs. log(Q) derived from the observed streamflow for 51 individual recession events selected over a period of 16 years. The straight line with a slope of 1 illustrates the upper boundary of the plot. This upper boundary is discussed in Section 5.5.

3.4. Estimation of ASNL

The 51 individual recession curves were separated into several ranges according to the logarithmic values of the minimum discharge at the end of each recession (Q_{min}). The number of ranges was chosen subjectively to make each range include at least seven recession curves with similar values of Q_{min}. As a result, five ranges were used, and are shown in Figure 4 and Table 1. The mean discharge of recessions of each range varies between 8054–58,899 m^3/d and encompasses a wide range of wetness conditions. Each plot in Figure 4 was indicated by two straight dash lines. The upper envelope was placed with a slope of 3. In keeping with Equation (5), this upper envelope represented the short time recession behavior. The lower envelope was placed visually by ignoring the data points with an inverse U shape. Thus, all of these initial data points were ignored during the following analysis.

The value of log (α) is taken as the intercept of the line of upper envelope with y-axis. Key values of each plot are summarized in Table 1. With the geometric means of k and f, and the mean of L estimated in Section 3.2, B can be obtained using Equation (6) as long as D is specified, and then the recession flow is calculated with Equation (2). The calculation process was repeated by increasing D from 0 with an interval of 1 m regularly. Only the values of D which led to the recession curve (dQ/dt vs. Q on a log–log scale) being located between two envelopes in Figure 4 were noted down. Hence the ranges of both D and B were obtained.

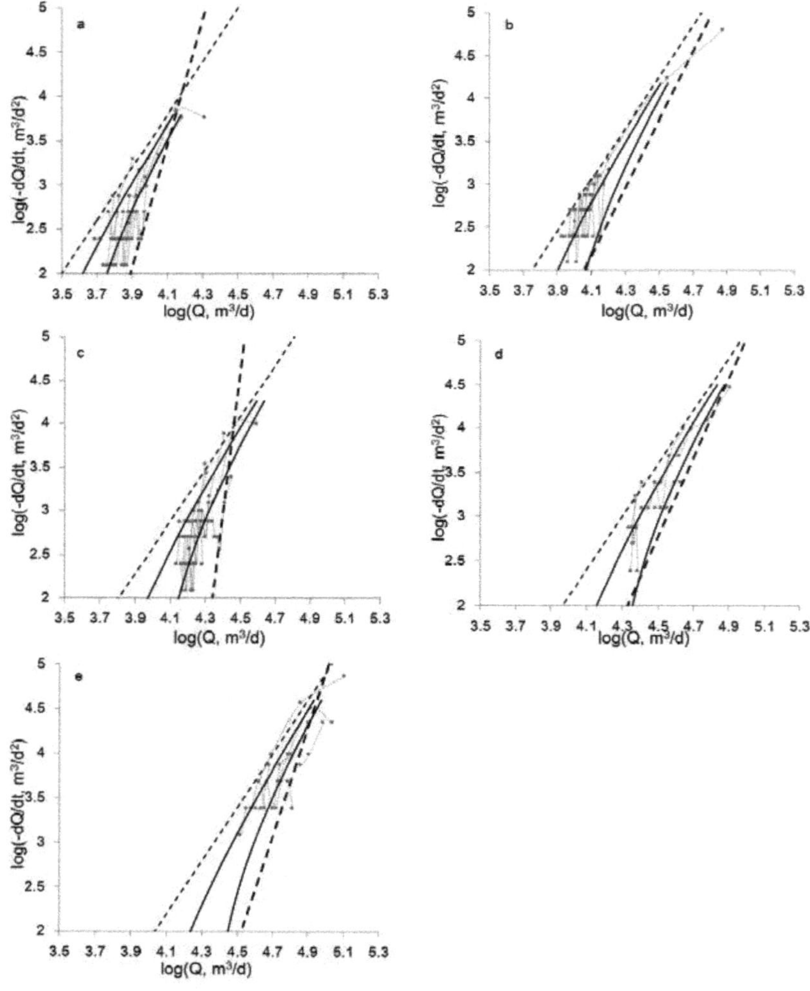

Figure 4. Separated plots of dQ/dt vs. Q, based on the range of Q_{min} on a log scale: (**a**) $\log(Q_{min}) < 3.9$, (**b**) $3.9 \leq \log(Q_{min}) < 4.1$, (**c**) $4.1 \leq \log(Q_{min}) < 4.3$, (**d**) $4.3 \leq \log(Q_{min}) < 4.5$, and (**e**) $\log(Q_{min}) \geq 4.5$. Gray dots illustrate the recession curves in each range. Two straight dash lines illustrate the upper and lower envelopes. Two black curves are obtained from the solution Equation (2) with the condition of D equal to 4 m (right curve), and 27 m (left curve).

4. Results

As a result, the preliminary range of D, 4–27 m, was estimated, which is not related to α. The two black curves shown in Figure 4 are based on the solutions of Equation (2) when D equals 4 m (right curve), and 27 m (left curve). We infer that the established range of D is only related to a dimensionless parameter, $-aL = \tan(i)L/2pD$, which represents the magnitude of the slope term in Equation (2) [22]. When D is large enough (more than 27 m in this case), $-aL$ has a small value, which leads to the catchment's behavior as a horizontal flow. Under this condition, the concave shape of the early recession curve disappears. On the other hand, for a small value of D (less than 4 m in this case), $-aL$ is large, which lead the catchment to behave as a steeper aquifer. In this case, the duration of early recession is very short, less than one day; it therefore cannot be shown in a log–log plot with a time interval of one day.

In fact, Equation (6) illustrates a negative power law relationship between B and D. Figure 5 shows the curves of B vs. D for all of the five different ranges illustrated by Figure 4. The range of ASNL estimated by Equation (6) is shown in Table 1. Thus, a maximum range of stream network length, 1.6–173.5 km, can be obtained. With several additional assumptions, a more reasonable range of ASNL might be concluded. For example, as stated in Section 2, the thickness of alluvium in Sagehen Creek catchment is assumed to range between 0 and 10 m. If we take the saturated aquifer depth as 10 m in the wettest conditions (illustrated by curve e in Figure 5), ASNL would be estimated at 43.9 km. It is reasonable to expect the shrinkage of the stream network and decrease of the saturated aquifer's depth during the drying period of the catchment. The reasonable range of ASNL should thus be located in the gray zone shown in Figure 5. Moreover, with the assumption that only stream channels with high orders are active in dry conditions (illustrated by curve a in Figure 5), ASNL is expected to equal the length of the high order stream based on the result in Section 3.2, which is 9.8 km. As a result, the estimated ASNL is between 9.8 km and 43.9 km. With the illustrated dash line in Figure 5, each value of ASNL related to five different ranges can be further estimated. The results are shown in Table 1.

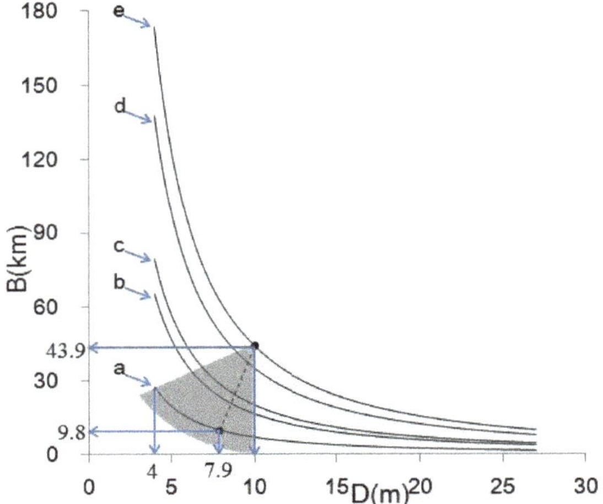

Figure 5. B vs. D curves calculated using Equation (6). The five black curves are related to five ranges, respectively. The gray zone illustrates the possible range of active stream network. The dash line in the gray zone illustrates the estimated ASNL according to the assumptions stated in Section 4.

Table 1. Key values of each plot in Figure 4a–e and estimated ASNL

Parameters	a	b	c	d	e
Range of log(Q_{min}, m^3/d)	<3.9	3.9 – 4.1	4.1 – 4.3	4.3 – 4.5	≥4.5
Number of recession curves	13	11	13	7	7
Mean Q (m^3/d)	8054	13445	18520	33041	58899
α (d/m^6)	3.16×10^{-9}	5.62×10^{-10}	3.80×10^{-10}	1.26×10^{-10}	7.94×10^{-11}
Range of ASNL (km)	1.6–27.5	3.7–65.2	4.5–79.3	7.9–137.8	9.9–173.5
ASNL (B) estimated from the dash line in Figure 5 (km)	9.8	20.7	24.3	37.1	43.9

5. Discussion

5.1. Comparison with Prior Results

Godsey and Kirchner [14] calculated dynamic stream length based on a field survey in Sagehen Creek. They found the connected flowing network length had a range of 12.4–32.5 km. while the stream flow discharge had a range of 4080–47,872 m^3/d. Consistent with their results, we revealed a range of 9.8–43.9 km when the average stream flow discharge ranges between 8054–58,899 m^3/d. Another estimation can be obtained from The National Hydrography Dataset (NHD) (http://nhd.usgs.gov/) maps. The NHD was developed by the USGS to support hydrography research [31]. Based on the NHD of Sagehen Creek, we calculated the lengths of the perennial, intermittent, and ephemeral stream, which are 6.8, 3.7, and 62.4 km respectively. The total stream network length is 72.9 km, which is about twice the estimated ASNL in wet conditions found in our research. One possibility for the discrepancy is that the length of stream in the NHD was mapped in high resolution (The map scale of the high-resolution NHD is 1:24,000), so that some minor channels were also mapped. These minor channels are mostly active only during rainfall events and transit surface runoff. We estimated ASNL based only on the recession flow discharging mostly from catchment storage, and ignored the influence of the surface runoff. Thus the minor stream network should shrink, and was not taken into account in our research. Another possibility is that the saturated depth in wet conditions may be smaller than 10 m. For example, when the saturated depth is taken as 7 m in wet conditions, a channel length of 74.9 km can be estimated based on the curve in Figure 5. However, without more detailed information on the dynamics of the groundwater table for Sagehen Creek, it is not easy to quantify the saturated depth accurately.

5.2. Uncertainty of the Parameters

As stated in Section 3.1, the analytical solution for an ideal rectangle sloping aquifer was used to perform the estimation of ASNL. It is obvious that a real catchment, such as Sagehen Creek, is far more complicated than a simple ideal catchment. Thus the catchment(s) average parameters, such as i, D, k, f, etc., should be obtained using an appropriate method. To account for the spatial variability of the catchment properties, these parameters are usually calculated with a reasonable range instead of a unique value. Thus, the uncertainty of the parameters in Equations (2)–(6) will introduce inaccuracy to the result. As shown in Section 5.1, the estimated ASNL changes significantly when the saturated depth value changes.

To reveal the influence of parameter uncertainty on the estimated result, sensitivity analysis of k and f was further carried out. Based on the parameters calibration result using GSFLOW in Sagehen Creek, the analysis range of k and f was taken as 0.026–0.39 m/d and 0.08–0.15, respectively. As in shown in Figure 6, the sensitivity of the results to f seems relatively limited, while the variation of k has obvious influence on the estimated results. The method proposed by Brutsaert and Nieber [1] is focused on the estimation of catchment averaged hydrogeological properties; thus, we consider the geometric means of k and f to be more appropriate in use as input for the estimation than their maximum or minimum values.

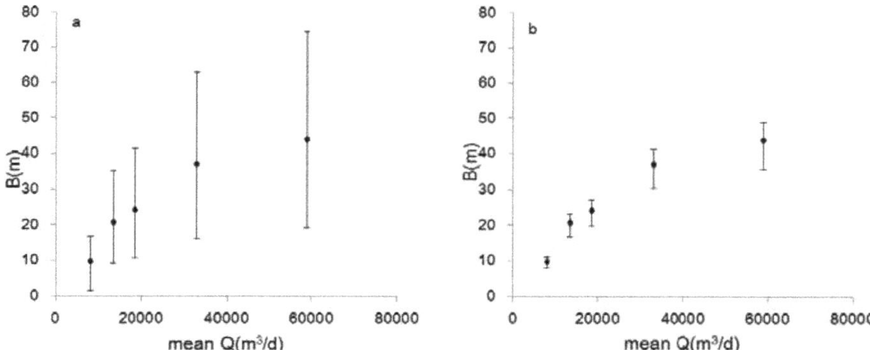

Figure 6. Sensitivity analysis of estimated ASNL as influenced by variations in k and f: (**a**) the error bars illustrate the influence of k with a range of 0.026–0.39 m/d; (**b**) the error bars illustrate the influence of f with a range of 0.08–0.15.

Consistent with prior research, and inherent to our research, is the assumption that there is no change in k and f in different recession events for a given catchment. As shown in Table 1, α increases by almost 40 times from wet to dry conditions. This would also imply the significant decrease of the product (kf), based on the negative relationship between α and kf shown by Equation (6). However this is less realistic. When the catchment dries and saturated areas shrink, the hydraulic response is dominated by riparian areas [32], and thus both k and f should increase generally. As a result, the most probable reason for the increase of α is the shrinking of ASNL when the watershed dries up.

5.3. Advantages and Limitations

The presented approach shows how to estimate ASNL by using recession flow analysis. This research moves our knowledge of recession flow forward. That is, the information embedded in observed recession streamflows can be used not only to estimate hydrogeology parameters, but also to estimate the total length of an active stream, especially in a hilly catchment. The method mainly relies on measured streamflow at the outlet of the catchment, thus it can be easily applied to other catchments.

However, the limitations are also obvious. Since we used observed streamflow at the gauge of catchment outlet, only those active stream channels connecting to the outlet can be estimated. Godsey and Kirchner [14] showed many disconnected stream channels based on their field survey. These disconnected stream channels cannot be estimated by the present approach. Another limitation of this study is that only one hilly catchment, Sagehen Creek, was tested. The accuracy of the method may be influenced by the catchment characteristics, such as topography, geology, climate etc. For example, the stream network would likely be less dynamic in a flat and rainy catchment, so that ASNL may be taken as static instead of dynamic and recession flow characteristics should have less relationship to channel length. More complete and reliable conclusions about the present approach will be reached only after broad applications in different catchments.

5.4. Precautions and Requirements

In this study, net infiltration flux (AI-AET) calculated by GSFLOW was used to select recession stream flows. Hydrologic fluxes are generally not known, but would be determined from measurements or calculated using some acceptable method. Thus, before the application of the present approach to a given catchment, hydrologic fluxes such as actual evapotranspiration and infiltration should be estimated based on some other method a priori. It is also necessary to have some preliminary understanding of catchment parameters such as hydraulic conductivity, specific yield, aquifer thickness, and slope angle of the aquifer. Generally, the more accurate these parameters are, the less the uncertainty of the estimated ASNL.

5.5. The Behavior of Short Time Recession

As shown in Figure 3, a cloud of $-dQ/dt$ vs. Q points is composed of a bunch of recession curves. The upper boundary of these points illustrates short time recession flow behavior between different recession events. An upper boundary with a slope of 1 has also been observed in other studies [5,7,8,33,34]. While it has not been discussed widely, Mendoza et al. [5] hypothesized that this 1:1 line might represent some maximum physical limit of the contributing aquifer on otherwise random-like recession flow behavior. We infer that this upper boundary with a slope of 1 reflects the variation of catchment parameters between different recession events. Based on Equations (4) and (6), the log form of Q and dQ/dt can be shown as

$$\log(Q) = -\frac{1}{2}\log(2t) - \frac{1}{2}\log\alpha \tag{9}$$

$$\log\left(-\frac{dQ}{dt}\right) = -\frac{3}{2}\log(2t) - \frac{1}{2}\log\alpha \tag{10}$$

Thus

$$\log\left(-\frac{dQ}{dt}\right) = \log(Q) - \log(2t) \tag{11}$$

As in shown by Equations (9) and (10), although the change of certain parameters (B, k, f, D, or i) between different recession events would lead to variations of α and the shape of recession curves, the variation of α has an equivalent influence on the magnitudes of Q and dQ/dt in a log–log scale. That is, as indicated by Equation (11), those data points with the same recession time (t) between different recession events would be located on a straight line with a slope of 1 and a y-intercept of $-\log(2t)$. It should be noted that this behavior is reflected only by the very early recession data, since Equations (4) and (6) are only valid for short recession times.

6. Concluding Remarks

Varying ASNL is a common phenomenon, especially in hilly headwater catchments. Accurate estimation of ASNL is valuable for hydrological modeling and is essential to hydro ecosystem maintenance and water resource sustainability. To address the difficulty in direct measurement of changes in the active stream network, this research links ASNL to recession flow dynamic. This approach shows how to estimate ASNL by using analysis of recession flow. To our knowledge, this has never been carried out before. In this first application to the Sagehen Creek catchment, the estimated range of ASNL is consistent with prior results. More complete and reliable conclusions about the present approach will be reached after broad applications in different catchments.

Acknowledgments: This research was partially supported by National Key Research and Development Program of China (No. 2016YFC040281002, 2016YFC0402808, and 2016YFC0402701), National Nature Science Foundation of China (No. 51409161), Natural Science Foundation of Jiangsu Province (Grants No. BK20140080), the Fundamental Research Funds for the Central Universities of China (No. 2015B28514), and the Priority Academic Program Development of Jiangsu Higher Education Institutions.

Author Contributions: Wei Li served as lead and corresponding author, and designed the program of research; Ke Zhang performed the analyses; Yuqiao Long provided editorial improvements to the paper; Li Feng analyzed the background of recession flow and improved the introduction.

Conflicts of Interest: The authors declare no conflict of interest.

References

1. Brutsaert, W.; Nieber, J.L. Regionalized Drought Flow Hydrographs from a Mature Glaciated Plateau. *Water Resour. Res.* **1977**, *13*, 637–643. [CrossRef]

2. Rupp, D.E.; Selker, J.S. On the use of the Boussinesq equation for interpreting recession hydrographs from sloping aquifers. *Water Resour. Res.* **2006**, *42*. [CrossRef]
3. Bogaart, P.W.; Rupp, D.E.; Selker, J.S.; van der Velde, Y. Late-time drainage from a sloping Boussinesq aquifer. *Water Resour. Res.* **2013**, *49*, 7498–7507. [CrossRef]
4. Pauritsch, M.; Birk, S.; Wagner, T.; Hergarten, S.; Winkler, G. Analytical approximations of discharge recessions for steeply sloping aquifers in alpine catchments. *Water Resour. Res.* **2015**, *51*, 8729–8740. [CrossRef]
5. Mendoza, G.F.; Steenhuis, T.S.; Walter, M.T.; Parlange, J.-Y. Estimating basin-wide hydraulic parameters of a semi-arid mountainous watershed by recession-flow analysis. *J. Hydrol.* **2003**, *279*, 57–69. [CrossRef]
6. Zhang, L.; Chen, Y.D.; Hickel, K.; Shao, Q. Analysis of low-flow characteristics for catchments in Dongjiang Basin, China. *Hydrogeol. J.* **2008**, *17*, 631–640. [CrossRef]
7. Oyarzún, R.; Godoy, R.; Núñez, J.; Fairley, J.P.; Oyarzún, J.; Maturana, H.; Freixas, G. Recession flow analysis as a suitable tool for hydrogeological parameter determination in steep, arid basins. *J. Arid Environ.* **2014**, *105*, 1–11. [CrossRef]
8. Santos, R.M.B.; Sanches Fernandes, L.F.; Moura, J.P.; Pereira, M.G.; Pachecoe, F.A.L. The impact of climate change, human interference, scale and modeling uncertainties on the estimation of aquifer properties and river flow components. *J. Hydrol.* **2014**, *519*, 1297–1314. [CrossRef]
9. Vannier, O.; Braud, I.; Anquetin, S. Regional estimation of catchment-scale soil properties by means of streamflow recession analysis for use in distributed hydrological models. *Hydrol. Process.* **2014**, *28*, 6276–6291. [CrossRef]
10. Bart, R.; Hope, A. Inter-seasonal variability in baseflow recession rates: The role of aquifer antecedent storage in central California watersheds. *J. Hydrol.* **2014**, *519*, 205–213. [CrossRef]
11. Biswal, B.; Nagesh Kumar, D. Study of dynamic behaviour of recession curves. *Hydrol. Process.* **2014**, *28*, 784–792. [CrossRef]
12. Patnaik, S.; Biswal, B.; Nagesh Kumar, D.; Sivakumaer, B. Effect of catchment characteristics on the relationship between past discharge and the power law recession coefficient. *J. Hydrol.* **2015**, *528*, 321–328. [CrossRef]
13. Shaw, S.B.; McHardy, T.M.; Riha, S.J. Evaluating the influence of watershed moisture storage on variations in base flow recession rates during prolonged rain-free periods in medium-sized catchments in New York and Illinois, USA. *Water Resour. Res.* **2013**, *49*, 6022–6028. [CrossRef]
14. Godsey, S.E.; Kirchner, J.W. Dynamic, discontinuous stream networks: Hydrologically driven variations in active drainage density, flowing channels and stream order. *Hydrol. Process.* **2014**, *28*, 5791–5803. [CrossRef]
15. Shaw, S.B. Investigating the linkage between streamflow recession rates and channel network contraction in a mesoscale catchment in New York state. *Hydrol. Process.* **2016**, *30*, 479–492. [CrossRef]
16. Biswal, B.; Marani, M. Geomorphological origin of recession curves. *Geophys. Res. Lett.* **2010**, *37*, L24403. [CrossRef]
17. Mutzner, R.; Bertuzzo, E.; Tarolli, P.; Weijs, S.V.; Nicotina, L.; Ceola, S.; Tomasic, N.; Rodriguez-Iturbe, I.; Parlange, M.B.; Rinaldo, A. Geomorphic signatures on Brutsaert base flow recession analysis. *Water Resour. Res.* **2013**, *49*, 5462–5472. [CrossRef]
18. Biswal, B.; Nagesh Kumar, D. What mainly controls recession flows in river basins? *Adv. Water Resour.* **2014**, *65*, 25–33. [CrossRef]
19. Biswal, B.; Marani, M. 'Universal' recession curves and their geomorphological interpretation. *Adv. Water Resour.* **2014**, *65*, 34–42. [CrossRef]
20. Markstrom, S.L.; Niswonger, R.G.; Regan, R.S.; Prudic, D.E.; Barlow, P.M. *GSFLOW-Coupled Ground-Water and Surface-Water FLOW Model Based on the Integration of the Precipitation-Runoff Modeling System (PRMS) and the Modular Ground-Water Flow Model (MODFLOW-2005)*; U.S. Geological Survey: Reston, VA, USA, 2008; p. 240.
21. Manning, A.H.; Clark, J.F.; Diaz, S.H.; Rademacher, L.K.; Earman, S.; Plummer, L.N. Evolution of groundwater age in a mountain watershed over a period of thirteen years. *J. Hydrol.* **2012**, *460–461*, 13–28. [CrossRef]
22. Brutsaert, W. The unit response of groundwater outflow from a hillslope. *Water Resour. Res.* **1994**, *30*, 2759–2763. [CrossRef]

23. Brutsaert, W.; Lopez, J.P. Basin-scale geohydrologic drought flow features of riparian aquifers in the Southern Great Plains. *Water Resour. Res.* **1998**, *34*, 233–240. [CrossRef]
24. Chen, B.; Krajewski, W.F. Recession analysis across scales: The impact of both random and nonrandom spatial variability on aggregated hydrologic response. *J. Hydrol.* **2015**, *523*, 97–106. [CrossRef]
25. Brutsaert, W. Long-term groundwater storage trends estimated from streamflow records: Climatic perspective. *Water Resour. Res.* **2008**, *44*. [CrossRef]
26. Lyon, S.W.; Giesler, R.; Humborg, C. Estimation of permafrost thawing rates in a sub-arctic catchment using recession flow analysis. *Hydrol. Earth Syst. Sci.* **2009**, *13*, 595–604. [CrossRef]
27. Kirchner, J.W. Catchments as simple dynamical systems: Catchment characterization, rainfall-runoff modeling, and doing hydrology backward. *Water Resour. Res.* **2009**, *45*. [CrossRef]
28. Stoelzle, M.; Stahl, K.; Weiler, M. Are streamflow recession characteristics really characteristic? *Hydrol. Earth Syst. Sci.* **2013**, *17*, 817–828. [CrossRef]
29. Shaw, S.B.; Riha, S.J. Examining individual recession events instead of a data cloud: Using a modified interpretation of dQ/dt–Q streamflow recession in glaciated watersheds to better inform models of low flow. *J. Hydrol.* **2012**, *434–435*, 46–54. [CrossRef]
30. Pauwels, V.R.N.; Troch, P.A. Estimation of aquifer lower layer hydraulic conductivity values through base flow hydrograph rising limb analysis. *Water Resour. Res.* **2010**, *46*. [CrossRef]
31. Simley, J.D.; Carswell, J.W. *The National Map—Hydrography*; U.S. Geological Survey: Reston, VA, USA, 2009.
32. Troch, P.A.; Berne, A.; Bogaart, P.; Harman, C.; Hillberts, A.G.J.; Lyon, S.W.; Paniconi, C.; Pauwels, V.R.N.; Rupp, D.E.; Selker, J.S.; et al. The importance of hydraulic groundwater theory in catchment hydrology: The legacy of Wilfried Brutsaert and Jean-Yves Parlange. *Water Resour. Res.* **2013**, *49*, 5099–5116. [CrossRef]
33. Malvicini, C.F.; Steenhuis, T.S.; Walter, M.T.; Parlange, J.-Y.; Walter, M.F. Evaluation of spring flow in the uplands of Matalom, Leyte, Philippines. *Adv. Water Resour.* **2005**, *28*, 1083–1090. [CrossRef]
34. Rupp, D.E.; Selker, J.S. Information, artifacts, and noise in dQ/dt−Q recession analysis. *Adv. Water Resour.* **2006**, *29*, 154–160. [CrossRef]

Article

An Eco-Hydrological Model-Based Assessment of the Impacts of Soil and Water Conservation Management in the Jinghe River Basin, China

Hui Peng [1,2,*], Yangwen Jia [2,*], Christina Tague [3] and Peter Slaughter [3]

1. Key Laboratory of Marine Environment and Ecology, Ministry of Education, Ocean University of China, Qingdao 266100, China
2. State Key Laboratory of Simulation and Regulation of River Basin Water Cycle (SKL-WAC), China Institute of Water Resources and Hydropower Research (IWHR), Beijing 100038, China
3. Donald Bren School of Environmental Science and Management, University of California at Santa Barbara, Santa Barbara, CA 93117, USA; ctague@bren.ucsb.edu (C.T.); Peter@bren.ucsb.edu (P.S.)

* Authors to whom correspondence should be addressed; pengh@ouc.edu.cn (H.P.); Jiayw@iwhr.com (Y.J.); Tel./Fax: +86-0532-66786568 (H.P.); +86-010-68785616 (Y.J.).

Academic Editor: Xuan Yu
Received: 21 September 2015; Accepted: 3 November 2015; Published: 11 November 2015

Abstract: Many soil and water conservation (SWC) measures have been applied in the Jinghe River Basin to decrease soil erosion and restore degraded vegetation cover. Analysis of historical streamflow records suggests that SWC measures may have led to declines in streamflow, although climate and human water use may have contributed to observed changes. This paper presents an application of a watershed-scale, physically-based eco-hydrological model—the Regional Hydro-Ecological Simulation System (RHESSys)—in the Jinghe River Basin to study the impacts of SWC measures on streamflow. Several extensions to the watershed-scale RHESSys model were made in this paper to support the model application at larger scales (>10,000 km^2) of the Loess Plateau. The extensions include the implementation of in-stream routing, reservoir sub-models and representation of soil and water construction engineering (SWCE). Field observation data, literature values and remote sensing data were used to calibrate and verify the model parameters. Three scenarios were simulated and the results were compared to quantify both vegetation recovery and SWCE impacts on streamflow. Three scenarios respectively represent no SWC, vegetation recovery only and both vegetation recovery and SWCE. The model results demonstrate that the SWC decreased annual streamflow by 8% (0.1 billion m^3), with the largest decrease occurring in the 2000s. Model estimates also suggest that SWCE has greater impacts than vegetation recovery. Our study provides a useful tool for SWC planning and management in this region.

Keywords: eco-hydrological model; soil and water conservation; the Jinghe River Basin

1. Introduction

The Loess Plateau, known for its highly erodible soil and fragile ecosystem, covers an area of approximately 640,000 km^2 in the upper and middle reaches of China's Yellow River. The plateau generally has a semi-arid climate, with extensive monsoonal influence. Centuries of deforestation, intensive agriculture and highly erodible soils have resulted in degenerated ecosystems and intense soil and water loss [1]. In order to support ecosystem recovery and reduce soil and water loss, the Chinese government instituted a variety of soil and water conservation (SWC) measures from the 1950s, such as vegetation management (including plantation and ecosystem recovery) and soil and water conservation engineering (SWCE, such as terrace farmland and check dams). These SWC measures have been proved to be useful in erosion control and sediment reduction [2,3].

It is generally believed that these SWC measures have impacts on the hydrological cycle [2,4]. Previous studies in the Loess Plateau region indicated that the SWC (including vegetation recovery, terrace land, check dams, *etc.*) could reduce and delay surface runoff, and decrease soil erosion [5,6]. Observations suggested that sediments were reduced by 2.2×10^8 t·year^{-1} from the Loess Plateau to the Yellow River from 1970 to 1996 [7]. Substantial reductions in streamflow, in the decades following SWC, have also been reported in the Loess Plateau. Xu *et al.*, [8] estimated that runoff from the Loess Plateau to the Yellow River has decreased by 1.0×10^9 m^3·year^{-1} since the 1950s based on observation data [8]. He *et al.*, [4] estimated that the annual streamflow of the Yellow River was reduced by 60% in the treated period (1973–2006) compared with a contrast period (1957–1973) based on parametric and non-parametric Mann-Kendall tests of trends in streamflow data. Although these previous studies showed declines in streamflow in the Loess Plateau, the observed changes in streamflow may also be impacted by climate and human water use [9]. In this paper, an eco-hydrological model is used to disentangle the impacts of climate and human water use from SWC measures, and to link specific SWC measures with hydrographic changes.

Eco-hydrological models that simulate interactions between vegetation dynamics and the hydrological cycle can serve as useful tools for evaluating the impact of both climate trends and human modifications on the landscape [10]. Given the vegetation and hydrological modifications from SWC described above, the Loess Plateau is a logical setting for the implementation of such models. Previous studies have used traditional hydrological models to evaluate streamflow changes in this region [11–17]. These models, however, did not account for vegetation dynamics, including growth following planting and interactions among vegetation properties (such percent cover, leaf area or total biomass), climate and hydrology. These interactions are crucial for the evaluation of SCW impacts because vegetation recovery is a key part in the SCW and vegetation recovery is highly dependent on climate and hydrology [18–20]. In addition, most previous efforts were conducted at small watershed scales and were not easily extrapolated to the larger basin scale. In this study, we examine the impacts for the full Jinghe River basin (over 45,000 km^2).

This paper presents an analysis of SWC impacts on streamflow of the Jinghe River Basin in the Loess Plateau by using a modified version of a fully coupled model of vegetation growth and hydrology, RHESSys. Modifications include the addition of sub-models of in-stream routing, and reservoir operation. Existing model data structures and parameters were used to represent different SWCE structures in RHESSys. To quantify the relative contributions of different SWC measures to streamflow changes, the model was used to estimate the streamflow of the Jinghe River during 1980–2010 given two different management scenarios: SWC including vegetation recovery and SWCE. The implications of the results for future SWC planning and water resource management are discussed.

2. Materials and Methods

2.1. Site Description

The Jinghe River is a 455-km long tributary of the Yellow River. The Jinghe basin (area of 45,421 km^2) is located in the middle of the Loess Plateau (106°20′ E–108°48′ E, 34°24′ N–37°48′ N), and 4.3% of the area is mountainous, 41.7% is loess tableland and broken plateau areas, and 48.8% is loess hilly and gully regions (Figure 1). The basin has deep layers (50–80 m) of loess, composed mainly of fine sand, silt and clay, with silt accounting for up to 50%. The loess has high porosity and is prone to landslides. This basin is one of the most highly eroded areas of the Loess Plateau.

Land use types within the basin include farm land (41.6%), forest (10.2%), grassland (46.5%), water (0.4%), bare land (0.1%), and residential and industrial land (1.31%). The natural vegetation of the basin is of a temperate forest-steppe transition type. Due to historical development processes, human activities have destroyed much of the forest, and now degraded grassland covers most of the area. The growing season of the vegetation is between April and October.

The basin spans seven cities in the provinces of Ningxia, Gansu and Shanxi. There are approximately 6 million people living in the middle and downstream portions of the basin. Water scarcity and soil desertification have resulted in small levels of agriculture and animal husbandry. The industrial structure of the basin is resource-oriented, with energy, heavy and chemical industries accounting for more than 90% of the total industrial production.

The basin has a typical continental climate and is located between the semi-humid and semi-arid temperate zones. It is dry in winter and spring, and storms are frequent in summer and autumn. The average air temperature is 8 °C. Annual precipitation ranges between 350 and 600 mm, with substantial inter-annual variation. Summer precipitation accounts for more than 50% of annual precipitation. Both temperature and precipitation decrease from south to north within the basin.

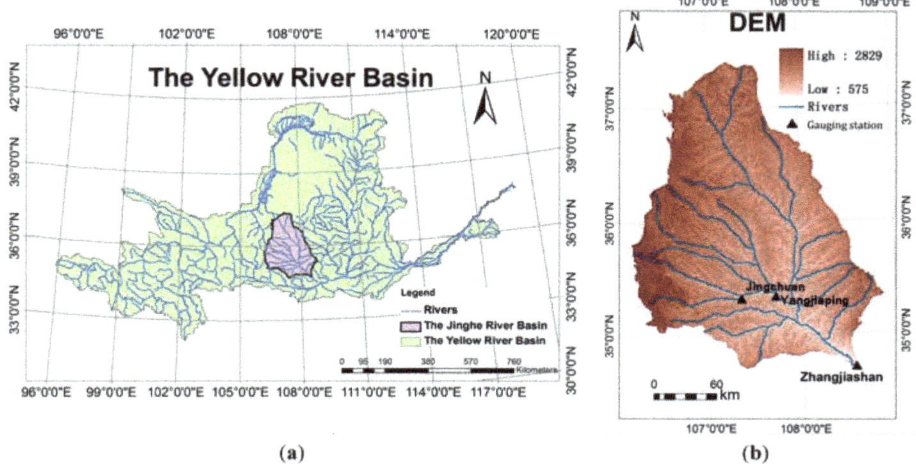

Figure 1. Digital map of the study area. (**a**): The yellow river basin; (**b**): DEM.

2.2. RHESSys Model

The Regional Hydro-Ecological Simulation System (RHESSys) is a biophysically based eco-hydrological watershed-scale model used to simulate water, carbon and nitrogen cycling and transport. RHESSys represents watersheds as a spatially nested hierarchical structure with a range of meteorological, hydrological and ecological processes associated with different levels of hierarchy [21]. Most vertical hydro-ecological processes are computed in the finest spatial resolution objects, patches and strata, where strata are vertical vegetation layers above the patch. Meteorological inputs and incoming radiation are organized with zones and drainage between patches organized within hillslopes that drain to stream reaches. All of the spatial resolution objects (patches, strata, zones and hillslopes) are generated in a Geographic Information System (GIS)-Geographic Resources Analysis Support System (GRASS-GIS).

The simulation of hydrological processes in RHESSys reflects water storage and transport vertically between the surface canopy/litter and subsurface soil layers and laterally between the simulation units. The soil profile is a simple three-layer soil, including a rooting zone, an unsaturated layer and a saturated layer. RHESSys also calculates storage and evaporation in litter, surface depressions, canopies and snow, and transpiration from overstory and understory canopies. The ecosystem carbon and nitrogen simulation process of RHESSys includes estimates of canopy photosynthesis, respiration and the allocation of net primary productivity to growth, plant turnover, and soil organic matter accumulation and decomposition.

RHESSys has been used for a wide variety of hydrological and ecosystem biogeochemical cycling applications [22–26]. A full description of the RHESSys process representation is given in [21].

2.3. Model Modification

Previous applications of the RHESSys model have focused on small watersheds (<1000 km^2) [21,27–29]. Several improvements were made to the model in order to apply RHESSys in the Jinghe River Basin, which include in-stream routing, reservoir operation, approaches to parameterize SWCE and a landscape partitioning strategy that supported computational efficiency while maintaining the representation of key components of landscape heterogeneity.

2.3.1. In-Stream Routing

Previous versions of RHESSys assume that any water entering a stream reach leaves the basin within a daily time step. For larger basins, such as the Jinghe River basin, travel time within the stream may exceed one day. To resolve this limitation issue, a nonlinear kinematic wave stream routing module was added to RHESSys. The kinematic wave model is defined by a continuity equation, momentum equation and Manning equation [30]. The numerical solution scheme—Newton's method—is used to solve these equations.

Continuity:

$$\frac{\partial Q}{\partial x} + \frac{\partial A}{\partial t} = q \tag{1}$$

Momentum:

$$S_0 = S_f \tag{2}$$

Manning Equation:

$$Q = \frac{1.49}{n} R^{\frac{2}{3}} \cdot S_0{}^{\frac{1}{2}} \tag{3}$$

where Q is the flow rate, A is the cross area of the stream, q is the lateral input, x is the stream length, t is the time interval, S_0 is the gravity force term, S_f is the friction force term, R is the hydraulic radius, and n is the Manning coefficient.

Inputs required by the stream routing model include the parameters listed in Table 1 and the stream network topology (e.g., identification of upstream and downstream reaches for each reach). A GRASS-GIS [31] program was developed to automatically construct a table that organizes the stream routing information as RHESSys inputs. This GRASS-GIS program requires spatial raster data to define the stream network, stream cross section information (top width, bottom width and height), Manning roughness and patches. The patch is the smallest spatial modeling unit in RHESSys.

The stream routing sub-model is triggered at the end of each time step (hourly or daily). Any lateral flow (surface or subsurface) from a RHESSys patch containing a stream is included in the stream routing. All improvements to RHESSys in this paper are available on the RHESSys github site. Version 5.14.8 was used for this paper.

Table 1. The format of the stream routing information table.

Item	Information
Overall reaches	Number of stream reaches
Reach	Reach ID, bottom width, top width, max height, slope, Manning roughness, length, Numbers of intersecting units, Numbers of upstream reaches, Number of downstream reaches
Intersecting unit	Patch ID, Zone ID, Hill ID
Upstream reach	Reach ID
Downstream reach	Reach ID

2.3.2. Reservoirs Sub-Model

To calculate the reservoir impacts on flow, a simple reservoir operation sub-model was added to RHESSys, which is called by the in-stream routing model introduced above. The reservoir operation sub-model is based on user-supplied monthly maximum volume requirements and minimum monthly outflow requirements of reservoirs. The outflow of the reservoir is computed as follows:

$$Q_{out} = \begin{cases} Q_{min} & Q_{in} - Q_{min} + V < V_{max} \\ Q_{in} + V - V_{max} & Q_{in} - Q_{min} + V > V_{max} \end{cases} \quad (4)$$

where Q_{out} is the outflow of the reservoir, Q_{min} is the required minimum outflow, V is the storage of the reservoir at the beginning of the time step, V_{max} is the storage capability of the reservoir for one month, and Q_{in} is the inflow of the reservoir. The outflow of the reservoir must satisfy the requirement of the minimum outflow requirement first, followed by the monthly volume limitation. User-prescribed human water use is subtracted from the reservoir volume. A reservoir information input file was added to the RHESSys input files, which records the operation information of each reservoir.

2.3.3. SWCE

In this study, two kinds of SWCE were included in the simulation: terrace land and check dams. The impacts of terrace land are simulated based on the ratio terrace land area to the total area for each patch, where patches are the fine-scale simulation unit in the model (average patch size is 1 km^2). These ratios are calculated from the terrace land area data available at the county scale from the water conservancy statistical yearbooks of each county. An increased surface detention storage capacity is used in the RHESSys model as a parameter to define the impact of terraced land on runoff. The relationship between terrace land ratio and detention storage capacity was based on the literature [32]. A reduced slope associated with terraced land is also implicitly included through the use of slope in the RHESSys subsurface and surface routing computation.

The hydrological impacts of check dams are simulated at the hillslope scale (e.g., area draining either side of a stream reach). Information on the annual new check dam deposition area of every county is available from water conservancy statistical yearbooks. To estimate the number of check dams in a hillslope, this annual new check dam deposition area was divided by the mean annual deposition area of one check dam. The mean annual deposition area of a typical check dam was 0.0029 km^2 and the drainage area of one check dam was 1.04 km^2 according to local survey data. Hillslope-scale runoff generated within the drainage area is assumed to be the inflow of the check dam. Check dams are simulated using the reservoir sub-model described above, assuming Q_{min} to be zero. Thus, the check dam is assumed be an overflow dam with a maximum storage (160,000 m^3). The surface evaporation of water stored in the check dam is calculated using the Penman Method. The water surface area is computed as follows:

$$A_{water} = \sqrt{2VB/i} \tag{5}$$

where V is the reservior storage, m^3; B is the reservior width, m; i is the bottom slope.

RHESSys allows the user to change the spatial resolution and shape of the modeling units [10]. It needs to consider both computational efficiency and spatial resolution in spatial units partitioning. Based on a preliminary simulation to evaluate changes in streamflow with resolution, patches were based on a 1-km resolution Digital Elevation Map (DEM). The hillslope was partitioned by GRASS-GIS watershed generation, with a threshold drainage area of 100 km^2. The basin was created by the GRASS-GIS basin boundary program based on the outlet hydrological measurement station in the Jinghe River Basin.

2.4. Data

Meteorological data collected from 8 climate stations and 13 precipitation stations were used to generate climate data input for each zone (RHESSys meteorology spatial object). Spatial patterns were generated through interpolation using a standard Theissen polygon approach. Meteorological data were retrieved from the China Meteorological Data Sharing Service System, including daily precipitation, maximum, minimum, and mean daily air temperature from 1956 to 2010. Daily precipitation data from 1956 to 2010 were obtained from the Yellow River Water Conservancy Commission (YRWCC). YRWCC also provided monthly stream flow data from the Zhangjiashan hydrological station, which is located at the outlet of the Jinger River Basin.

The 30 Arc-Second DEM data were retrieved from the website of the U.S. Geological Survey (USGS) [33]. Slope, aspect, stream and basin boundaries were created based on these DEM data in GRASS-GIS. Soil type and characters were obtained from the National Second Soil Survey Data and Soil Types of China.

Land use/cover data for 1980, 1985, 1997, 2000 and 2005 (scale 1:100,000) were used in the model, which were derived from Landsat TM data and revised with survey data of land use in the yearly reports from administrative districts. Land use was reclassified into 3 categories: undeveloped, urban and agriculture land. Undeveloped land includes forest, grassland and wasteland. In the agriculture land, irrigation and fertilization input are included. The input values for irrigation and fertilization are 0.1 m^3 water/m^2/year and 16.05 g NH_4/m^2/year according to local survey data. The vegetation type map was also obtained from the land use/cover data with categories of forest, grass and crop.

The stream cross-section data (bottom width, top width, depth, Manning roughness) for the main stream were available from 13 hydrological stations. For the upstream reach, bottom width, top width, and depth were calculated based on the relationship between the drainage area and cross section [32], and Manning roughness was calculated using the method in reference [34]. Reservoir operation information was obtained from water resource survey data. Human water use data were obtained from the Second Countrywide Comprehensive Water Resources Planning in China. SWCE information was obtained from the water conservancy statistical yearbooks of each county.

Net Primary Production Yearly L4 Global 1-km products and Leaf Area Index—Fraction of Photosynthetically Active Radiation 8-Day L4 Global 1-km products of the Moderate Resolution Imaging Spectroradiometer (MODIS) were retrieved from U.S. Geological Survey (USGS) website, maintained by the Land Processes Distributed Active Archive Center (LP DAAC) of the National Aeronautics and Space Administration (NASA) at the USGS/Earth Resources Observation and Science (EROS) Center, Sioux Falls, South Dakota, 2011.

2.5. Soil and Water Conservation Measures

The soil and water conservation measures included in the model can be divided into two types: vegetation restoration and SWCE. Vegetation restoration is defined by changes in vegetation type (forest, grass and farmland), which are obtained from remote sensing data. The vegetation type would be updated based on Landsat TM data of 1980, 1985, 1997, 2000 and 2005. These updates were used to reflect vegetation changes that occurred as a result of restoration measures. The RHESSys vegetation growth model was used to calculate vegetation biomass and associated characteristics such as leaf area index and net primary productivity. The ratios of the vegetation area in the whole basin are shown in Figure 2.

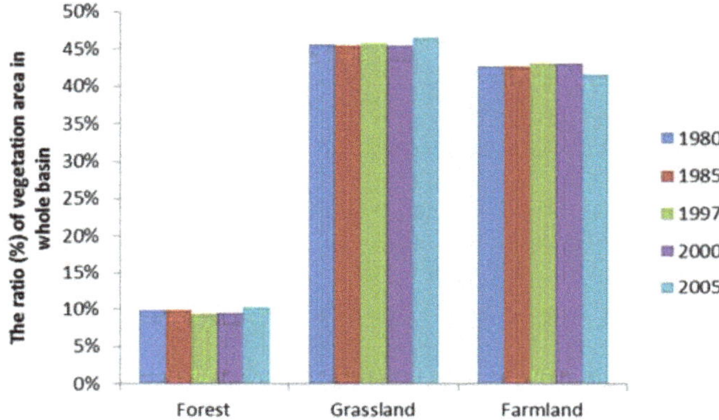

Figure 2. The ratio of the vegetation area in whole basin.

With the approaches for SWCE simulation described in Section 2.3, two kinds of SWCE were included in the simulation: terrace land and check dams. Terrace land and check dams were obtained from water conservancy statistical yearbooks of each county in the basin, which were updated for years 1980, 1985, 1997, 2000 and 2005. Figure 3 shows the terrace land area and check dam control area for the whole basin. The terrace land area ratio in each county was used to generate a raster map of the terrace land ratio for all patches. The ratio of terrace land was divided into 6 categories to match different detention storage capacity parameters (Table 2). The annual new check dam deposition area ratios were calculated in each county and the result was written into a raster file.

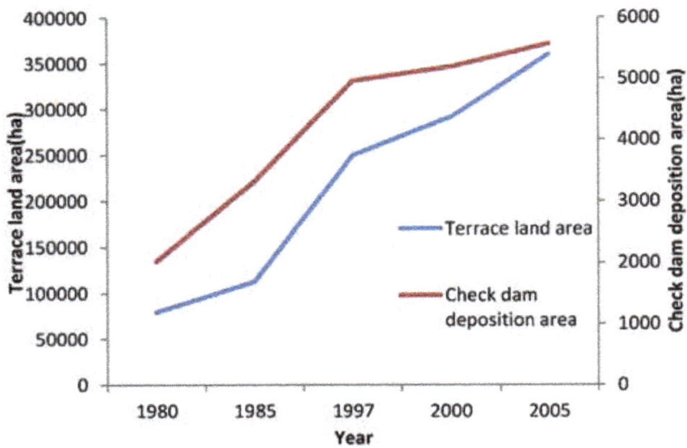

Figure 3. Total terrace land area and check dam control area of the basin.

Table 2. Relationship between detention storage capacity and terrace land area ratio of the whole area.

Categories	1	2	3	4	5	6
Terrace land area ratio	<0.04	0.04–0.06	0.06–0.08	0.08–0.12	0.12–0.2	0.25
Detention storage capacity (m)	0.005	0.0065	0.009	0.0125	0.02	0.03

2.6. Model Parameterization and Calibration

Vegetation parameters are available from the RHESSys parameter database and literature [35]. Some of the parameters were modified based on previous field work in the basin [36].

Drainage parameters are mostly obtained from the RHESSys parameter database, which links these parameters with soil texture. Soil depth, grading composition and porosity were obtained from the National Second Soil Survey Data. These parameters are linked to the classification and are distributed spatially according to the 1:10,000,000 soil classification map in China [32]. Calibrated drainage parameters include: saturated hydraulic conductivity (K), decay of saturated hydraulic conductivity with depth (m), pore size index (pore_size_index), air entry pressure (psi_air_entry), percentage of infiltrated water that percolates to the deeper groundwater store (gw1), and drainage rate of the deep groundwater store (gw2).

A spin-up RHESSys simulation of 2200 years was run prior to calibration and model analysis to stabilize the soil carbon and nitrogen pools. Fifty-four years of climate data (1957–2010) were repeated to create weather sequences for this spin-up run. Following spin-up, drainage parameters in the model were calibrated by comparing observed and modeled streamflow. A Monte Carlo approach was used to sample parameter sets of m, K, pore_size_index, psi_air_entry, gw1 and gw2 values. By comparing observed and simulated streamflow, an acceptable parameter set was selected to use in the model analysis. The calibrations included 250 simulations of a 5-year period, 1991–1995. The performance metrics used to evaluate the model performance include the error in mean annual streamflow estimates and the Nash-Sutcliffe efficiency (NSE) [37] between observed and simulated monthly streamflow of the Zhangjiashan gauging station. The parameter set selected from the calibration was able to capture major hydrologic trends, based on the Nash–Sutcliffe efficiency for monthly streamflow (0.74) (Figure 4). The error in estimating mean monthly streamflow for the calibration period was 9.2%.

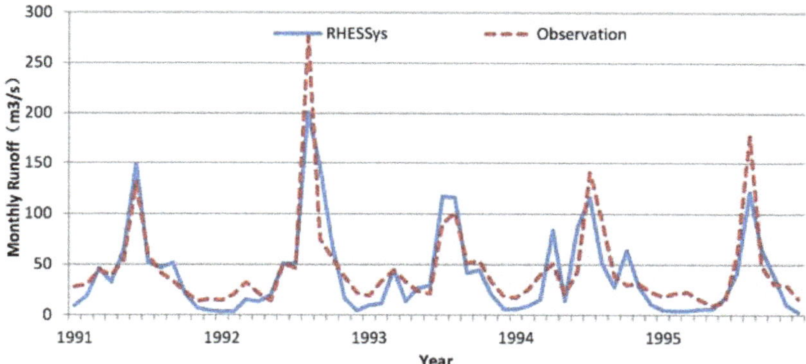

Figure 4. Calibration results of monthly discharge at the Zhangjiashan gauging station.

2.7. Model Verification Method

The model verification includes two parts: hydrological verification and ecological verification. In the hydrological verification, a single parameter set based on the performance metrics outlined in the calibration was selected and used for subsequent model validation. The global parameter uncertainty analysis, such as Generalized Likelihood Uncertainty Estimation (GLUE), is not included in this study because it is too computationally intensive in such a large basin running at relatively fine scales. In ecological verification, RHESSys estimations of net primary productivity (NPP) and leaf area index LAI were compared with Moderate Resolution Imaging Radiometer (MODIS) NPP and LAI products. Specifically, RHESSys estimates of annual average NPP for the years 2000–2010, and LAI distribution for 22 August 2009 were compared with MODIS NPP and LAI products. The basin mean values of the MODIS NPP products were calculated in ArcGIS software.

2.8. Model Scenarios

To evaluate the SWC impacts on streamflow, several different RHESSys model scenarios were compared, which separate the vegetation changes and SWCE (terrace land and check dams) impacts. Three scenarios were simulated in the simulation from 1980 to 2010: Scenario 1 had no prescribed vegetation changes (vegetation area and vegetation types are kept the same as 1980; however, vegetation changes associated with climate variation that are simulated by the model were included), and also assumed no new SWCE following 1980; Scenario 2 prescribed vegetation changes using the actual vegetation data from remote sensing-based land use/cover but did not include new SWCE; Scenario 3 included both prescribed vegetation changes and conservation practices based on SWCE data, which are the closest to historical simulations. Thus, the streamflow differences between scenario 1 and scenario 2 illustrate the impacts from vegetation management, while the difference between scenario 2 and scenario 3 illustrate the additional impacts associated with SWCE.

To isolate other influences of streamflow, the climate and human water use data were the same in the three scenarios. The data were the actual historical data as described above. Therefore, differences in model estimates between the 3 scenarios can be attributed to the impact of management practices on streamflow. The results of the scenario analysis were used to estimate the impact of SWCE, vegetation change and their combination on streamflow.

3. Results and Discussion

3.1. Model Verification

Monthly streamflow data from 1996 to 2000 were used for validation. For the validation period, simulated mean annual streamflow differed from observed values by only 0.4% and the Nash-Sutcliffe Efficiency was 0.70 for monthly streamflow (Figure 5). These results suggested that the model did a reasonable job of capturing the major hydrological processes.

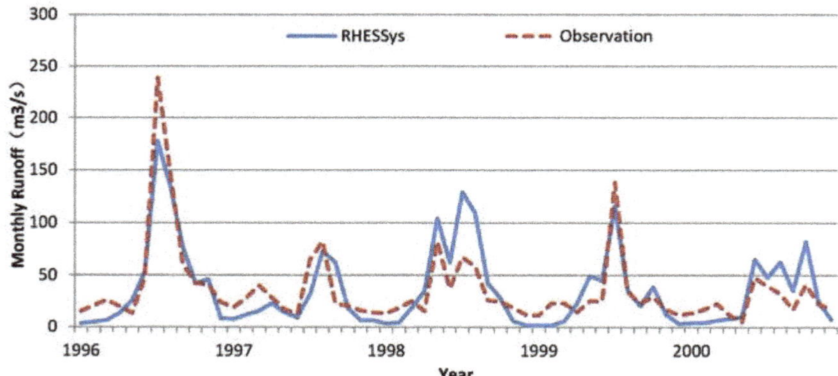

Figure 5. Verification results of monthly discharge at the Zhangjiashan gauging station.

The model estimates of annual average NPP for the years 2000–2010 were compared with Terra/MODIS Net Primary Production Yearly L4 Global 1-km products. The NPP results of RHESSys matched well with those derived from the MODIS products (Figure 6) (mean error is less than 5%). Model estimates of spatial distribution of LAI for 20 August 2009 were compared with Terra/MODIS Leaf Area Index—Fraction of Photosynthetically Active Radiation 8-Day L4 Global 1-km products. The LAI estimates from RHESSys are similar to the MODIS products: MODIS LAI ranges from 0 to 6.8; RHESSys LAI from 0 to 7.7. Spatial patterns are also similar (Figure 7), although the MODIS product showed a greater spatial variation in LAI than the RHESSys. These simulations using the RHESSSys spatial resolution were often coarser than the resolution of the MODIS products.

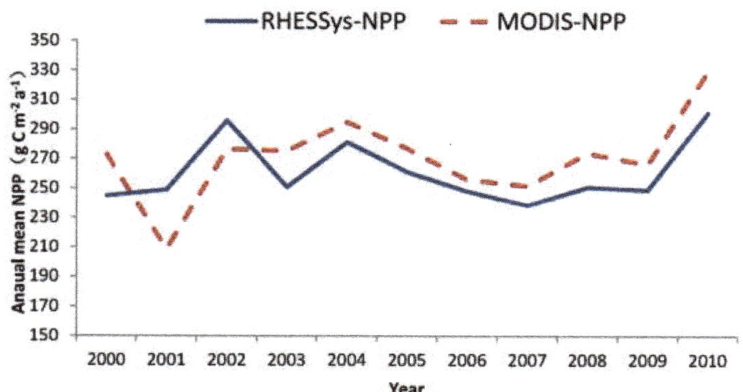

Figure 6. Comparison of annual mean NPP of the Jinghe River Basin from the RHESSys results and MODIS products.

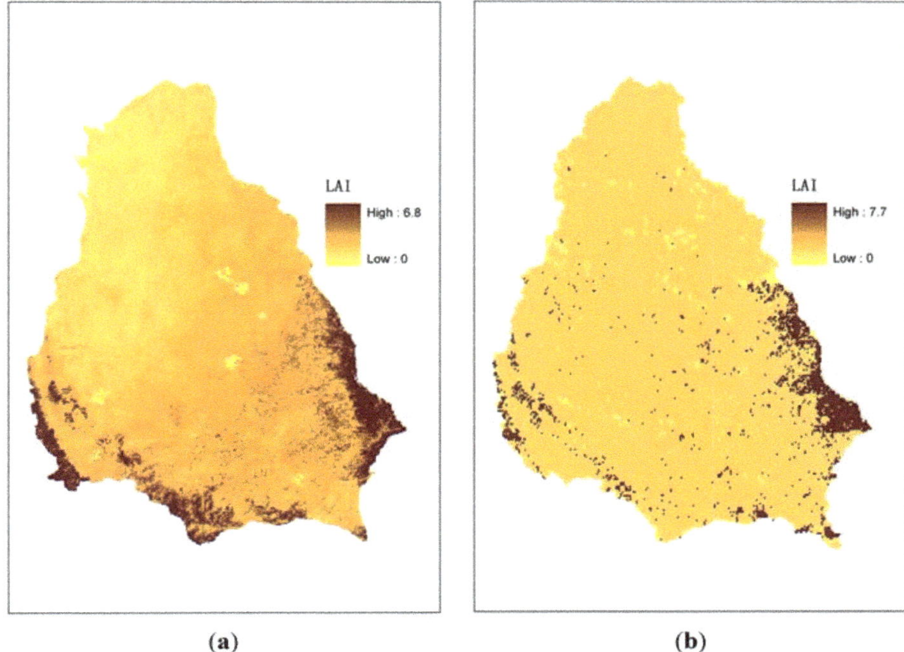

Figure 7. Comparison of LAI distribution between the MODIS products (**a**) and RHESSys results (**b**) on 20 August 2009.

3.2. Streamflow Decrease and Other Impacts

Before the analysis of SWC impacts on streamflow, the streamflow change in the recent decades was evaluated. The Mann-Kendall trend test was used to evaluate the temporal trend of annual mean streamflow at the Zhangjiashan gauging station. The results showed that annual mean streamflow decreased from 1980 to 2010 with a significant change trend of 0.80 m^3/s per year (with 95% confidence interval, Figure 8).

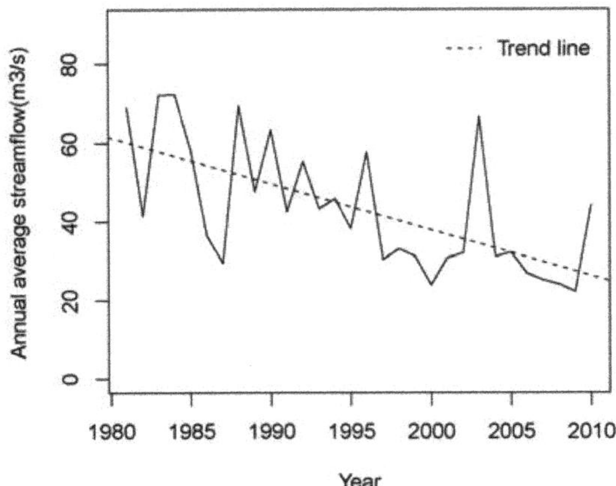

Figure 8. Annual mean discharge of the Jinghe River (Zhangjiashan gauging station) and its change trend.

The climate change trends of the Jinghe River Basin in the recent decades were evaluated. The Mann-Kendall trend test was used to evaluate the change trend of annual precipitation. The results showed that the annual precipitation from 1980 to 2010 has a non-significant change trend. The Mann-Kendall trend test was also used to analyze the annual mean value of daily maximum and minimum temperature, which both showed a significant increasing trend of 0.7 °C/decade from 1980 to 2010 with a 95% confidence coefficient. The temperature increase may cause evapotranspiration (ET) increases; however, ET has a non-significant change trend. The impacts of climate are difficult to evaluate. The human water use increased significantly from the 1980s to 2000s. The annual human water use ranged from 0.38 to 0.66 billion m^3. To isolate this impact on streamflow, Scenario 1 was used in the simulation.

3.3. Impacts of Vegetation Changes

First, analysis was performed for the changes in vegetation type and the associated ET of vegetation patches in the recent decades from scenario 3 (historical simulation). According to Figure 2, the forest area decreased in the 1990s compared with the 1980s, and then increased in 2005. Grassland area changed little from the 1980s to the 1990s, and then increased in 2005. Farmland increase in the 1990s compared with the 1980s, and then decreased in 2005. These results demonstrated that the impact of efforts to recover natural vegetation and reclamation both happened in the 1990s, and there is some evidence of a turning point around 2000 after which the natural vegetation (forest and grassland) started to increase and the farmland started to decrease.

The vegetation area changes suggested that some forest areas and grasslands were destroyed in the 1990s. Farmland increases in the 1990s imply that some forest areas and grasslands became farmland. But there is a turning point around 2000, when the area of forest and grassland increased and farmland decreased. In 1999, a state-funded project, "Grain for Green" was launched in western China for soil erosion control and vegetation improvement by converting slope cropland into grassland or forest [2]. The analysis of land cover data implied that this project had a small but visible impact on vegetation patterns, specifically an increase in forest.

The estimated total ET of the vegetation patches was averaged by decade (Figure 9), and the results showed that in scenario 3, the ET in the 1990s was the highest (21.2 billion m^3), followed by the 1980s (20.9 billion m^3), whereas the ET in the 2000s was the lowest (20.3 billion m^3). Similarly, the ET of farmland in the 1990s is the largest, followed by the 1980s, but lowest in the 2000s. The ET of grassland in the 1990s is the largest, followed by the 1980s, and the ET of grassland in the 2000s was the lowest. The ET of forest in the 1980s is the largest, followed by the 2000s, and the ET of forest in the 1990s was the lowest. The ET results of scenario 1 showed the change trends were similar to scenario 3, but with differences (Figure 9).

The ET of vegetation patches in scenario 3 reflected changes in the distribution of land cover type between forest, grassland and farm land. A comparison between results of scenario 3 and scenario 1 shows the combined impact of SWC measures (Figure 9). The forest area decreased in the 1990s and increased after 2000, as did the contribution of forest ET in scenario 3. The ET of farmland increased in the 1990s and decreased in the 2000s in scenario 3, which is the same trend as the farmland area change. The differences between scenario 3 and scenario 1 reinforce this argument. The only disagreement between ET and the vegetation area change trend was the grassland in the 2000s in scenario 3. The grassland area increased in the 2000s, but the contribution of ET from grassland decreased in the 2000s. Since soil and other factors were constant in the model, this suggests that climate differences played an important role. The contribution of ET from grassland in scenario 3 was larger than that in scenario 1 in the 2000s, which showed the grassland area increase caused the ET increase. All of the results demonstrate the vegetation area influences the ET. Although they had the lowest rainfall in the 1990s, the total ET of the vegetation patches was the highest. The total ET of the vegetation patches increased in the 1990s and decreased in the 2000s.

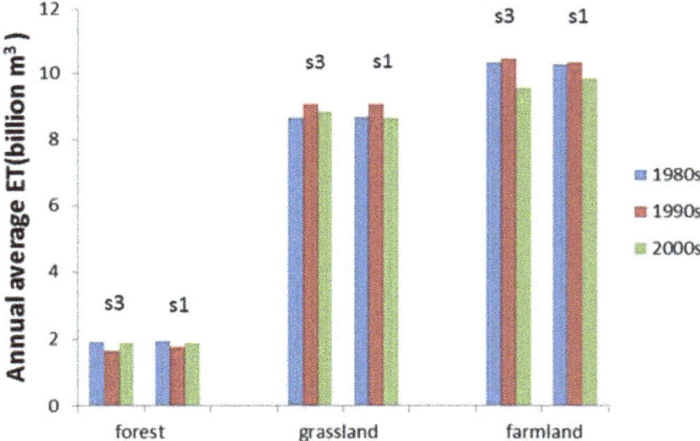

Figure 9. The ET of every vegetation type in scenario 3.

Streamflow in every decade was compared across the 3 scenarios (Figure 10). Scenario 1 is the baseline simulation with no vegetation change or SWCE included (e.g., vegetation from 1980 was used). Vegetation type change was included in scenario 2. Therefore, the streamflow differences between scenario 1 and scenario 2 were the streamflow changes caused by the vegetation changes. The streamflow results show that the vegetation change from the 1980s to the 2000s decreased the annual average streamflow by about 0.027 billion m^3 (1.8%). The annual average decrease is 0.002 billion m^3 in the 1980s (0.1%), 0.043 billion m^3 in the 1990s (3.3%), and 0.036 billion m^3 in the 2000s (2.7%, Table 3).

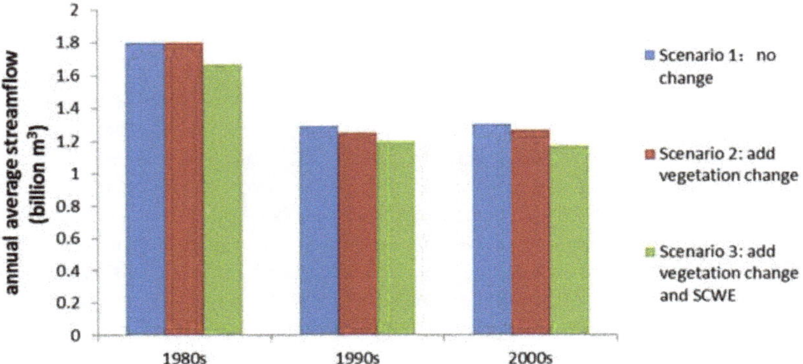

Figure 10. Simulated annual average streamflow of the Jinghe River in every decade.

Table 3. Annual average streamflow simulation results and SWC impacts on annual average streamflow (unit: billion m^3).

Item	1981–1990	1991–2000	2001–2010	1981–2010
Vegetation change impacts (Scenario 1–Scenario 2)	0.002 (0.1%)	0.043 (3.3%)	0.036 (2.7%)	0.027 (1.8%)
SWCE impacts (Scenario 2–Scenario 3)	0.132 (7.3%)	0.054 (4.2%)	0.100 (7.7%)	0.095 (6.5%)
Total SWC impacts (Scenario 1–Scenario 3)	0.134 (7.4%)	0.097 (7.5%)	0.136 (10.4%)	0.122 (8.3%)

The results showed that the vegetation changes from 1980 led to a small streamflow decrease. In addition to the recovery of vegetation resulting from SWC, other activities like farmland increase could also result in vegetation changes, which means that vegetation recovery from SWC is not the only reason for the streamflow change discussed above. It is difficult to distinguish the vegetation recovery from SWC from other vegetation changes in the vegetation map. Future studies may focus on the impacts of vegetation recovery from SWC only when better methods of distinguishing vegetation recovery have been found.

3.4. SWCE Impacts

The SWCE including the terrace land and check dams had more substantial impacts on streamflow estimates. The survey data from water conservancy statistical yearbooks of each county showed that the area of the terrace land and check dams increased dramatically in the past decades.

The streamflow changes caused by SWCE were calculated by scenario analysis using our eco-hydrological model. The estimated streamflow differences between scenario 2 and scenario 3 are streamflow changes caused by SWCE. The simulated streamflow shows that SWCE decreased annual streamflow by 0.095 billion m^3 (6.5%). The annual decrease is 0.132 billion m^3 in the 1980s (7.3%), 0.054 billion m^3 in the 1990s (4.2%), and 0.100 billion m^3 in the 2000s (7.7%, Table 3).

These results showed that SWCE rather than vegetation changes were likely to be the greater source of reduced streamflow. Further differences between scenario 2 and 3 showed only the impact of SWCE, since other input data were the same for both scenarios. The substantially greater streamflow reductions in scenario 3 therefore reflect SWCE rather than vegetation changes. These results highlight how the combination of check dams and terrace land could increase ET losses and infiltration associated with a larger detention storage capacity. Similarly, the check dams could block the streamflow and increase the water storage in the stream reaches and the associated ET losses. Field measurements in the Loess Plateau showed that the terrace land can reduce 73.5%–94.5% of the surface runoff [38] at a hillslope scale. In the larger-scale study, these local scale reductions translate into a basin-wide reduction of close to 8% (for the 2000s).

3.5. Integrated Impacts

Since scenario 3 includes both components of SWC (vegetation change and engineered structures) while scenario 1 included neither, the differences between them are the impacts of both soil and conservation measurements. The model results showed that the soil and conservation decreased annual streamflow by 0.122 billion m^3, which is 8.3% of the annual streamflow of scenario 1. The annual decrease was 0.134 billion m^3 in the 1980s, 0.097 billion m^3 in the 1990s, and 0.136 billion m^3 in the 2000s, which are 7.4%, 7.5% and 10.4% of the annual streamflow of scenario 1, respectively (Table 3). The largest impacts occurred in the 2000s, suggesting that the SWC measures after 1999 aggravated the impacts on streamflow. These results were consistent with the shifts in SWC policy during the same period.

The SWCE had larger impacts on streamflow than the vegetation changes. The SWCE decreased annual streamflow by 0.095 billion m^3 from 1981 to 2010, which accounts for 78% of the total impacts (Table 3). The vegetation change decreased annual streamflow by 0.027 billion m^3 from 1981 to 2010 and accounts for 22% of the total impacts. These results suggested that efforts to reduce the impacts of soil and water conservation on streamflow in this region should focus on the design of SWCE (check dams and terraces) to reduce evaporative losses rather than on vegetation changes.

Streamflow in scenario 1, which isolated the impact of climate and human water use, also changed with time. There was a substantial decline in streamflow (27%) between the 1980s and 2000s. The magnitude of the impact of climate and human water use on streamflow is greater than the impact of SWCE and the vegetation change. Scenarios 2 and 3 demonstrated that the effects of SWCE and vegetation change were in the same direction (declines in streamflow) and thus aggravated the impacts of climate and human water use.

Although this model-based study confirmed that SWC is likely responsible for declines in annual streamflow, it is important to acknowledge that SWC has positive impacts on erosion control. Liu *et al.*, [39] found that the sediment in the Jinghe River decreased 12.1% from 1960 to 2000 and attributed this to SWC in this basin. Erosion control and surface water resources always have tradeoffs that must be considered when proposing soil and water conservation measures.

4. Conclusions

This paper presents an application of a biophysically-based eco-hydrological model—Regional Hydro-Ecological Simulation System (RHESSys)—In the Jinghe River Basin to study the impacts of SWC measures including vegetation recovery and engineering construction in the Loess Plateau. Several improvements have been made to the model, including adding in-stream routing and reservoir operation sub-models. Field observation data, literature values and remote sensing data were used to calibrate the model parameters. The hydrological and ecological verifications showed that the model offers a reasonable representation of the eco-hydrological dynamic processes in this basin. Three scenarios were developed to compare the impacts of vegetation changes with SWCE on streamflow at the Jinghe River Basin scale. Scenarios were based on remote sensing and planning documents that summarized SWC deployment over the past 3 decades. Model estimates suggested that SWC decreased annual streamflow by 8.3% (0.122 billion m^3/year), with the largest decreases occurring in the 2000s, which is consistent with the intensification of SWC measures after 1999. Engineering changes such as terraces and check dams (SWCE) account for approximately 78% of the total impacts, with the remainder attributed to vegetation recovery. There is also a decline in streamflow associated with climate and human water use, which is accentuated by the impact of SWC. These results suggested that efforts to reduce water loss from SWC should focus on reducing evaporative losses from SWCE rather than vegetation changes. More generally, this study demonstrated the feasibility of applying a mechanistically coupled eco-hydrological model at the scale of the Jinghe River Basin and provided a new tool to support the management of SWC and water resources in the Loess Plateau.

Acknowledgments: This study received financial support from the National Basic Research Program of China (973 Program) (2015CB452701), the Key Study Project of the National Natural Science Foundation of China (51379215, 50939006), the Creative Research Group Fund Project of the National Natural Science Foundation of China (51021006), the National 11 Five-Year Scientific and Technical Support Program of China (2010BAC69B02-02) and the Fundamental Research Funds for Central Universities of the Ministry of Education of China (201413060).

Author Contributions: Hui Peng carried out the model modification, calculation, result analysis and drafted the manuscript. Yangwen Jia conceived and guided this study. Christina Tague guided the model calculation, designed the model modification and contributed to result analysis and discussion. Peter Slaughter developed the program in GRASS-GIS software. All authors read and approved the final manuscript.

Conflicts of Interest: The authors declare no conflict of interest.

References

1. Shi, H.; Shao, M. Soil and water loss from the Loess Plateau in China. *J. Arid Environ.* **2000**, *45*, 9–20. [CrossRef]
2. Chen, L.; Wei, W.; Fu, B.; Lü, Y. Soil and water conservation on the Loess Plateau in China: Review and perspective. *Prog. Phys. Geogr.* **2007**, *31*, 389–403. [CrossRef]
3. Fu, B. Soil erosion and its control in the loess plateau of China. *Soil Use Manag.* **1989**, *5*, 76–82. [CrossRef]
4. Dou, L.; Huang, M.; Hong, Y. Statistical Assessment of the Impact of Conservation Measures on Streamflow Responses in a Watershed of the Loess Plateau, China. *Water Resour. Manag.* **2009**, *23*, 1935–1949. [CrossRef]
5. He, X.; Li, Z.; Hao, M.; Tang, K.; Zheng, F. Down-scale analysis for water scarcity in response to soil-water conservation on Loess Plateau of China. *Agric. Ecosyst. Environ.* **2003**, *94*, 355–361.
6. Huang, M.; Zhang, L. Hydrological responses to conservation practices in a catchment of the Loess Plateau, China. *Hydrol. Process.* **2004**, *18*, 1885–1898. [CrossRef]
7. Ran, D.; Liu, B.; Wang, H. Analysis on sediment reduction of the Yellow River through soil and water conservation measures. *Soil Water Conserv. China* **2002**, *10*, 35–36.
8. Xu, J.; Li, X.; Wang, Z. Analysis on ecological water consumption of soil and water conservation measures in the Loess Plateau. *Yellow River* **2003**, *25*, 21–22.
9. Wang, Q.; Fan, X.; Qin, Z.; Wang, M. Change trends of temperature and precipitation in the Loess Plateau region of China, 1961–2010. *Glob. Planet. Chang.* **2012**, *92*, 138–147. [CrossRef]
10. Band, L.E.; Tague, C.L.; Brun, S.E.; Tenenbaum, D.E.; Fernandes, R.A. Modelling Watersheds as Spatial Object Hierarchies: Structure and Dynamics. *Trans. GIS* **2000**, *4*, 181–196. [CrossRef]
11. He, H.; Zhou, J.; Zhang, W. Modelling the impacts of environmental changes on hydrological regimes in the Hei River Watershed, China. *Glob. Planet. Chang.* **2008**, *61*, 175–193. [CrossRef]
12. Jaskierniak, D. *Modelling the Effects of Forest Regeneration on Streamflow Using Forest Growth Models*; University of Tasmania: Hobart, Australia, 2011.
13. Li, Z.; Liu, W.; Zhang, X.; Zheng, F. Impacts of land use change and climate variability on hydrology in an agricultural catchment on the Loess Plateau of China. *J. Hydrol.* **2009**, *377*, 35–42. [CrossRef]
14. O'Loughlin, E.M.; Short, D.L.; Dawes, W.R. Modelling the Hydrological Response of Catchments to Land Use Change. In Proceedings of the Hydrology and Water Resources Symposium 1989: Comparisons in Austral Hydrology, Canberra, Australia, 28–30 November 1989; pp. 335–340.
15. Sun, G.; Zhou, G.; Zhang, Z.; Wei, X.; McNulty, S.G.; Vose, J.M. Potential water yield reduction due to forestation across China. *J. Hydrol.* **2006**, *328*, 548–558. [CrossRef]
16. Zhang, L.; Dawes, W.R.; Hatton, T.J. Modelling hydrologic processes using a biophysically based model—Application of WAVES to FIFE and HAPEX-MOBILHY. *J. Hydrol.* **1996**, *185*, 147–169. [CrossRef]
17. Zhang, X.P.; Zhang, L.; McVicar, T.R.; van Niel, T.G.; Li, L.T.; Li, R.; Yang, Q.; Wei, L. Modelling the impact of afforestation on average annual streamflow in the Loess Plateau, China. *Hydrol. Process.* **2008**, *22*, 1996–2004. [CrossRef]

18. Chapin, F.S., III; Mooney, H.A.; Matson, P. *Principles of Terrestrial Ecosystem Ecology*; Springer: Berlin, Germany; Heidelberg, Germany, 2002.
19. Khurana, E.; Singh, J.S. Ecology of seed and seedling growth for conservation and restoration of tropical dry forest: A review. *Environ. Conserv.* **2001**, *28*, 39–52. [CrossRef]
20. Sullivan, C.Y.; Eastin, J.D. Plant physiological responses to water stress. *Agric. Meteorol.* **1974**, *14*, 113–127. [CrossRef]
21. Tague, C.L.; Band, L.E. RHESSys: Regional Hydro-Ecologic Simulation System—An Object-Oriented Approach to Spatially Distributed Modeling of Carbon, Water, and Nutrient Cycling. *Earth Interact.* **2004**, *8*, 1–42. [CrossRef]
22. Band, L.E.; Tague, C.L.; Groffman, P.; Belt, K. Forest ecosystem processes at the watershed scale: Hydrological and ecological controls of nitrogen export. *Hydrol. Process.* **2001**, *15*, 2013–2028. [CrossRef]
23. Zierl, B.; Bugmann, H.; Tague, C.L. Water and carbon fluxes of European ecosystems: An evaluation of the ecohydrological model RHESSys. *Hydrol. Process.* **2007**, *21*, 3328–3339. [CrossRef]
24. Christensen, L.; Tague, C.L.; Baron, J.S. Spatial patterns of simulated transpiration response to climate variability in a snow dominated mountain ecosystem. *Hydrol. Process.* **2008**, *22*, 3576–3588. [CrossRef]
25. Tague, C.; Grant, G.; Farrell, M.; Choate, J.; Jefferson, A. Deep groundwater mediates streamflow response to climate warming in the Oregon Cascades. *Clim. Chang.* **2008**, *86*, 189–210. [CrossRef]
26. Tague, C. Modeling hydrologic controls on denitrification: Sensitivity to parameter uncertainty and landscape representation. *Biogeochemistry* **2009**, *93*, 79–90. [CrossRef]
27. Claessens, L.; Tague, C.L. Transport-based method for estimating in-stream nitrogen uptake at ambient concentration from nutrient addition experiments. *Limnol. Oceanogr. Methods* **2009**, *7*, 811–822. [CrossRef]
28. Tague, C. Application of the RHESSys model to a California semiarid shrubland watershed. *J. Am. Water Resour. Assoc.* **2004**, *40*, 575–589. [CrossRef]
29. Tague, C.L.; Band, L.E. Evaluating explicit and implicit routing for watershed hydro-ecological models of forest hydrology at the small catchment scale. *Hydrol. Process.* **2001**, *15*, 1415–1439. [CrossRef]
30. Chou, V.T.; Maidment, D.R.; Mays, L.W. *Applied Hydrology*; McGraw-Hill: New York, NY, USA, 1988.
31. Neteler, M.; Bowman, M.H.; Landa, M.; Metz, M. GRASS GIS: A multi-purpose open source GIS. *Environ. Model. Softw.* **2012**, *31*, 124–130. [CrossRef]
32. Jia, Y.; Wang, H.; Zhou, Z.; Qiu, Y.; Luo, X.; Wang, J.; Yan, D.; Qin, D. Development of the WEP-L distributed hydrological model and dynamic assessment of water resources in the Yellow River basin. *J. Hydrol.* **2006**, *331*, 606–629. [CrossRef]
33. U.S. Geological Survey. Global 30 Arc-Second Elevation (GTOPO30). Available online: https://lta.cr.usgs.gov/GTOPO30 (accessed on 3 April 2015).
34. Wang, G.; Li, W. *Reasonability Analysis of Hydrologic Design Results*; Huanghe Water Conservancy Press: Zhengzhou, China, 2002.
35. White, M.A.; Thornton, P.E.; Running, S.W.; Nemani, R.R. Parameterization and Sensitivity Analysis of the BIOME–BGC Terrestrial Ecosystem Model: Net Primary Production Controls. *Earth Interact.* **2000**, *4*, 1–85. [CrossRef]
36. Peng, H.; Jia, Y.; Qiu, Y.; Niu, C.; Ding, X. Assessing climate change impacts on the ecohydrology of the Jinghe River basin in the Loess Plateau, China. *Hydrol. Sci. J.* **2013**, *58*, 651–670. [CrossRef]
37. Nash, J.E.; Sutcliffe, J.V. River flow forecasting through conceptual models part I—A discussion of principles. *J. Hydrol.* **1970**, *10*, 282–290. [CrossRef]
38. Xiong, Y.; Wang, H.; Bai, Z.; Tian, Y. Preliminary study on benefit indexes of runoff and sediment reduction by terraced field, forest land and grass land. *Soil Water Conserv. China* **1996**, *8*, 10–13.
39. Liu, G.; Yu, P.; Wang, Y.; Tu, X.; Xiong, W.; Xu, L. Spatial-temporal variation of annual sediment yield during 1960–2000 in the Jinghe Basin of Loess Plateau in China. *Sci. Soil Water Conserv.* **2011**, *9*, 1–7.

Article

Hydrologic Alteration Associated with Dam Construction in a Medium-Sized Coastal Watershed of Southeast China

Zhenyu Zhang [1,2], Yaling Huang [1,2] and Jinliang Huang [1,2,*]

1 Coastal and Ocean Management Institute, Xiamen University, Xiamen 361005, China; tczzy2007@gmail.com (Z.Z.); hyl@stu.xmu.edu.cn (Y.H.)
2 Fujian Provincial Key Laboratory of Coastal Ecology and Environmental Studies, Xiamen University, Xiamen 361005, China
* Correspondence: jlhuang@xmu.edu.cn; Tel.: +86-218-3833

Academic Editor: Christopher J. Duffy
Received: 9 April 2016; Accepted: 8 July 2016; Published: 26 July 2016

Abstract: Sustainable water resource management requires dams operations that provide environmental flow to support the downstream riverine ecosystem. However, relatively little is known about the hydrologic impact of small and medium dams in the smaller basin in China. Flow duration curve, indicators of hydrologic alteration andrange of variability approach were coupled in this study to evaluate the pre- and post-impact hydrologic regimes associated with dam construction using 44 years (1967–2010) of hydrologic data in the Jiulong River Watershed (JRW), a medium-sized coastal watershed of Southeast China, which suffered from intensive cascade damming. Results showed that the daily streamflow decreased in higher flow while daily streamflow increased in lower flow in both two reaches of the JRW. The dams in the North River tended to store more water while the dams in the West River tended to release more water. The mean daily streamflow increased during July to January while decreased during February to May after dam construction in both two reaches of the JRW. After dam construction, the monthly streamflow changed more significantly and higher variability of monthly streamflow exhibited in the West River than in the North River. The homeogenizing variability of monthly streamflow was observed in both two reaches of the JRW. The earlier occurrence time of extreme low streamflow event and later occurrence time of extreme high streamflow event exhibited after dams construction. The extreme low and high streamfow both decreased in the North River while both increased in the West River of the JRW. All of the indicators especially for the low pulse count (101.8%) and the low pulse duration (−62.1%) changed significantly in the North River. The high pulse count decreased by 37.1% in the West River and the count of low pulse increased abnormally in the North River. The high pulse duration in the post-impact period increased in the two reaches of JRW. The rise rate decreased by 26.9% and 61.0%,and number of reversals increased by 40.7% and 46.4% in the North River and West River, respectively. Suitable ranges of streamflow regime in terms of magnitude, rate, and frequency were further identified for environmental flow management in the North River and West River. This research advances our understanding of hydrologic impact of small and medium dams in the medium-sized basin in China.

Keywords: damming; hydrologic alteration; eco-hydrology; environmental flow management; coastal watershed

---

1. Introduction

Rivers play an important role in the development of human society by providing goods and services for human beings, by which the streamflow regime in turn has been altered for thousands of years due to various human activities [1–3]. By constructing large numbers of dams, human can utilize and control rivers by changing natural streamflow variability to suit human needs [4]. As a result, the past decades have witnessed the great alteration of streamflow regime in the watersheds throughout the world for their extensive dam construction [5–8]. Identifying the environmental impacts caused by hydraulic engineering facilities (e.g., dams) has therefore become an essential component in water resources planning and management [9–12].

The construction of large modern dams produced a dramatic change in the magnitude of hydrologic, geomorphologic and ecologic impacts on rivers [12,13]. Water development, mostly related to dams and diversions, contributed to the declines of more threatened and endangered species than any other resources-related activity [14]. Previous studies show that dam regulation generally had stronger effects on hydrologic regime than other disturbances by reducing the hydrologic variability of river systems [5,15,16]. Obviously, hydrologic regime alteration is responsible for the ecological system change in the rivers [1,17,18]. Dam construction has great impacts on hydrology, therefore, it is of scientific importance to evaluate the hydrologic alteration induced by dam construction.

Many attempts have been made to explore the hydrologic consequences associated with dam construction in recent decades [13,16,19–21]. More than 170 hydrologic metrics (e.g., average flow, flood frequency, peak discharge) have been developed to elaborate the different components of streamflow regime and their contribution to ecological consequence in the river ecological system in the past decade [22]. However, studies on streamflow-related disturbances are mainly on high-steamflow and low-streamflow events [1], which just partially characterize streamflow change. The full range of natural streamflow needs to be identified for its necessarity in evaluating ecosystem health of rivers [12,23–26]. The method of indicators of hydrologic alteration (IHA) was developed by Richter et al. [27] because of their close relationship to ecological functioning as well as for their ability to reflect human induced changes to streamflow regimes for a wide range of disturbances [28]. The IHA method was employed widely to assess the hydrologic change of the dam construction in many rivers worldwide such as the Great Plains, Illinois River, Southeastern US, Yellow River, Yangtze River, and Huaihe River [8,13,16,24,26,29]. The results prove that it is possible to identify the hydrologic change by dam construction over a full range. Moreover, flow duration curve (FDC) was developed to evaluate the overall impact of the streamflow regulation [30] and further applied to effectively determine whether human activities including dams construction can modify the pattern of the ecodeficit and ecosuplus of streamflow [31,32].

Implementation of environmental flows is one of the measures taken to restore or to maintain good ecological statusof rivers [33]. Estimates suggest that by 2050 many countries will face water scarcity, placing increasing pressures on face the water-dependent ecosystems of rivers and estuaries [34]. Obviously, maintaining natural streamflow variability has become an essential principle for environment flow management [35,36]. So far, most of the studies only focus on the minimum release rule so as to maximize human benefits such as water supply or hydropower generation [37,38]. However, provision of a single minimum streamflow cannot protect the biodiversity of a river, which requires the full range of natural flows [24,39–41]. Therefore, to satisfy such strategy of environmental flow management, reservoir planners and operators should seek to minimize the degree of natural flow regime alteration along the regulated river [42]. Based on the 33 indicators of IHA, Richter et al. [27] introduced a useful approach referred to as Range of Variability Approach (RVA) to quantitatively evaluate the degree of hydrologic alteration induced by human disturbance. This method has been demonstrated as a practical and effective way to identify the reasonable range of streamflow regime for environmental flow management [24,41,42].

The hydrologic regime is an indispensable dynamic of the ecosystem change in the watershed, which requires reasonable strategies to protect the magnitude, frequency, duration, timing and rate of change in the streamflow [26,43,44]. The impacts of the dams construction on hydrologic regime show regional difference, since the effects of dams on magnitude, frequency and timing of streamflow change with the types, operations, storage capacity of dams [16,27,45,46]. For example, the groundwater levels rise significantly below ground surface after dam construction in the Tarim River [47], while the water level in the river decreased appreciably in time after dams construction in Yangtze River, Yellow River, Huaihe River [24,26,48]. Moreover, it has been reported that the pattern of the monthly streamflow change due to damming is location-dependent [7,13,49,50]. In China, the accelerating development of economy increases the demand for energy and water resource, thus raising the need for the hydraulic engineering facilities, such as dams and reservoirs [51]. So far, most studies have focused on the impacts of large dams in large basins of China including Yangtze River and Yellow River [12,26,48,52,53]. However, relatively little is known about the hydrologic impact of small and medium dams in the smaller basins. More attentions should be paid to the small and medium dams because of their abundance (with more than 800,000 throughout the world) and their vital roles in maintaining local aquatic ecosystem health and water security [54–56].

The Jiulong River Watershed (JRW), a medium-sized coastal watershed in Southeast China, suffered from intensive human activities with over 13,500 hydraulic engineering facilities including over 120 small or medium dams along the mainstream and major tributaries. Our previous study partly characterized the hydrologic impact of cascade dam in JRW [57]. However, we need more attempts to fully delineate the hydrologic alteration associated with for watershed management. The objectives of this study are: (1) to evaluate the full range of streamflow regime change induced by dam construction; (2) to identify the suitable range of streamflow regime for environmental flow management in the JRW.

2. Materials and Methods

2.1. Study Area

The Jiulong River Watershed (JRW), covering approximately 14,700 km^2 in the eastern coastal area of China (116°46′55″–118°02′17″ E, 24°23′53″–25°53′38″ N) (Figure 1). Two main tributaries, namely, the North River and West River reaches, meet in Zhangzhou, which produces an annual flow of 12 billion m^3 into the Jiulong River estuary and Xiamen-Kinmen coast. The JRW plays an extremely important role in the region economic and ecological health. Water resources in the JRW have been highly developed and supply great demand to many stakeholders, like water supply, irrigation, hydropower and industry. More than ten million residents from Xiamen, Zhangzhou and Longyan use the Jiulong River as their source of water for residential, industrial and agricultural activities. The construction of large dams along the mainstream and major tributaries of JRW greatly altered the natural streamflow regime of the river over the last several decades.

Our previous study showed that the earliest changes in streamflow regime associated with dam construction in the JRW were detected in 1992 [3,57]. As shown in Figure 2, there is distinct difference in terms of flashiness index (the ratio of absolute day-to-day fluctuations of streamflow relative to total flow in a year) between pre-impact period (namely, 1967–1991) and post-impact period (namely, 1992–2010).

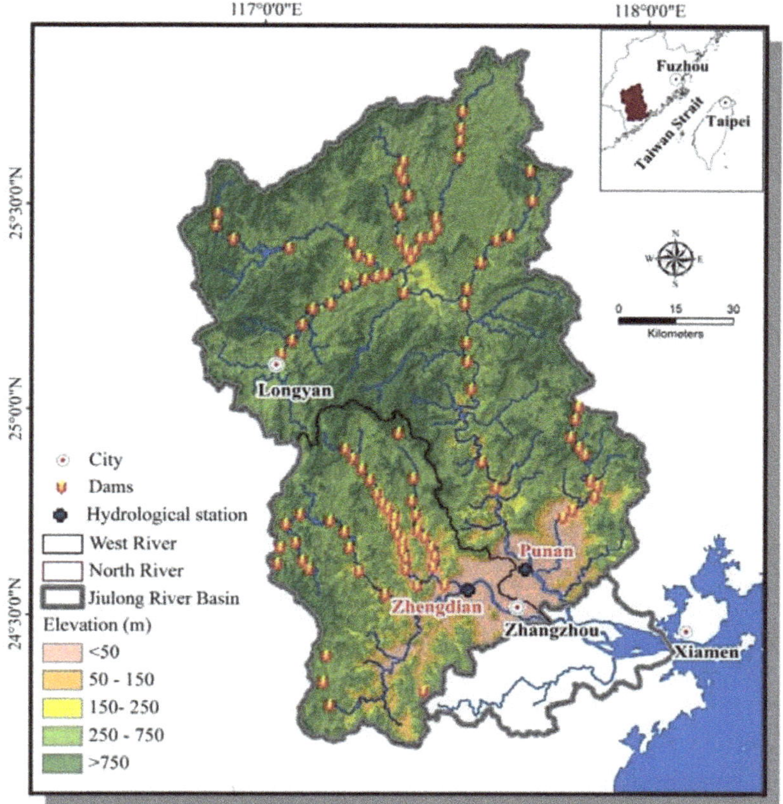

Figure 1. Study area.

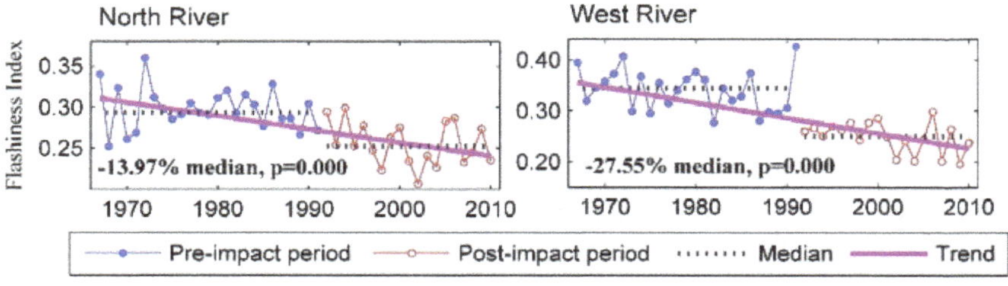

Figure 2. Temporal trend of flashiness index during 1968–2010 (modified from Huang et al., 2013).

2.2. Data Source

In this study, daily streamflow data during 1967–2010 for two downstream hydrologic stations (Punan and Zhengdian) in two reaches, namely, North River and West River were used to evaluate the effect of dam regulation on downstream streamflow in the JRW. A basic description of the two hydrologic stations is shown in Table 1.

Table 1. The streamflow for two gauging station in the JRW.

Station	Longitude	Latitude	Discharge Area (km^2)	Length (km)	Number of Dams Upstream from Stations	Pre-Impact	Post-Impact
Punan (North River)	117.67° E	24.61° N	9640	274	87	1967–1999	2000–2010
Zhengdian (West River)	117.53° E	24.56° N	3940	172	37	1967–1994	1995–2010

2.3. Method

2.3.1. Flow Duration Curve Analysis (FDC)

FDC is constructed from streamflow data over a time interval of interest and to provide a measure of the percentage of time duration that streamflow equals to or exceeds a given value. An annual FDC reflects the variability of daily streamflow during a typical period in a year. The FDC plots can be calculated by the following formula:

$$\mathrm{Pi} = i/(n+1) \tag{1}$$

where n is the number of the days of streamflow and i is the rank.

2.3.2. Indicators of Hydrologic Alteration (IHA)

IHA was used in this study to assess the hydrologic shifts associated with dam construction in a full range. The IHA computes 32 indices that describe the hydrologic regime in the watershed through the frequency, magnitude, duration, timing and rate of the streamflow (Figure 3). The IHA indicators can be divided into five groups: monthly streamflow indices, extreme flow indices, timing indices, high-flow and low-flow indices, and rising and falling indices.

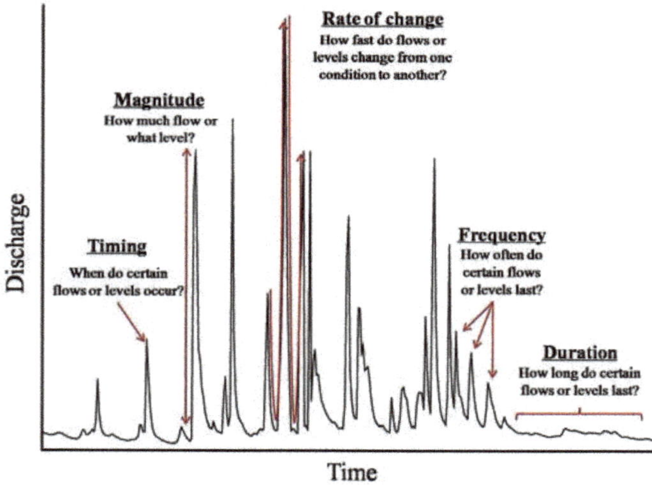

Figure 3. Quantified hydrologic regime using IHA method.

2.3.3. Range of Variability Approach (RVA)

RVA was used in this study to quantitatively evaluate the degree of hydrologic alteration induced by dam construction. In an RVA analysis, the full range of pre-impact data for each parameter is divided into three different categories. The low-level category contains all values less than or equal to the 33th percentile; the middle-level category contains all values falling in the range of the 34th to 67th percentile; and the high-level category contains all values greater than the 67th percentile. A Hydrologic Alteration (HA) factor is calculated for each of the three categories as following formula:

$$\text{HA} = (\text{observed frequency} - \text{expected frequency})/\text{expected frequency} \tag{2}$$

In this study, we divided the absolute ranges of HA factor into three classes: little or no alteration (⩽33%), moderate alteration (33%–67%) and high alteration (⩾67%) according to Richter et al. (1998) [58].

The coefficient of dispersion was a commonly used indicator to evaluate the variability of daily streamflow. It is calculated as the following formula:

$$\text{The coefficient of dispersion} = (\text{75th percentile} - \text{25th percentile})/\text{50th percentile} \tag{3}$$

2.3.4. Method for Measuring Environmental Flow

RVA was used in this study to identify the reasonable range of streamflow regime for environmental flow management. The basic consideration using RVA method for environmental flow identification here is that the preferred environmental flow regime in terms of magnitude, rate, and frequency should maintain streamflow variability as natural as possible in order to sustain the majority of riverine ecological functions. The variability of streamflow magnitude (i.e., daily streamflow) and the three indicators on the rate and frequency of the daily streamflow (i.e., rise rate, fall rate and number of reversals) involved in the IHA indicators can effectively represent the environment change induced by hydropower operations in the river system. In practice, the suitable range of the magnitude, rate and frequency should be identified for environmental flow associated with dams regulation.

The Tennant method (also known as Montana method), by which 20% of the daily average flow was used as the minimum ecological and environmental flow [59,60], was performed in this study to calculate the environmental flow in JRW in order to make comparison with the values from RVA method.

2.3.5. Statistical Analysis

The Kolmogorov–Smirnov (K–S) goodness of fit test was first used to test for normality ofthe distribution of the indicators representing the regime of streamflow. The *t*-test was then used to determine if means for each of the indicators during pre-impact period were statistically different from one another during post-impact period where significance was defined as $p < 0.05$.

3. Result

3.1. Overall Hydrologic Impact Assessment

The effect of regulation by dams on downstream flow regime was assessed using FDC in the North River and West River. As shown in Figure 4, the variability of daily streamflow exhibited anoverall decreasing trend for both two reaches of JRW. In the North River, the daily streamflow slightly decreased in lower percentiles (i.e., higher flows), while daily streamflow slightly increased in the higher percentiles (i.e., lower flows). Comparatively, daily streamflow slightly increased in the higher percentiles (i.e., the lower flows) in the West River. The crossing point of the two curves was calculated to be 136 days and 116 days in the North River and West River, respectively. The dams in the North River stored approximately 0.42 billion m^3 water during a 136-day higher flows period while release only 0.06 billion m^3 water in the subsequent 229 days in lower flow. In the West River, the dams stored only 0.06 billion m^3 water during a 116-day higher flow period while release 0.13 billion m^3 water in the subsequent 249 days in lower flow. This phenomenon indicated that the dams in the North River tend to store more water while the dams in the West River tend to release more water.

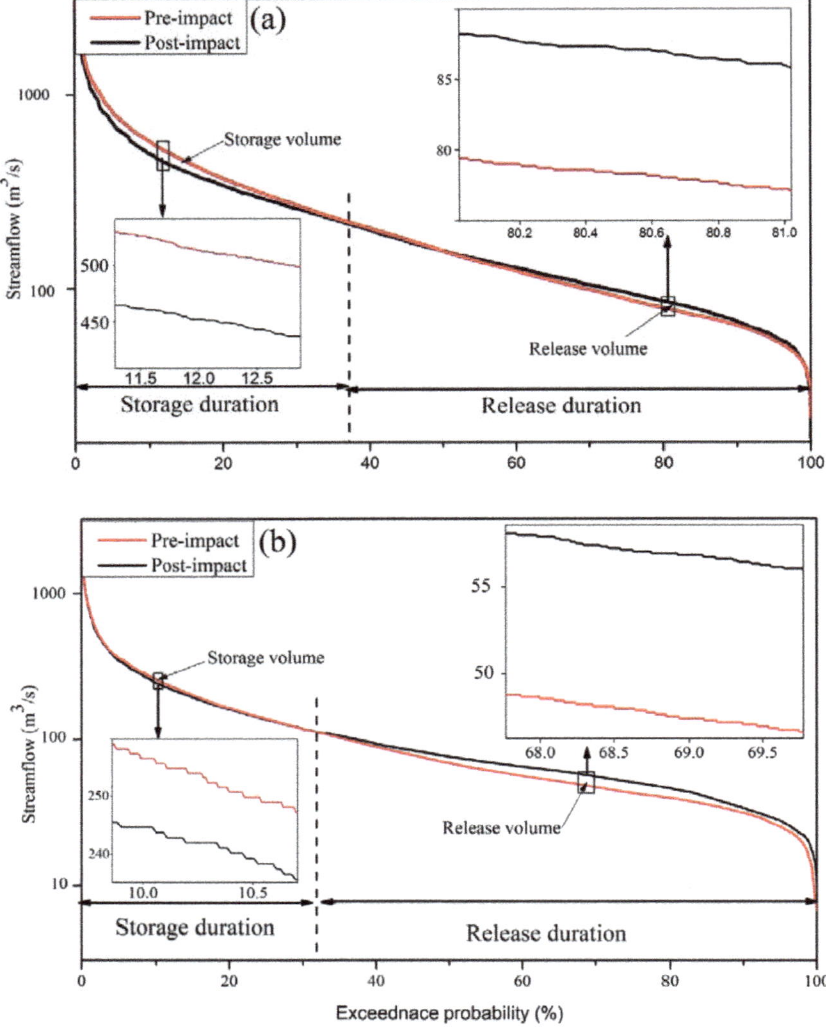

Figure 4. The flow duration curve change in JRW (**a**) North River reach; (**b**) West River reach.

3.2. *Hydrologic Regime Changein a Full Range*

3.2.1. Magnitude of Monthly Streamflow Regime

Themagnitude of monthly streamflow alteration was identified using IHA in the North River and West River. Similar variability patterns were observed in both two reaches of the JRW, namely, the monthly streamflow during July to January increased while decreased during February to May (Table 2). In addition, the monthly streamflow changed more significantly in the West River than in the North River. The monthly streamflow in January, August and December increased significantly in the West River while the monthly streamflow decreased significantly in May in the North River (Table 2).

Table 2. Monthly streamflow change due to dam construction in the JRW (%).

Two Reaches	Jan.	Feb.	Mar.	April	May	Jun.	Jul.	Aug.	Sep.	Oct.	Nov.	Dec.
North River	10.5	−12.1	−37.7	−10.1	−33.4 *	6.4	21.6	17.4	10.7	6.2	4.8	10.7
West River	18.7 *	−13.3	−14.1	−19.1	−23.2	−8	30.3	54.2 *	25.3	28	11.1	20.9 *

Note: * $p < 0.05$.

Figure 5 shows the homogenizing variability of monthly streamflow in the JRW. The coefficient of dispersion increased in the period with high variability of streamflow (e.g., March in the North River, Figure 5a) while decreased in the period with low variability of streamflow in the JRW (e.g., Jan in the North River, Figure 5b). Particularly, the variability of monthly streamflow tend to decrease with the coefficient of dispersion more than 0.7 whereas it increased with the coefficient of dispersion less than 0.7 in both two reaches of the JRW. Interestingly, the variability of monthly streamflow in the flood season (e.g., in July and August) in the West River increased after dam construction, which might be related to the increasing frequency of extreme weather events.

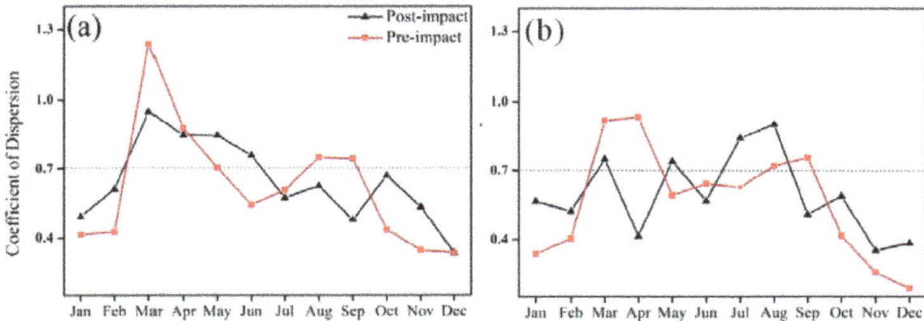

Figure 5. The variability of monthly streamflow in the JRW (**a**) North River reach; (**b**) West River reach.

Variability patterns of streamflow magnitude were delineated using RVA method and are presented in Figure 6. The similar variability patterns of monthly streamflow exhibited in the North River and West River, namely, the frequency of the monthly streamflow with high-level category increased during August to January (Figure 6) in both two reaches of JRW. Obviously, after dam construction, the monthly streamflow changed more intensively in the West River than in the North River. Particularly, the frequency of the monthly streamflow with high-level category all increased during August to January in the West River.

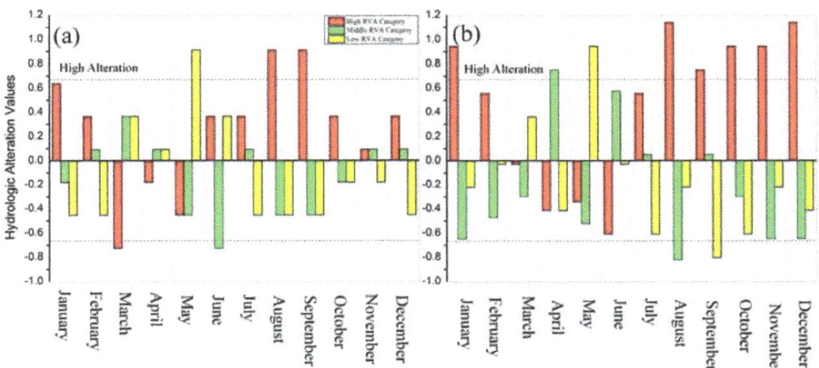

Figure 6. Hydrologic alteration of monthly streamflow in the JRW based on RVA method. (**a**): North River reach; (**b**) West River reach.

3.2.2. Magnitude, Duration and Timing of Extreme Streamflow Regime

The occurrence time of the annual extreme streamflow in the North River was approximately one month earlier than that in the West River (Table 3). The minimum and maximum streamflow occurred in early February and early July in the North River, while the minimum and maximum streamflow were in late February and early August in the West River. The Julian dates of minimum streamflow in the post-impact period for both two reaches of JRW were earlier than those in the pre-impact period, while the Julian date of maximum streamflow showed the opposite results (Table 3). In particular, the extreme low streamflow event occurred 14 days earlier than before in the West River and the extreme high streamfow event occurred 18 day later than before in the North River.

Table 3. The median of Julian date of the extreme streamflow.

Two Reaches	Pre-Impact Min Q	Post-Impact Min Q	Pre-Impact Max Q	Post-Impact Max Q
NorthRiver	34.0	33.0	171.0	189.0
West River	64.5	50.5	211.5	215.0

The baseflow index (BFI) increased by 18.8% and 17.7% in the West River and North River reaches, respectively. Given that BFI is the ratio of the 7-day minimum streamflow to the annual streamflow, the BFI increased more than the 7-day minimum streamflow, which might indicate a potential discharge decreasing trend in both two reaches of JRW (Table 4). Most of indicators of extreme streamflow decreased in the post-impact period in the North River. In contrast, most of the indicators of the extreme streamflow increased in the West River.

Table 4. The change of magnitude and duration of annual extreme streamflow condition in the JRW (%).

Two Reaches	1-Day Min	3-Day Min	7-Day Min	30-Day Min	90-Day Min	1-Day Max	3-Day Max	7-Day Max	30-Day Max	90-Day Max	BFI
North River	−20.3	−5.9	7.2	10.4	−0.4	−9.6	−7.4	−6.6	−7.5	−8.2	18.8 *
West River	8	11.7	13.7	12.7	7.1	6	−2.1	−0.2	8	9.8	17.7

Note: * $p < 0.05$.

Most of the indicators change little (HA ⩽ 33%) or moderately (HA = 33%–67%) in the JRW (Figure 7). The frequency of 30-day minimum streamflow which falled into high-level RVA category (HA ⩾ 67%) increased intensively in the North River and West River, indicating the increased frequency of minimum monthly streamflow with high-level category in both two reaches of JRW. The frequency of 30-day maximum streamflow which falled into high RVA category (HA ⩾ 67%) increased intensively in the West River, suggesting the increased frequency of maximum monthly streamflow with high-level category in the West River.

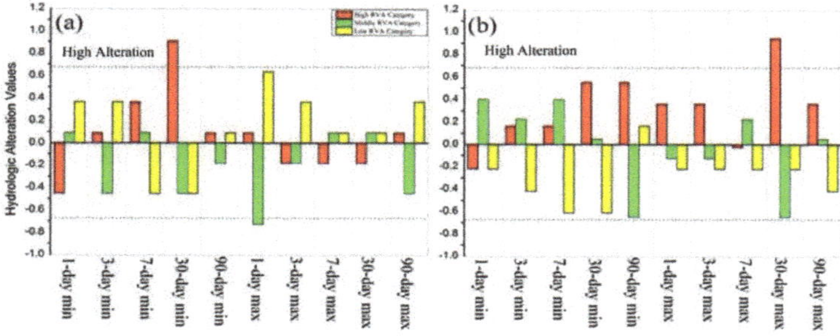

Figure 7. Hydrologic alteration of annual extreme streamflow in the JRW based on RVA method. (**a**) North River reach; (**b**) West River reach.

3.2.3. Frequency and Duration of High and Low Pulses

The frequency and duration of high and low pulses are quantified by the count and duration of the low pulse and high pulse, respectively. Except for the low pulse count, the other three indicators in Table 5 displayed the same trend after dams construction in the JRW. The frequency and duration of high and low pulses changed significantly in the last 44 years especially during the post-impact period in both two reaches of JRW. All of the indicators, especially for the low pulse count (101.8%) and the low pulse duration (−62.1%), changed significantly in the North River. The high pulse count decreased by 37.1%in the West River (Table 5). Most of the indicators increased significantly in low or middle level categories whereas decreased intensively in high-level category in the North river and West River, implying that the dams might effectively attenuate the high flow pulse event in both two reaches of JRW (Figure 8).

Table 5. The change of the frequency and duration of high and low pulses in the JRW (%).

Two Reaches	Low Pulse Count	Low Pulse Duration	High Pulse Count	High Pulse Duration
North River	101.8 **	−62.1 **	−21.4 *	19.2 *
West River	−15.5	−29.0	−37.1 **	6.7

Notes: ** $p < 0.005$; * $p < 0.05$.

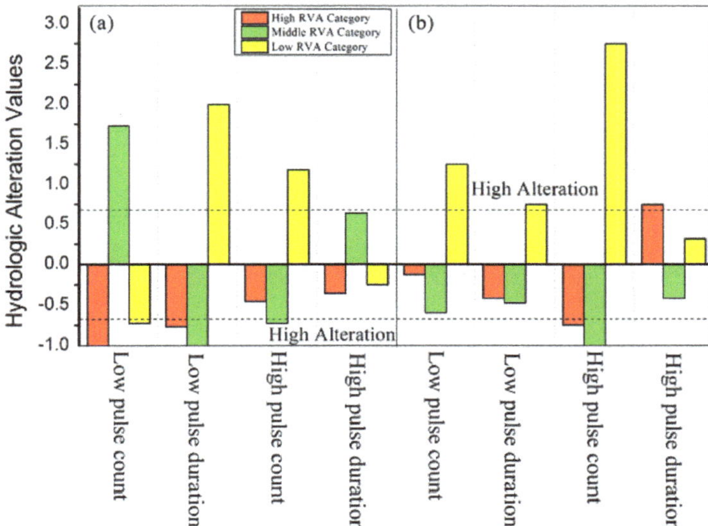

Figure 8. Hydrologic alteration frequency and duration of high and low pulses in the JRW based on RVA method. (**a**) North River reach; (**b**) West River reach.

3.2.4. Rate and Frequency of Streamflow Regime Change

The rate and frequency of the daily streamflow regime were quantified by the rise rate, fall rate and number of reversals (Table 6, Figure 9). The rate and frequency of streamflow regime changed intensively in both two reaches of JRW. The three indicators on the rate and frequency of the daily streamflow (i.e., rise rate, fall rate and number of reversals) revealed the similar trend in the post-impact period based on the observed data in the two reaches of JRW. The rise rate decreased by 26.9% and 61.0% in the North River and West River, respectively. The number of reversals increased by 40.7% and 46.4% in the North River and West River, respectively. The fall rate increased by 28.3% in the North River while increased by 0.8% in the West River (Table 6).

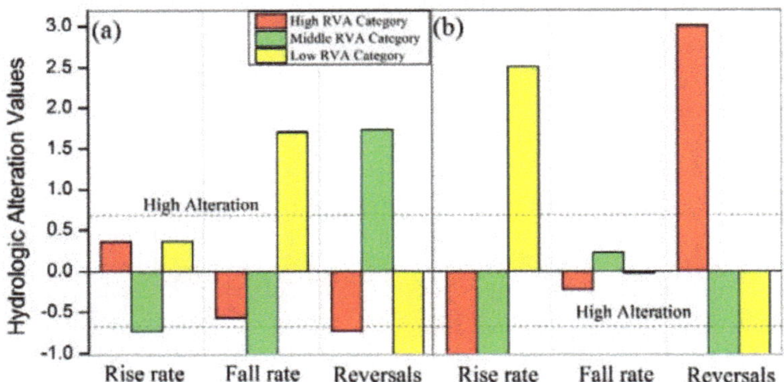

Figure 9. Hydrologic alteration of rate and frequency of the daily streamflow in the JRW based on RVA method. (**a**) North River reach; (**b**) West River reach.

Table 6. The change of the rate and frequency of the daily streamflow in the JRW (%).

Two Reaches	Rise Rate	Fall Rate	Number of Reversals
North River	−26.9 *	28.3 *	40.7 **
West River	−61.0 **	0.8	46.4 **

Notes: ** $p < 0.005$; * $p < 0.05$.

3.3. The Possible Suitable Range of Streamflow Regime

The suitable ranges of the daily streamflow are meaningful to be developed for environmental flow management in the JRW. Using Tennant method, the minimum environmental flow were 51.3 $m^3/s \cdot d$ and 24.5 $m^3/s \cdot d$ in the North River and West River. On the other hand, the suitable ranges of the daily streamflow were identified for each month using the RVA method (Figure 10). As shown in Figure 10, the monthly streamflow in August was higher than the natural streamflow boundary (i.e., RVA boundary), and the monthly streamflow in May was lower than the natural streamflow boundary in the North River. The similar result was found in the West River where the monthly streamflow in August, September, November, and December were higher than the natural streamflow boundary.

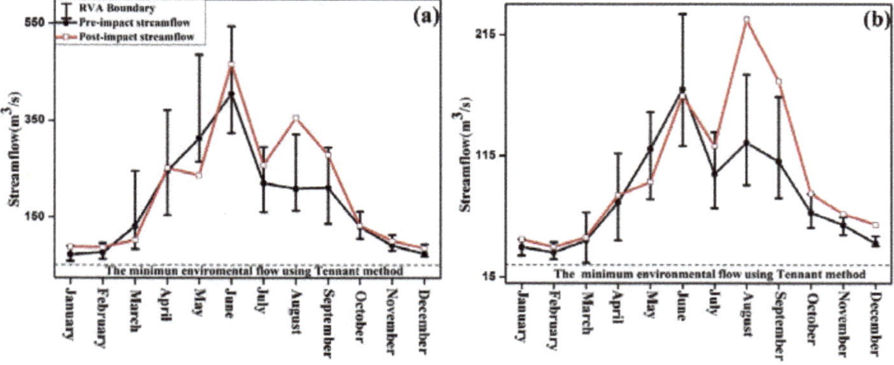

Figure 10. The suitable ranges of the monthly streamflow in the JRW. (**a**) North River reach; (**b**) West River reach.

The rate and frequency of streamflow changed largely due to dams construction in the JRW (Figure 9). From the perspective of the practical regulation, the suitable ranges for rise rate, fall rate and number of reversal in the JRW were identified based on the RVA method and the results are presented in Table 7. The fall rate in the North River and the rise rate in the West River were higher than the RVA boundary. Moreover, the number of reversal was higher than the RVA boundary in both reaches of the JRW.

Table 7. The suitable ranges of the rate and frequency in the JRW.

Hydrologic Parameters	North River			West River		
	RVA Range	**Pre-Impact**	**Post-Impact**	**RVA Range**	**Pre-Impact**	**Post-Impact**
Rise rate	16.0~33.5	24.0	17.6	10.1~19.8	15.2	5.9
Fall rate	−17.5~−12.2	−15.0	−19.2	−8.4~−5.5	−7.1	−7.2
Number of Reversals	103.0~137.0	119.9	168.7	98.3~108.8	106.7	156.2

Note: Lower and upper targets of RVA range are the 25th and 75th percentile value of the pre-impact hydrologic parameters including rise rate, fall rate, and number of reversals.

4. Discussion

4.1. Overall Impact of Streamflow Regulation on Flow Regimes

The overall streamflow regime change can be delineated through FDCs, which has been widely examined [30–32]. Our results revealed that the annual variability of daily streamflow decreased in the post-impact period in the North River and West River of JRW. The daily streamflow decreased in higher flow resulting from water storage regulated by dams while daily streamflow increased in lower flow resulting from water release regulated by dams in both two reaches of the JRW. The similar results also were found in the Han River and Yangtze River [8,31]. The dams tend to store more streamflow in the North River. For one thing, there are more dams and more length of river in the North River, which might slow down the downstream streamflow. For another, the downstream reservoirs or dams in the North River are designed to supply local residents with water for drinking, agricultural and industrial uses, which would make the downstream streamflow regulated.

4.2. Magnitude of Monthly Streamflow Regime

The effect of the dams on the monthly streamflow has been widely reported in the literature [13,16,29,61]. In this study, the mean daily streamflow increased during July to January while decreased during February to May after dam construction in the two reaches of JRW. This is similar with the previous findings in Huaihe River, Yangtze River and Langcang River [8,12,24]. Compared to the North River, the daily streamflow displayed a higher variability of monthly streamflow in the West River. This may be attributable to the smaller areas of watershed and less dams in the West River (Table 1). The effects of dam construction on water storage and release are greater for smallwatersheds than for large watersheds [62]. Generally, the impact of damconstruction in the JRW on monthly streamflow regime alteration is relatively complicated when combined with magnitude and inter-annual variability of monthly streamflow.

4.3. Magnitude, Duration and Timing of Extreme Streamflow Regime

Generally, baseflow is the most sensitive indicator to the streamflow regime change associated with dam construction [1,63–66]. The baseflow index (BFI) tends to increase after dam construction [3,8]. The increased BFI is the combined results of the increase in the 7-day minimum streamflow and the decrease in the annual streamflow due to dam construction [8]. Our study revealed that BFI value increased by 18.8 and 17.7%, respectively, in the post-impact period in the North River and West River. This suggests the general hydrologic regime alteration in the JRW due to dam construction. Our study also shows the potential discharge decreasing after construction in the JRW. The reduced discharged may be attributed to the result of increased evaporation or infiltration and loss to groundwater [13]. Our prior study shows that the 10-year average streamflowduring 2001–2010 in the North River and West Riverreaches decreased by 9.2% and 6.7%, respectively, compared to the average annual streamflow during 1967–2000, and the most important driving force for streamflow regime change in the JRW is related to population growth and economic development [3]. The potential discharge decrease after dam construction in the JRW might be related to annual streamflow decrease in recent years and the increasing water demanding in the watershed. Climatic variability in term of precipitation changes had limited effects on this situation, since that there is slightly increasing tendency of the annual precipitation in the West River while the precipitation in the North Riverseemed not to have changed in the past 50 years, as observed in our prior study [3].

The occurrence time of annual extreme condition can effectively reflect the seasonal variation of the hydrologic conditions [29]. The variation of the extreme events changed with geographic areas and climate patterns. Our study indicates the earlier occurrence time of extreme low streamflow event and later occurrence time of extreme high streamflow event after dams construction in the JRW, which is consistent with these asonal variation of the hydrologic conditions in other watersheds in South China [8,24,67]. This means the mode of dams regulation on streamflow regime in the JRW was similar to those in South China.

Previous studies in other watersheds show that minimum streamflow increased and maximum streamflow decreased after dam construction [5,46,68]. However, our study suggests the extreme low and high streamfow both decreased in the North River while both increased in the West River of the JRW. This is the results caused by the seasonal variation of streamflow regime changed in the JRW. The variability of the streamflow increased in February while decreased in July in the North River (Figure 4a), thus the minimum streamflow and maximum streamflow decreased in such two months in the North River. The situation was different in the West River, the variability of the streamflow decreased in March while increased in August (Figure 4b), therefore, the minimum streamflow and maximum streamflow increased in such two months respectively.

4.4. Pulse, Rate and Frequency Change

In response to theoperation of dams and reservoirs, the number of low pulse and high pulse will decrease [13,45,69]. Our study showed the count of low pulse in the North River increased abnormally (Table 5). This might be related to the increasing water demanding in the North River in recent years. As the water source for approximate 10 million resident in Xiamen, Zhangzhou and Longyan, large number of irrigation facilities was established, thereby inducing the low pulse count increased in the North River [3]. Similar observations were obtained in Huaihe River Basin and Tarim River Basinof China [24,47].

Due to the limited capacity of reservoirs, the duration of high pulse will change [29,70]. Our study also showed the increasing high pulse duration in the post-impact period in the JRW. This might be the result of the water storage and release regulated by the dams in JRW. The count of the high extreme high pulse decreased when water storage was operated by dams. This means the dams can store and attenuate all high flow pulse event. However, the dams tend to release the water when the water level exceeds the limited capacity of reservoir in the flood season.

The rate and frequency of streamflow can provide a measure of the rate and frequency of intra-annual environmental change and the decreased rise rate of hydrographs and increased in reversals after dams construction was widely observed [45,46]. Our study revealed that the rise rate decreased by 26.9% and 61.0% and number of reversals increased by 40.7% and 46.4% in the North River and West River, respectively. This is a byproduct of hydropower generation, wherein water is stored in the reservoir until sufficient head is attained to generate power efficiently, at which time the flow is rapidly released through the dam tubbiness [4]. Furthermore,the decreased rise rate and increased fall rate in the JRW also suggested the rate changing from high flow to low flow slowed down and the rate change from low flow to high flow speeded up, implying that the streamflow peak might be delayed (Table 3) and the variability of streamflow changed (Figure 4). Similar results were also found in Huaihe River, Yellow River, Taiwan and Great Plains [13,24,67,70].

4.5. Feasible Streamflow Regime in JRW

Streamflow regime is a primary determinant of structure and function of an aquatic and water quality in streams [3,24,26]. The magnitude of streamflow will influence the available habitat for organisms and the water quality in downstream. Too low streamflow will induce degrading water quality while too high streamflow will increase water level causing lost habitats. In order to protect native biodiversity and evolutionary potential of aquatic, riparian and wetland systems, the natural flow paradigm emphasizes the need to maintain or restore the range natural intra-annual and inter-annual variation of hydrologic regimes [24,39–41]. Our study revealed that the minimum flow requirement using Tennant method were close to the lower target identified using RVA method (Figure 10). Obviously, Tennant method only considers the lowest streamflow (i.e., the lowest streamflow in dry season) in the river but rarely considered the effective habitat quality at varying flow. Our study quantified reasonable range of the streamflow regime using RVA method in order to maintain the natural streamflow regime in the JRW.

Dams may not homogenize all river systems, but may move them outside the bounds of normal river function [16]. Our results suggest a suitable streamflow framework to generalize seasonal patterns inhydrologic alterations due to dam regulation. One the one hand, the suitable ranges in the flood season (e.g., in August) for both reaches of the JRW should be guaranteed by the reasonable regulation from downstream dams. On the other hand, more attention should be paid to the streamflow release or storage in the average season (e.g., May) and dry season (e.g., December) in the North River and West River, respectively.

Our study shows that the three indicators on the rate and frequency of the daily streamflow (i.e., rise rate, fall rate and number of reversals) were informative to delineate the critical role of dam construction on streamflow change. In this study, the fall rate in the North River and the rise rate in the West River were higher than the RVA boundary, which suggests that the dams in the JRW should release more water to get the natural targets of the rise rate and fall rate, so as to maintain the natural streamflow regime in the JRW. Moreover, the mean number of the reversal increased significantly and was higher than the RVA boundary, we therefore suggest that the dams in the JRW should decrease the frequency of store-release streamflow.

5. Conclusions

Flow duration curve analysis, indicators of hydrologic alteration, and range of variability approach were coupled in this study to evaluate the streamflow regime change induced by dam constructionin a full range in Jiulong River Watershed (JRW). The dailystreamflow decreased in higherflow resulting fromwater storage regulated by dams whiledaily streamflow increased in lower flow resulting fromwater release regulated by dams in both two reaches of the JRW. The dams in the North River tend to store more water while the dams in the West River tend to release more water. The mean dailystreamflow increased during July to January while decreased during February to May after dam construction in the two reaches of JRW. After dam construction, the monthly streamflow changed more significantly and higher variability of monthly streamflow was observed in the West River than in the North River. The homeogenizing variability of monthly streamflow exhibited in both two reaches of JRW. The earlier occurrence time of extreme low streamflow event and later occurrence time of extreme high streamflow event after dams construction. The extreme low and high streamfow both decreased in the North River while both increased in the West River of the JRW. All of the indicators especially for the low pulse count (101.8%) and the low pulse duration (−62.1%) changed significantly in the North River. The high pulse count decreased by 37.1% in the West River and the count of low pulse increased abnormally in the North River. The high pulse duration in the post-impact period increased in both two reaches of the JRW. The rise rate decreased by 26.9% and 61.0%,and number of reversals increased by 40.7% and 46.4% in the North River and West River, respectively. The fall rate increased by 28.3% in the North River.

Reasonable range of streamflow regime in terms of magnitude, rate, and frequency was identified using RVA method to sustain environmental flow management. More attention should be paid to the streamflow release or storage regulated by dams in May, August and December in the JRW. The dams in the JRW should release more water and decrease the frequency of store-releases treamflow. This research advances our understanding of hydrologic impact of small and medium dams in the medium-sized basin in China.

Acknowledgments: This study was supported by the Natural National Science Foundation of China (Grant No. 41471154), the Fundamental Research Funds for the Xiamen Universities (Grant No. 20720150129), and the National Science and Technology Support Program (Grant No. 2013BAC06B01). Anonymous reviewers supplied constructive feedback that helped to improve this paper.

Author Contributions: Jinliang Huang and Zhenyu Zhang conceived and designed the experiments; Zhenyu Zhang performed the experiments; Zhenyu Zhang and Yaling Huang analyzed the data; Zhenyu Zhang and Yaling Huang contributed reagents/materials/analysis tools; Zhenyu Zhang and Jinliang Huang wrote the paper.

Conflicts of Interest: The authors declare no conflict of interest.

References

1. Poff, N.L.; Allan, J.D.; Bain, M.B.; Karr, J.R.; Prestegaard, K.L.; Richter, B.D.; Spark, R.E.; Stromberg, J.C. The natural flow regime: A paradigm for river conservation and restoration. *BioScience* **1997**, *47*, 769–784. [CrossRef]
2. Ripl, W. Water: The bloodstream of the biosphere. *Philos. Trans. R. Soc. B* **2003**, *358*, 1921–1934. [CrossRef] [PubMed]
3. Huang, J.L.; Zhang, Z.Y.; Feng, Y.; Hong, H.S. Hydrologic response to climate change and human activities in a subtropical coastal watershed of southeast China. *Reg. Environ. Chang.* **2013**, *13*, 1195–1210. [CrossRef]
4. Richter, B.D.; Andrew, T.; Warner, J.L.; Meyer, K. A collaborative and adaptive process for developing environmental flow recommendations. *River Res. Appl.* **2006**, *22*, 297–318. [CrossRef]
5. Poff, N.L.; Olden, J.D.; Merritt, D.M.; Pepin, D.M. Homogenization of regional river dynamics by dams and global biodiversity implications. *PNAS* **2007**, *104*, 5732–5737. [CrossRef] [PubMed]
6. Li, Z.W.; Zhang, Y.K. Multi-scale entropy analysis of Mississippi River flow. *Stoch. Environ. Res. Risk Assess.* **2008**, *22*, 507–512. [CrossRef]
7. Chen, Y.Q.; Yang, T.; Xu, C.Y.; Zhang, Q.; Chen, X.; Hao, Z.C. Hydrologic alteration along the middle and upper East River (Dongjiang) basin, South China: A visually enhanced mining on the results of RVA method. *Stoch. Environ. Res. Risk Assess.* **2010**, *24*, 9–18. [CrossRef]
8. Gao, B.; Yang, D.; Zhao, T.T.G.; Yang, H.B. Changes in the eco-flow metrics of the upper Yangtze River from 1961 to 2008. *J. Hydrol.* **2012**, *448–449*, 30–38. [CrossRef]
9. Flug, M.; Seiz, H.L.H.; Scott, J.F. Multicriteria decision analysis applied to Glen Canyon dam. *J. Water Resour. Plan. Manag.* **2000**, *126*, 270–276. [CrossRef]
10. Cowell, C.M.; Stoudt, R.T. Dam-induced modifications to upper Allegheny River streamflow patterns and their biodiversity implications. *J. Am. Water Resour. Assoc.* **2002**, *38*, 187–196. [CrossRef]
11. Shiau, J.T.; Wu, F.C. Assessment of hydrologic alterations caused by Chi-Chi diversion weir in Chou-Shui Creek, Taiwan: Opportunities for restoring natural flow conditions. *River Res. Appl.* **2004**, *20*, 401–412. [CrossRef]
12. Zhao, Q.H.; Liu, S.L.; Deng, L.; Dong, S.K.; Wang, C.; Yang, Z.F.; Yang, J.J. Landscape change and hydrologic alteration associated with dam construction. *Int. J. Appl. Earth Obs. Geoinf.* **2012**, *16*, 17–26. [CrossRef]
13. Costigan, K.H.; Daniels, M.D. Damming the prairie: Human alteration of Great Plains river regimes. *J. Hydrol.* **2012**, *444–445*, 90–99. [CrossRef]
14. Losos, E.; Hayes, J.; Phillips, A.; Wilcove, D.; Alkire, C. Taxpayer-subsidized resource extraction harm species: Double jeopardy. *Bioscience* **1995**, *45*, 446–455. [CrossRef]

15. Trush, W.J.; McBain, S.M.; Leopold, L.B. Attributes of an alluvial river and their relation to water policy and management. *PNAS* **2000**, *97*, 11858–11863. [CrossRef] [PubMed]
16. McManamay, R.Y.; Orth, D.J.; Dolloff, C.A. Revisiting the homogenization of dammed rivers in the southeastern US. *J. Hydrol.* **2012**, *424–425*, 217–237. [CrossRef]
17. Bunn, S.E.; Arthington, A.H. Basic principles and ecological consequences of altered flow regimes for aquatic biodiversity. *Environ. Manag.* **2002**, *30*, 492–507. [CrossRef]
18. Moyle, P.B.; Mount, J.F. Homogenous rivers, homogeneous fauna. *PNAS* **2007**, *104*, 5711–5712. [CrossRef] [PubMed]
19. Jansson, R.; Nilsson, C.; Renofalt, B. Fragmentation of riparian floras in rivers with multiple dams. *Ecology* **2000**, *81*, 899–903. [CrossRef]
20. Chovanec, A.; Schiemer, F.; Waibacher, H.; Spolwind, R. Rehabilitation of a heavily modified river section of the Danube in Vienna (Austria): Biological assessment of landscape linkages on different scales. *Int. Rev. Hydrobiol.* **2002**, *87*, 183–195. [CrossRef]
21. Tockner, K.; Stanford, J.A. Riverine flood plains: Present state and future trends. *Environ. Conserv.* **2002**, *29*, 308–330. [CrossRef]
22. Olden, J.D.; Poff, N.L. Redundancy and the choice of hydrologic indices for characterizing stremflow regimes. *River Res. Appl.* **2003**, *19*, 101–121. [CrossRef]
23. Dudgeon, D. River rehabilitation for conservation of fish biodiversity in monsoonal Asia. *Ecol. Soc.* **2005**, *10*, 15.
24. Hu, W.W.; Wang, G.X.; Deng, W.; Li, S.N. The influence of dams on ecohydrological conditions in the Huaihe River basin, China. *Ecol. Eng.* **2008**, *33*, 233–241. [CrossRef]
25. Zeilhofer, P.; de Moura, R.M. Hydrological changes in the northern Pantanal caused by the Manso dam: Impact analysis and suggestions for mitigation. *Ecol. Eng.* **2009**, *35*, 105–117. [CrossRef]
26. Yang, Z.; Yan, Y.; Liu, Q. Assessment of the flow regime alterations in the lower Yellow River, China. *Ecol. Inform.* **2012**, *10*, 56–64. [CrossRef]
27. Richter, B.D.; Baumgarter, J.V.; Powell, J.; Braun, D.P. A method for assessing hydrologic alteration within ecosystems. *Conserv. Biol.* **1996**, *10*, 1163–1174. [CrossRef]
28. Mathews, R.; Richter, B.D. Application of the indicators of hydrologic alteration software in environmental flow setting. *J. Am. Water Resour. Assoc.* **2007**, *43*, 1400–1413. [CrossRef]
29. Lian, Y.Q.; You, J.Y.; Sparks, R.; Demissie, M. Impact of human activities to hydrologic alterations on the Illiois River. *J. Hydrol. Eng.* **2012**, *17*, 537–546. [CrossRef]
30. Vogel, R.M.; Sieber, J.; Archfield, S.A.; Smith, M.P.; Apse, C.D.; Huber-Lee, A. Relations among storage, yield and instream flow. *Water Resour. Res.* **2007**, *43*, W05403. [CrossRef]
31. Kim, N.; Lee, J.; Kim, J. Assessment of flow regulation effects by dams in the Han River, Korea, on the downstream flow regimes using SWAT. *J. Water Resour. Plan. Manag.* **2012**, *138*, 24–35. [CrossRef]
32. Brown, A.E.; Western, A.W.; McMahon, T.A.; Zhang, L. Impact of forest cover changes on annual stramflow and flow duration curves. *J. Hydrol.* **2013**, *483*, 39–50. [CrossRef]
33. Acreman, M.C.; Ferguson, A.J.D. Environmental flows and the European Water Framework Directive. *Freshw. Biol.* **2010**, *55*, 32–48. [CrossRef]
34. Petts, G.E. Instream flow science for sustainable reiver management. *J. Am. Water Resour. Assoc.* **2009**, *45*, 1071–1086. [CrossRef]
35. Jowett, I.G.; Biggs, B.J.F. Application of the 'natural flow paradigm' in a New Zealand context. *River Res. Appl.* **2009**, *25*, 1126–1135. [CrossRef]
36. Poff, N.L.; Richter, B.; Arthington, A.; Bunn, S.E.; Naiman, R.J.; Kendy, E.; Acreman, M.; Apse, C.; Bledsoe, B.P.; Freeman, M.; et al. The ecological limits of hydrologic alteration (ELOHA): A new framework for developing regional environmental flow standards. *Freshw. Biol.* **2010**, *55*, 147–170. [CrossRef]
37. Halleraker, J.H.; Sundt, H.; Alfredsen, K.T.; Dangelmaier, G. Application of multiscale environmental flow methodologies as tools for optimized management of a Norwegian regulated nation salmon water course. *River Res. Appl.* **2007**, *23*, 493–510. [CrossRef]
38. Jager, H.I.; Smith, B.T. Sustainable reservoir operation: Can we generate hydropower and preserve ecosystem values? *River Res. Appl.* **2008**, *24*, 340–352. [CrossRef]
39. Richter, B.D.; Thomas, G.A. Restoring environmental flows by modifyingdam operations. *Ecol. Soc.* **2007**, *12*, 12.

40. Naiman, R.J.; Latterell, J.J.; Pettit, N.E.; Olden, J.D. Flow variability and thebiophysical vitality of river systems. *C. R. Geosci.* **2008**, *340*, 629–643. [CrossRef]
41. Yin, X.A.; Yang, Z.F.; Petts, G.E. Reservoir operating rules to sustain environmental flows in regulated rivers. *Water Resour. Res.* **2011**, *47*, W08509. [CrossRef]
42. Yin, X.A.; Yang, Z.F.; Petts, G.E. Optimizing environmental flows below dams. *River Res. Appl.* **2012**, *28*, 703–716. [CrossRef]
43. Arthington, A.H.; Bunn, S.E.; Poff, N.L.; Naiman, R.J. The challenge of providing environmental flow rules to sustain river systems. *Ecol. Appl.* **2006**, *16*, 1311–1318. [CrossRef]
44. Richter, B.D. Re-thinking environmental flows: From allocations and reserves to sustainability boundaries. *River Res. Appl.* **2010**, *26*, 1052–1063. [CrossRef]
45. Magilligan, F.J.; Nislow, K.H. Long-term changes in regional hydrologic regime following impoundment in a humid-climate watershed. *J. Am. Water Resour. Assoc.* **2001**, *37*, 1551–1569. [CrossRef]
46. Pyron, M.; Neumann, K. Hydrologic alterations in theWabash river watershed, USA. *River Res. Appl.* **2008**, *24*, 1175–1184. [CrossRef]
47. Zhang, X.Q.; Chen, Y.N.; Li, W.H.; Yu, Y.; Sun, Z.H. Restoration of the lower reaches of the Tarim River in China. *Reg. Environ. Chang.* **2013**, *13*, 1021–1029. [CrossRef]
48. Dai, Z.J.; Liu, J.T. Impact of large dams on downstream fluvial sedimentation: An example of the Three Gorges Dam (TGD) on the Changjiang (Yangtze River). *J. Hydrol.* **2013**, *480*, 10–18. [CrossRef]
49. Räsänen, T.A.; Koponen, J.; Lauri, H.; Kummu, M. Downstream hydrological impacts of hydropower development in the Upper Mekong Basin. *Water Resour. Manag.* **2012**, *26*, 3495–3513. [CrossRef]
50. Sun, Z.; Huang, Q.; Opp, C.; Hennig, T.; Marold, U. Impacts and Implications of major changes caused by the Three Gorges Dam in the middle reaches of the Yangtze River, China. *Water Resour. Manag.* **2012**, *26*, 3367–3378. [CrossRef]
51. Cai, B.M.; Zhang, B.; Bi, J.; Zhang, W. Energy's thirst for water in China. *Environ. Sci. Technol.* **2014**, *48*, 11760–11768. [CrossRef] [PubMed]
52. Zhang, Q.; Zhou, Y.; Singh, V.P.; Chen, X.H. The Influence of dam and lakes on the Yangtze River streamflow: Long-range correlation and complexity analyses. *Hydrol. Process.* **2012**, *26*, 436–444. [CrossRef]
53. Li, S.; Xiong, L.H.; Dong, L.H.; Zhang, J. Effects of Three Gorges reservoir on the hydrologic droughts at the downstream Yichang station during 2003–2011. *Hydrol. Process.* **2013**, *27*, 3891–3993. [CrossRef]
54. World Commissionon Dams (WCD). *Dams and Development: A New Framework for Decision-Making*; Report of the World Commissionon Dams; Earthscan Publishing: London, UK, 2000.
55. Deitch, M.J.; Merenlender, A.M.; Feirer, S. Cumulative effects of small reservoirs and streamflow in Northern coastal California catchments. *Water Resour. Manag.* **2013**, *27*, 5101–5118. [CrossRef]
56. World Wide Fund for Nature (WWF). WWF's Dams Initiative: Rivers at Risk. Available online: http://wwf.panda.org/what_we_do/footprint/water/dams_initiative/ (accessed on 1 July 2012).
57. Zhang, Z.; Huang, J.; Huang, Y.; Hong, H. Streamflow variability response to climate change and cascade dams development in a coastal China watershed. *Estuar. Coast. Shelf Sci.* **2015**, *166*, 209–217. [CrossRef]
58. Richter, B.D.; Baumgartner, J.V.; Braun, D.P.; Powell, J. A spatial assessment of hydrologic alteration within a river network. *Regul. Rivers Res. Manag.* **1998**, *14*, 329–340. [CrossRef]
59. Baxter, G. River utilization and the preservation of migratory fish life. *Proc. Inst. Civ. Eng.* **1961**, *18*, 225–244.
60. Tennant, D.L. Instream flow regimens for fish, wildlife, recreation and related environmental resources. *Fisheries* **1976**, *1*, 6–10. [CrossRef]
61. Majhi, I.; Yang, D.Q. Streamflow characteristics and changes in Kolyma basin in Siberia. *J. Hydrometeorol.* **2008**, *9*, 267–279. [CrossRef]
62. Matteau, M.; Assani, A.A.; Mesfioui, M. Application of multivariate statistical analysis methods to the dams hydrologic impact studies. *J. Hydrol.* **2009**, *371*, 120–128. [CrossRef]
63. Hirsch, R.M.; Walker, J.F.; Day, J.C.; Kallio, R. *The Influence of Man on Hydrological Systems*; Wolman, M.G., Riggs, H.C., Eds.; Surface Water Hydrology, Geological Society of America, Geological Society of America: Boulder, CO, USA, 1990.
64. Stanford, J.A.; Ward, J.V.; Liss, W.J.; Frissell, C.A.; Williams, R.N.; Lichatowich, J.A.; Coutant, C.C. A general protocol for restoration of regulated rivers. *Regul. Rivers Res. Manag.* **1996**, *12*, 391–413. [CrossRef]
65. Graf, W.L. Damage control: Restoring the physical integrity of America's rivers. *Ann. Assoc. Am. Geogr.* **2001**, *91*, 1–27. [CrossRef]

66. Baker, D.B.; Richards, R.P.; Loftus, T.T.; Kramer, J.W. A new flashiness index: Characteristics and applications to Midwestern rivers and streams. *J. Am. Water Resour. Assoc.* **2004**, *40*, 503–522. [CrossRef]
67. Shiau, J.T.; Wu, F.C. Feasible diversion and instream flow release using range of variability approach. *J. Water Res. Plan. Manag.* **2004**, *130*, 395–403. [CrossRef]
68. Magilligan, F.J.; Nislow, K.H. Changes in hydrologic regime by dams. *Geomorphology* **2005**, *71*, 61–78. [CrossRef]
69. Wang, H.; Yang, Z.; Satio, Y.; Liu, J.P.; Sun, X. Interannual and seasonal variation of the Huanghe (Yellow River) water discharge over the past 50 years: Connections to impacts from ENSO events and dams. *Glob. Planet. Chang.* **2006**, *50*, 212–225. [CrossRef]
70. Yang, T.; Zhang, Q.; Chen, Y.Q.; Tao, X.; Xu, C.Y.; Chen, X. A spatial assessment of hydrologic alteration caused by dam construction in the middle and lower Yellow River, China. *Hydrol. Process.* **2008**, *22*, 3829–3843. [CrossRef]

Article

Flooding in Delta Areas under Changing Climate: Response of Design Flood Level to Non-Stationarity in Both Inflow Floods and High Tides in South China

Yihan Tang [1,3], Qizhong Guo [2], Chengjia Su [1,3] and Xiaohong Chen [1,3,*]

1 Center for Water Resources and Environment, Sun Yat-sen University, Guangzhou 510275, China; yihan.olivia.tang@outlook.com (Y.T.); suchengj@mail2.sysu.edu.cn (C.S.)

2 Department of Civil and Environmental Engineering, Rutgers, The State University of New Jersey, Piscataway, NJ 08854, USA; qguo@soe.rutgers.edu

3 Guangdong Engineering Technology Research Center of Water Security Regulation and Control for Southern China, Sun Yat-sen University, Guangzhou 510275, China

* Correspondence: eescxh@mail.sysu.edu.cn; Tel.: +86-20-84114575

Received: 30 March 2017; Accepted: 22 June 2017; Published: 11 July 2017

Abstract: Climate change has led to non-stationarity in recorded floods all over the world. Although previous studies have widely discussed the design error caused by non-stationarity, most of them explored basins with closed catchment areas. The response of flood level to nonstationary inflow floods and high tidal levels in deltas with a dense river network has hardly been mentioned. Delta areas are extremely vulnerable to floods. To establish reliable standards for flood protection in delta areas, it is crucial to investigate the response of flood level to nonstationary inflow floods and high tidal levels. Pearl River Delta (PRD), the largest delta in South China, was selected as the study area. A theoretical framework was developed to quantify the response of flood level to nonstationary inflow floods and the tidal level. When the non-stationarity was ignored, error up to 18% was found in 100-year design inflow floods and up to 14% in 100-year design tidal level. Meanwhile, flood level in areas that were $\leq$22 km away from the outlets mainly responded to the nonstationary tidal level, and that $\geq$45 km to the nonstationary inflow floods. This study will support research on the non-stationarity of floods in delta areas.

Keywords: non-stationarity; flood level; delta flood; Pearl River Delta

---

1. Introduction

Since the early twentieth century, climate change has altered hydrological cycles all over the world. Flood records are no longer stationary [1–5]. In particular, delta areas have experienced intensified flooding due to the combined impacts of the inflow floods from the upstream watershed and the rising high tidal level induced by sea-level rise (SLR). Meanwhile, livable climate and convenient sea transportation have contributed to increasing population and economy in delta areas around the world [6,7]. Thus, delta zones have become extremely vulnerable to floods. It is important to understand the response of flood level to nonstationary inflow floods and high tidal levels in delta areas, especially ones with a dense river network.

There were some previous studies discussing the variation in design tidal levels under the impact of SLR. Trends in recording series, as well as the correlation between these trends and other climate factors, were explored using sophisticated extreme value analysis methodology [8]. To calculate reliable at-site frequency of the increasing tidal level, many previous researchers generated a stationary series before carrying out frequency analysis. Some restored the stationary time series by removing the increment caused by SLR [9–11], while some tried to generate stationary samples by duplicating the characteristics of historical tidal processes, which were assumed to be stationary [12]. Other researchers assumed that the rate of high tidal levels was time-dependent, and conducted frequency analysis on the recorded series with a time-correlated variable [8–13].

Design flood levels in delta areas were also analyzed by statistical methods in past studies. Joint distribution was the most widely used approach to statistically estimate flood levels in delta areas, which was simultaneously affected by multiple forcing factors like rainstorms, storm surges, upstream floods, and high tides. To take non-stationarity caused by environmental changes into concern, joint distribution incorporated the nonstationary characteristics of changing factors into marginal distribution functions. However, in this way, both the difficulty of parameter estimation and the uncertainty of results were significantly increased [14]. Another commonly used method for delta flood protection is to first calculate two design flood levels, under the impact of either the inflow floods or the high tidal level, and choose the higher one as the design value [15]. This method is simple and convenient, but it neglects the interconnection between inflow floods and high tides.

This research aims to estimate the response of flood level to the impact of inflow floods and downstream tidal level, when the time series of these two factors are both nonstationary. To achieve this purpose, Pearl River Delta (PRD) in South China was selected as the study area. This paper is arranged as follows. In Section 2, a brief introduction of the study area, dataset and methodology is provided. In Section 3, the response of the flood level to the nonstationary inflow floods and the high tidal level is estimated, and its spatial pattern is discussed. Conclusions are made in Section 4. The investigation in PRD can be used as reference for estimations of the flood level response in other delta areas with similarly complicated environmental conditions.

2. Study Area and Methods

2.1. Study Area and Dataset

Pearl River Delta, which covers an area of 39,380 km^2, is the largest delta in South China (Figure 1). It has a humid subtropical climate. The monsoon period in this area lasts from April to September. Due to the terrain altitude, the dense river network develops from the northwest to the southeast. Water flow in PRD comes from West River and North River, and all the water in the delta flows into Pearl River Estuary through multiple outlets. Based on its source of inflow floods water, PRD area can be further divided into a West River sub-delta, and a North River sub-delta. Floods in two upstream river basins mainly occur between June and August. Meanwhile, extreme tides in the estuary, induced by typhoons or storm surges, mainly occur around September [16].

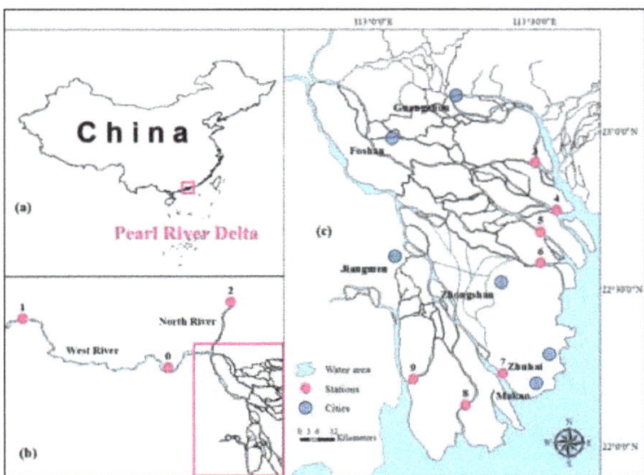

Figure 1. Map of Pearl River Delta and locations of gauge stations: (**a**) location of Pearl River Delta on the map of China; (**b**) gauge stations of upstream rivers; (**c**) a closer view of Pearl River Delta.

PRD is the megaregion where Macao is located [17]. In 2014, the population of cities in PRD surpassed that of Chicago and it was ranked in the top 30 in the world [18]. In 2015, the total income of only two cities in PRD, i.e., Guangzhou and Foshan, reached $20.6 million [19]. With its fast growth in economy and population, PRD is extremely vulnerable to flood risk.

In this research, recorded peak flood volumes in the flood season, i.e., from June to August, in Wuzhou and Shijiao were used. So were the high tidal levels in seven outlet stations (Figure 1, Table 1). All the data were collected during the period from 1958 to 2011.

Table 1. Gauge stations and trend detection.

River Basin	No.	Gauge Station	Channel Name	Mann–Kendall
West River	0	Gaoyao	-	-
	1	Wuzhou	-	Upward (z = 1.66)
North River	2	Shijiao	-	No (z = 0.38)
Pearl River Delta Outlets	3	Sanshakou	Humen	No (z = 1.43)
	4	Nansha	Jiaomen	No (z = −0.37)
	5	Wanqinshaxi	Hongqimen	No (z = 1.57)
	6	Hengmen	Hengmen	Upward (z = 2.96)
	7	Denglongshan	Modaomen	Upward (z = 1.96)
	8	Huangjin	Jitimen	Upward (z = 2.65)
	9	Xipaotai	Hutiaomen	Upward (z = 2.44)

Note: When $|Z| > 1.65$, the trend is 90% significant.

2.2. Methodology

In this research, a methodology was developed to estimate the impact of the nonstationary inflow floods and the high tidal level on flood level. First, a time-varying moments (TVM) model was applied to evaluate the design inflow floods from the upstream river basins and the high tidal levels around the delta estuaries. Then, the influence of the nonstationary inflow floods and the high tidal level on flood level was estimated through a one-dimensional hydrodynamic model using the TVM-estimated design inflow floods and the high tidal level as its inputs. Four scenarios were created by combining the nonstationary inflow floods and the high tidal level, and then the nonstationary water level in PRD was estimated under these scenarios. The whole schedule of this combined methodology is shown in the flowchart presented in Figure 2. All the parts that are considered nonstationary are highlighted in blue.

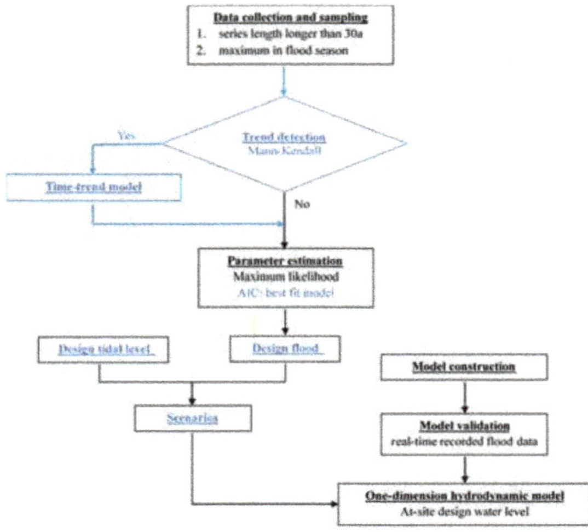

Figure 2. Flow chart of the methodology. Parts highlighted in blue considered non-stationarity in time series.

2.2.1. Time-Varying Moments Estimating Nonstationary Inflow Floods and High Tidal Levels

The time-varying moments method, which has been widely applied in the statistical estimation of nonstationary peak flood volume, was used in this investigation to calculate the nonstationary design value of inflow floods and high tidal level [20]. In this method, to incorporate the non-stationarity, the first two moments of the time series were initially assumed to have trends that are either linear (L) or parabolic (P) [21,22].

Four conditions were generated by assembling the L or P trend in the first two moments, labeled A to D [22]: (1) when only the first moment, i.e., the mean of the time series, is time-dependent, whether the trend is L or P, the condition is named A; (2) when only the second moment, i.e., the standard deviation of the time series, is time-dependent, whether the trend is L or P, the condition is named B; (3) when both the mean and the standard deviation have time-dependent trends in, and the standard deviation is a product of, the mean and a constant value, i.e., the coefficient of variation (Cv), the condition is named C; (4) when both the first two moments have trends, but they are not directly correlated, the condition is named D. By combining Scenarios A–D and two types of trends, i.e., L and P, eight time-trend models were generated. All models are listed in Table 2, where m' represents the mean of the nonstationary time series that is defined in relation to both the mean of a stationary time series, m, and time, t. Likewise, σ' denotes the nonstationary standard deviation and is considered to be associated with both the stationary standard deviation σ and time, t. In all functions, a_m, a_σ, b_m, and b_σ denote the constants.

Table 2. Time trend models of the first two moments.

Probability Distribution Function	Time Trend Model	Time-Varying Model	m	σ
P3	AL	P3-AL	$m' = m + a_m t$	$\sigma' = \sigma$
	BL	P3-BL	$m' = m$	$\sigma' = \sigma + a_\sigma t$
	CL	P3-CL	$m' = m + a_m t$	$\sigma' = C_v \cdot m$
	DL	P3-DL	$m' = m + a_m t$	$\sigma' = \sigma + a_\sigma t$
	AP	P3-AP	$m' = m + a_m t + b_m t^2$	$\sigma' = \sigma$
	BP	P3-BP	$m' = m$	$\sigma' = \sigma + a_\sigma t + b_\sigma t^2$
	CP	P3-CP	$m' = m + a_m t + b_m t^2$	$\sigma' = C_v \cdot m$
	DP	P3-DP	$m' = m + a_m t + b_m t^2$	$\sigma' = \sigma + a_\sigma t + b_\sigma t^2$

Note: P3 denotes the Pearson type III.

According to studies carried out in the 1990s, Pearson type III performs the best among all commonly used probability distribution functions for either the flood flow or the high tidal level in PRD [23–30]. Therefore, Pearson type III was selected as the probability density function (PDF) used in design value estimation in this research. Based on the eight time trend models, parameters of Pearson type III were all defined to be time-dependent variables. All these parameters were estimated by the maximum likelihood method, where t was used as a variable. The goodness of fit of PDFs, i.e., Pearson type III with parameters of different time trend models, was tested by the Akaike Information Criterion (AIC).

2.2.2. One-Dimensional River Network Hydrodynamic Model

The one-dimensional hydrodynamic model has been widely used in flood level estimation. It is comparatively more accurate in representing the cross-sectional area of channels and requires only a small amount of field data to set up the model. Also, it can be quickly set up, and the computation is fast [31]. Therefore, the one-dimensional hydrodynamic model was chosen as the basic tool to estimate the flood level in this research.

1. Model Setup and Input Data

The one-dimensional hydrodynamic model used in the present study was developed from the one created by the Center for Water Resources and Environment (2015), and the whole model was set up in Fortran [32]. This model adopted the Saint-Venant continuity equation and the longitudinal momentum balance equation as the governing equations. The implicitly weighted four-point Preissmann scheme [33] was used to solve these governing equations.

The model set the time-step as 10 min, and a space lag from 500 to 2500 m. In all, the model included 196 channels and 144 nodes (Figure 3). The upstream inflow from Gaoyao (West River) and Shijiao (North River), as well as the tidal levels in seven outlets (Figure 1, Table 1), were set as its boundaries, and the recorded data in these stations were used as the input to run the model.

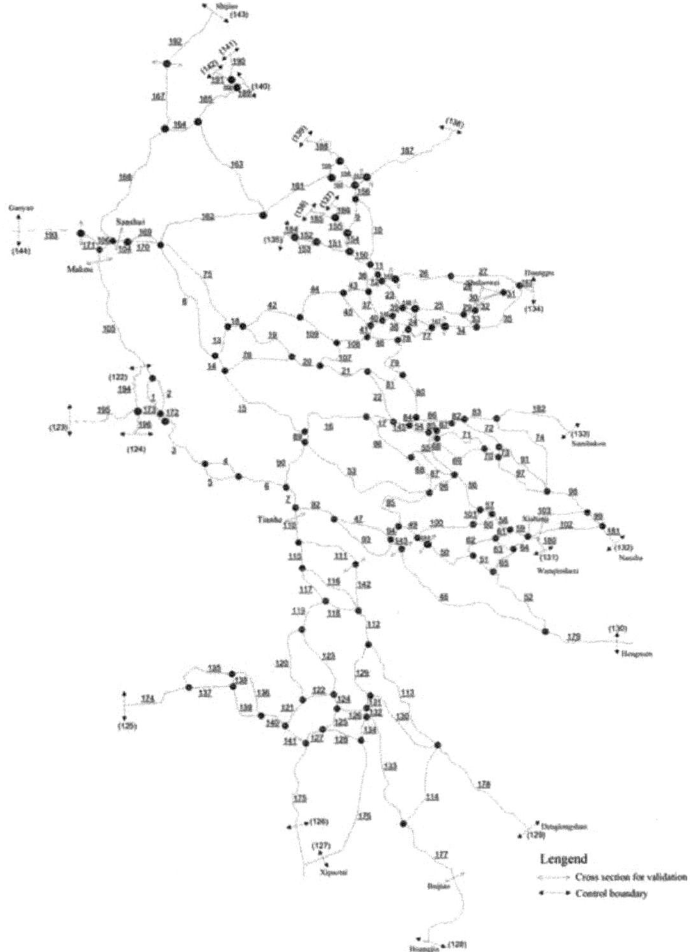

Figure 3. Overview of the simplified one-dimensional hydrodynamic model of PRD.

Conducting simultaneous field survey over a large basin like PRD region is extremely time-consuming and expensive. Meanwhile, although large-scale field measurement has been carried out by the Guangdong

Hydraulic Bureau every five years, the data are beyond our access. Owing to the limitations of survey data, this research applied the topographic data collected in the flood season of 1998 to develop a 1D model. All topographic maps were obtained from the Guangdong Hydraulic Bureau, in a measuring scale of 1:5000. The hourly flood flow and water level were simultaneously detected at 40 stations during a typical flood in PRD. Data collected from 12:00 p.m. on 25 June 1998 to 12:00 p.m. on 28 June 1998 at hydrological stations in Gaoyao and Shijiao were used to drive the model and calibrate the parameters. The fluvial water level at each sampling frequency was linearly interpolated to the model time-step.

2. Model Validation

The hydrodynamic model was validated before estimating the flood level at PRD stations. The coefficient of riverbed roughness was first set between 0.016 and 0.035 [33]. Next, after the model had been running for 30 days, the parameters were updated by the results obtained on day 30. The aforementioned procedures were repeated more than 10 times before the simulation results stabilized. The model parameters were finally set when the difference between the simulation results of day 1 and day 30 remained within 1%.

To assess the performance of the model, measured and model-estimated amplitudes for eight constituents were compared in six stations, i.e., Makou, Sanshui, Tianhe, Baijiao, Xiaheng, and Shaluowei, and the result was shown in Figure 4. Difference bands were both plotted at ±0.025 and ±0.05 m, with an overall RMSE of 6.87 cm. For each of these six stations, recorded and estimated flood levels were also compared. The average absolute errors in the six stations were below 0.05 m, with a minimum of 0.001 m at Baijiao. To further measure the model efficiency, an index (α) was applied, was calculated by the following equations:

$$r_i = \frac{Z' - Z_0}{Z_0} \times 100\% \ (i = 1, 2, 3, \ldots, 72) \tag{1}$$

$$A_i = \begin{cases} 0 & (5 \leq r_i) \\ 1 & (r_i < 5) \end{cases} \tag{2}$$

$$\alpha = \frac{\sum_{i=1}^{72} A_i}{72} \times 100\%, \tag{3}$$

where Z' is the estimated flood level; Z_0, the recorded flood level; and i, the number of hours. When $\alpha \geq 75\%$, the simulation of the model was predicted to be valid. The value of α is 95% in Makou, 83% in Sanshui, 90% in Tianhe, 92% in Baijiao, 81% in Xiaheng, and 92% in Shaluowei. All αs are above 80%, and the model has been proved to be valid.

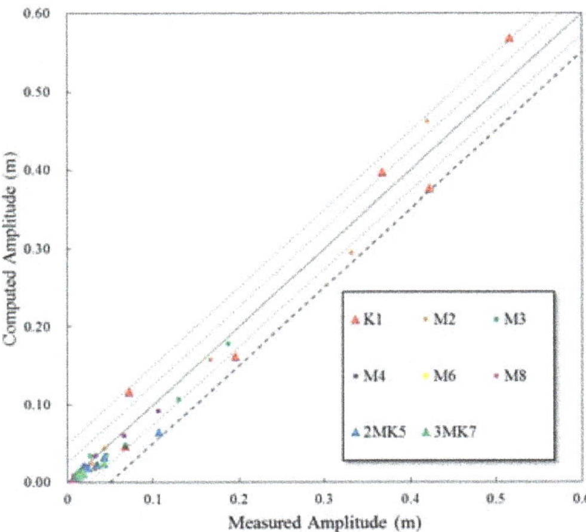

Figure 4. Comparison of harmonic constituent amplitudes, with difference bands located at 0.025 and 0.05 m.

3. Results and Discussion

3.1. Non-Stationarity in the Inflow Floods and the Downstream Tidal Level

3.1.1. Increasing Inflow Floods

Trends in the inflow floods of PRD were estimated by Mann–Kendall method [34,35]. As is shown in Table 1, inflow floods from West River had significantly increased since 1958. Since 1950, global warming and atmosphere circulation abnormities have led to fewer but more intense tropical cyclones in West River [36–38]. Meanwhile, the extreme precipitations (over 99 percentile) in West River have increased by more than 45 mm since 1959 [39]. As a result, the flood flow in lower West River became significantly higher in the flood season [40].

Based on the trend test results, design flood in West River was estimated by TVM, and CL was detected to be the best-fit time-trend model for the increasing inflow floods (Table 2). Thirty-year simple moving mean and standard deviation of annual peak flood flow series in West River was analyzed and displayed in Figure 5. Since the first two moments of the flood series both had linear trends, the results of AIC test were proven to be reliable.

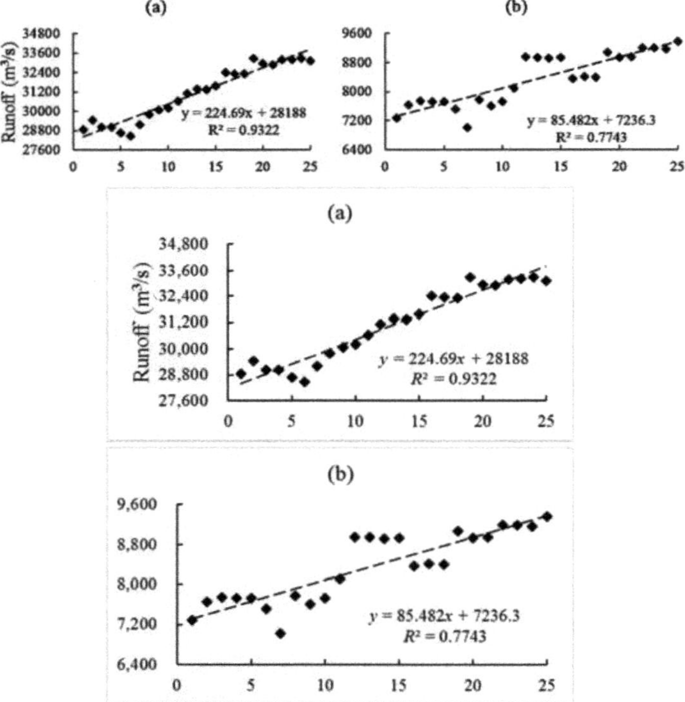

Figure 5. Trends in the first two moments of Wuzhou: (**a**) simple moving mean; (**b**) simple moving standard deviation.

The 100-, 50-, and 20-year design flood flow in West River in 2050 were then estimated. To compare the design volumes with and without taking non-stationarity into consideration, the relative difference E was calculated as follows:

$$E = \frac{X_T - X_0}{X_0} \times 100\%, \tag{4}$$

where X_T denotes the nonstationary design value, and X_0 denotes the stationary one. The gap between a 100-year and a 20-year X_T is ΔX_T, and the difference between a 100-year and a 20-year X_0 is ΔX_0. Both ΔX_T and ΔX_0 were calculated and compared. The greater the value of ΔX_T or ΔX_0, the steeper the slope of the upper tail (and the greater the pressure in flood protection. (The "upper tail" here specifically refers to the part of the flood frequency curve with an exceedance probability lying between 1% and 5%.)

Results showed that by ignoring the increasing tendency, the 100-year peak flood flow was underestimated by 17.51% (9825 m^3/s) (Table 3). Meanwhile, the relative underestimation (E) was augmented when the exceeding probability reached 5%, i.e., by 17.91% (9417 m^3/s) for the 50-year flood, and by 18.52% (8809 m^3/s) for the 20-year flood. In addition, ΔX_T (9553 m^3/s) was larger than ΔX_0 (8537 m^3/s), indicating that non-stationarity in the flood had contributed to a steeper upper tail.

Table 3. Comparison of design flood.

Station	Unit	T = 100			T = 50			T = 20		
		X_T	X_0	E (%)	X_T	X0	E (%)	X_T	X0	E (%)
Wu-zhou	m^3/s	65,926	56,101	17.51	61,990	52,573	17.91	56,373	47,564	18.52
Heng-men	m	2.93	2.68	9.33	2.83	2.57	10.12	2.69	2.42	11.16
Denglong-shan	m	3.00	2.63	14.07	2.88	2.50	15.20	2.70	2.31	16.88
Huang-jin	m	3.65	3.20	14.06	3.50	3.00	16.67	3.28	2.73	20.15
Xipao-tai	m	3.18	3.34	−4.79	3.06	3.12	−1.92	2.90	2.83	2.47

Note: T is the return period.

3.1.2. Varied High Tidal Level

High tidal levels in seven outlets were tested, and tidal levels in all outlets but Nansha had upward tendencies. Somehow, only the tendencies in Hengmen, Denglongshan, Huangjin, and Xipaotai reached a significance of 90% (Table 1). Rising sea-level rise at the Pearl River Estuary pushed up the high tidal levels, and severe sand excavation and channel regulation from the 1980s to the early 21st century around the upstream of Nansha neutralized the rising trend [41].

Nonstationary estimations were carried out in the six outlets that had significant trends. In each station, best-fit models was detected, and the nonstationary design value was compared with the stationary one (Tables 2 and 3). Results showed that, whether the return period was 100, 50, or 20 years, design levels were all underestimated in Hengmen, Denglongshan, and Huangjin, and the increase in the return period had led to a greater underestimation, which was displayed in the value of E. Meanwhile, the 100-, 50-, and 20-year Es ranged from 9.33% (0.25 m) to 11.16% (0.27 m) in Hengmen, 14.07% (0.37 m) to 16.88% (0.39 m) in Denglongshan, and 14.06% (0.45 m) to 20.15% (0.55 m) in Huangjin. Both X_T and Es decreased along the coastline from the southwest to the northeast, i.e., from Huangjin and Denglongshan to Hengmen. Furthermore, since ΔX_T reached 0.24 m in Hengmen, 0.30 m in Denglongshan, and 0.37 m in Huangjin, it is easy to conclude that the slope of upper tail was steeper in the southwest (Huangjin) than in the northeast (Hengmen).

The direction of the ocean current affected the impact of SLR on the tidal level. Based on this criteria, Wu [42] divided the PRD outlets into three groups. In the first group, to which Huangjin belonged, high tides in outlet were mainly influenced by the current from the Indian Ocean. In the second group, to which Denglongshan belonged, high tidal levels were affected by a substantial impact of ocean currents from both the Pacific Ocean and the Indian Ocean. In the third group, to which Hengmen belonged, high tidal levels were solely influenced by the Pacific Ocean current.

The surrounding environment of all these outlet stations affected the ocean current impact as well. Take Huangjin as an example: since it was located in a bay surrounded by tiny islands, the tidal vibration affected by the current impact was amplified. In Hengmen, conversely, the tidal vibration induced by the current effect was reduced, due to the land lying to the east of Pearl River Estuary [42,43].

3.2. *Response of the Design Flood Level to Nonstationary Inflow Floods and the High Tidal Level*

Design flood levels in 22 stations were estimated under six scenarios, combining nonstationary inflow floods and nonstationary high tidal level (Table 4, Figure 6). The design flood level estimated with these nonstationary inputs was called the *nonstationary flood level*, whereas the one without considering non-stationarity was named the *stationary flood level*. For all 22 stations, *nonstationary* and the *stationary flood levels* were compared under Scenarios 1 and 4.

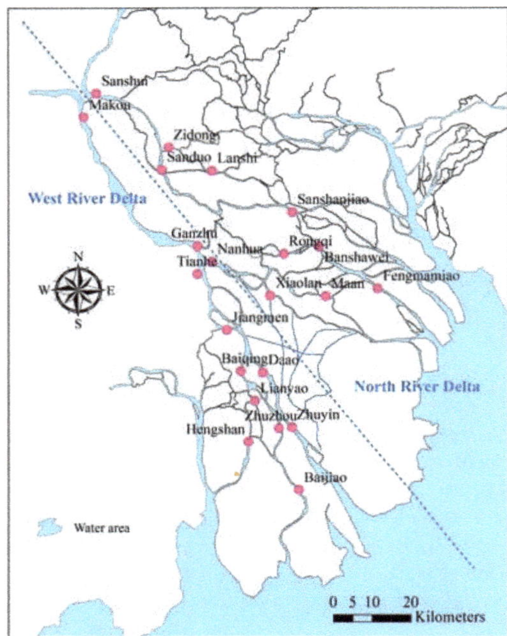

Figure 6. Locations of the 22 gauging stations in Pearl River Delta.

Table 4. Scenarios employed.

Scenario	Return Period	
	Flood Flow	Extreme Tidal Level
1	100a	100a
2	100a	50a
3	50a	100a
4	50a	50a

3.2.1. Response of Flood Level at a Single Spot

In 21 out of 22 stations, the *nonstationary flood level* was higher than the *stationary*. Only in Hengshan was the *stationary flood level* higher. This is because Hengshan was located just 18 km away from the outlet at Xipaotai, and its flood level was highly affected by the nonstationary design tidal level in Xipaotai. Due to the result in Table 1, the nonstationary design tidal level in Xipaotai was lower than the stationary one. Therefore, the *stationary flood level* in Hengshan was higher than the nonstationary one.

To further assess the response of flood level to the nonstationary inflow floods and high tidal level in PRD, the 22 stations were divided into three groups by the K-means clustering method, based on the absolute difference between their *nonstationary* and *stationary flood levels* (Table 5). Among the three groups, Group 1 had the highest mean value, and Group 3 the lowest. If a station belonged to the same group under both scenarios, it was marked in blue (Table 5).

Table 5. K-means clustering results.

Scenario Number	Group 1	Group 2	Group 3
1	Baiqing, Da'ao, Lianyao, Makou, Sanshui, Zhuyin, Zhuzhou	Baijiao, Ganzhu, Lanshi, Nanhua, Rongqi, Sanduo, Tianhe, Xiaolan, Zidong	Banshawei, Fengmamiao, Hengshan, Jiangmen, Ma'an, Sanshanjiao
4	Baijiao, Baiqing, Da'ao, Lianyao, Sanduo, Sanshui, Zhuzhou, Zhuyin, Zidong	Ganzhu, Lanshi, Makou, Ma'an, Rongqi, Xiaolan	Banshawei, Fengmamiao, Jiangmen, Hengshan, Nanhua, Sanshanjiao, Tianhe

Under both scenarios, five stations, i.e., Baiqing, Da'ao, Lianyao, Zhuyin, and Zhuzhou, belonged to the first group. These five stations were located between 25 and 42 km away from the delta outlets, and their flood levels were similarly affected by the high tidal levels. Meanwhile, the responses in Banshawei, Sanshanjiao, and Fengmamiao stations were the lowest under both scenarios. The distance between these stations and their outlets was within the area of 15–33 km. Outlets of these stations, i.e., Wanqingshaxi and Sanshanjiao, exhibited no significant trend in the high tidal level.

When the scenario changed from 1 to 4, stations located in the West River sub-delta, i.e., Makou, Baijiao, Jiangmen, Nanhua, and Tianhe, fell from groups with higher mean values to ones with lower mean values. Meanwile, the stations in the North River sub-delta, i.e., Zidong and Sanduo stations, jumped from the group with lower mean (Group 2) to the group with a higher one (Group 1). Alteration in geometries in the upstream river channel increased the flood flow distribution ratio in the North River sub-delta. Under Scenario 1, the flow distribution ratio in Sanshui, i.e., the income of the North River sub-delta, was 0.7% (300 m^3/s), higher than under Scenario 4 (Table 3). In other words, the increase in the upstream flood flow in the North River sub-delta was greater under Scenario 4, while the increment of income flood flow in the West River sub-delta dropped by 300 m^3/s.

3.2.2. Response of the Flood Level along the River Channel

The flood level in each station along the same river channel was connected into a flood line, and the 100-year nonstationary flood line was compared with the stationary one in two representative river channels (Figures 6 and 7). Makou–Zhuyin channel was located in the West River sub-delta, where there were five stations along the way, i.e., Makou, Tianhe, Jiangmen, Da'ao, and Zhuyin. Sanshui–Sanshanjiao channel was located in the North River sub-delta, and Sanshui, Sanduo, and Sanshanjiao were located along the river channel from the delta income to the outlet.

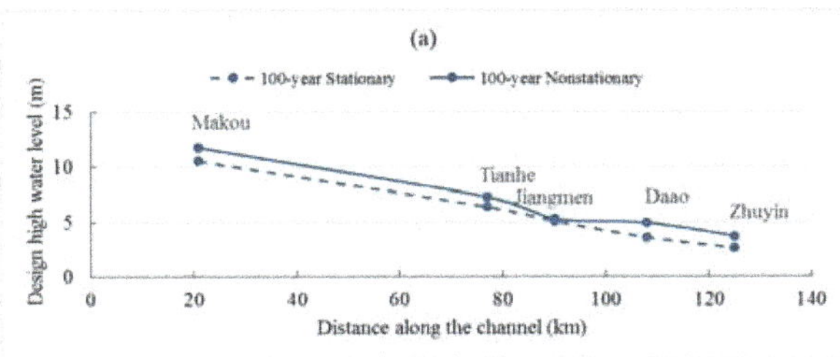

Figure 7. *Cont.*

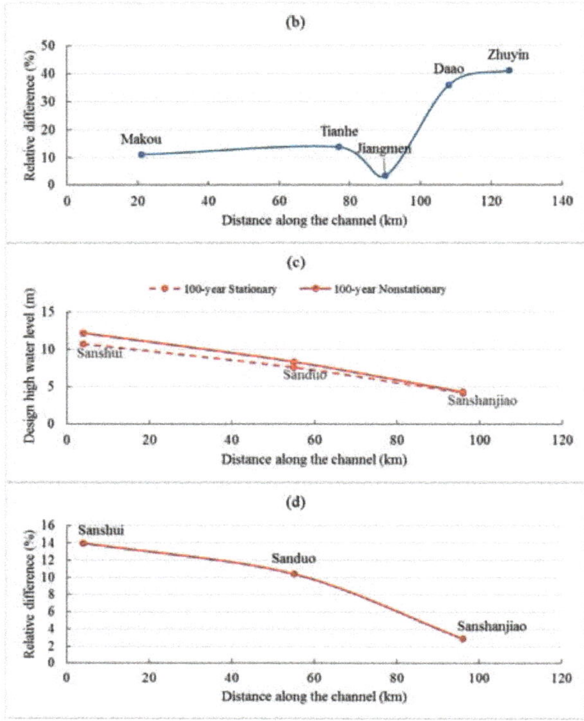

Figure 7. Flood line difference in river channels: (**a**) difference in Makou–Zhuyin Channel; (**b**) relative difference in Makou–Zhuyin Channel; (**c**) difference in Sanshui–Sanshanjiao Channel; (**d**) relative difference in Sanshui–Sanshanjiao Channel.

In either channel, the nonstationary flood line was higher than the stationary one, and the absolute gap between two flood lines gradually narrowed from the delta incomes to the outlets (Figure 7a,b). The relative difference E in the two channels varied dramatically in the spot 60 km away from the outlet (Figure 7c,d). When it came around the area that was 60 km away from the outlet, E kept dropping along the Sanshui–Sanshanjiao channel, while it increased sharply in Makou–Zhuyin channel. In Sanshui–Sanshanjiao channel, since the high tidal level in the outlet displayed no significant upward trend (Table 1), the increase in the design flood line mainly responded to the upstream floods' increment. Thus, the impact of the nonstationary inflow floods decreased smoothly along the channel solely due to the reduction in the elevation. On the contrary, however, the increment of the 100-year tidal level reached 14.07% in the outlet of Makou–Zhuyin channel (Table 3). Affected by the interactive impact of increasing inflow floods and high tidal level, the relative difference between flood lines increased in the area that was 60 km away from the outlet.

Both the absolute and the relative difference between the nonstationary and stationary flood line declined sharply in Jiangmen, caused by the flow distribution ratio variation in its upstream, i.e., the node where Tianhe and Nanhua parted. A large portion of the flood flow (45.53%) from Makou went through Tianhe, located 13 km away from Jiangmen. However, the width–depth ratio in Tianhe was only half that in Makou [44]. Therefore, the relative gap between *nonstationary* and *stationary flood level* in Tianhe increased. Nevertheless, the width–depth ratio in Jiangmen is 2.4 times that in Tianhe [44]. Thus, the relative gap between the nonstationary and the stationary design level in Jiangmen was narrowed abruptly from Tianhe.

3.3. Leading Factors of Non-Stationarity and the Flood Level Response

Nonstationary flood levels under different scenarios were further calculated and compared. To compare the response of the flood level to nonstationary inflow floods and that to the nonstationary tidal level, index X was developed and calculated by the following equation:

$$X = \frac{|Z_1 - Z_3|/Z_1}{|Z_1 - Z_2|/Z_1} = \frac{|Z_1 - Z_3|}{|Z_1 - Z_2|}, \tag{5}$$

where Z_i (i = 1, 2, 3) is the nonstationary design flood level simulated under scenario i. When $X \leq 0.5$, the flood level in PRD mainly responded to the nonstationary high tide, whereas when $X \geq 5$, the flood level mainly responded to the nonstationary inflow floods.

In general, X in each station was positively correlated to the distance between this station and the delta outlet (Figure 8). Xs of stations that were less than 22 km away from the PRD outlet were all below 0.5. Meanwhile, when the distances between stations and the delta outlet were more than 45 km, all Xs were above 5. It can be deducted that in PRD, in stations that are more than 45 km away from the outlets, the flood level was mainly influenced by the nonstationary inflow floods. Furthermore, flood levels in stations that are located within 22 km of the delta outlets were strongly affected by the nonstationary high tidal level.

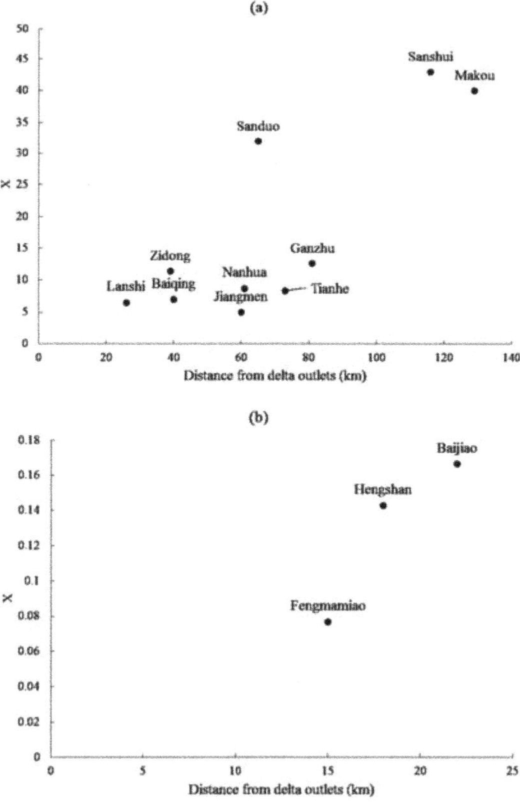

Figure 8. *Cont.*

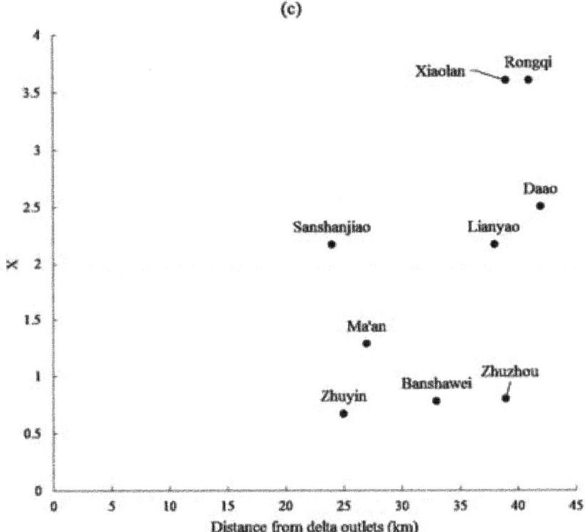

Figure 8. Xs of stations in Pearl River Delta: (**a**) stations with $X \geq 5$; (**b**) stations with $X \leq 0.5$; (**c**) stations with $0.5 < X < 5$.

As for the flood level in stations lying between 22–45 km from delta outlets, their response to either nonstationary inflow floods did not increase when they became farther away from the delta outlets. Take Zhuzhou, for example; though it is located 39 km away from the outlet, its X was the same as that of Banshawei, which was 33 km away from the outlet. Since the impact of extreme tides decreased when the distance between the station and delta outlets increased, X in Zhuzhou was supposed to be higher than that in Banshawei. On the other hand, however, Zhuzhou was much farther away from the entrance of PRD than Banshawei (Figure 8). Therefore, the impact of upstream change decreased to a greater extent in Zhuzhou than in Banshawei. Moreover, the outlet of Zhuzhou had a more severe upward tendency than Banshawei (Table 1). Thus, the non-stationarity of the high tidal level in the outlet of Zhuzhou exerted a more severe impact on the flood level in Zhuzhou.

Although Rongqi was located 2 km farther from the delta outlet than Xiaolan [44], X in Rongqi was the same as in Xiaolan (Figure 8). Simply according to the distance between stations and the delta outlets, X in Rongqi was supposed to be higher than in Xiaolan. However, the estimated X was inconsistent with our assumption. This was due to the impact of a dense river network and the directions of the development of river channels [37,44]. Rongqi was located on a river way almost parallel to the latitude. Since the channel was in accordance with the direction of the reduction of the elevation, flow speed in this channel slowed down. Meanwhile, the region where Rongqi was located had a higher density of river networks than other spots in PRD, and the complexity of river channels further weakened the impact of nonstationary high tidal levels, whose effect decreased from the outlets to the incomes. Based on Xs in Zhuyin and Sanshanjiao, it was concluded that the estimated water level in Sanshanjiao was more affected by non-stationarity in the upstream flood flow. Somehow, their distance to the delta outlets was almost the same, i.e., 25 km for Zhuyin and 24 km for Sanshanjiao. As was mentioned in Section 3.1.1, there was an increase in the upstream flow in the North River sub-delta with the rising input flood flow in the West River. Meanwhile, Sanshanjiao was closer to the income station than Zhuyin. The impact of nonstationary inflow floods was thus maintained. Likewise, although the distances from Lanshi and Ma'an to their outlets were both 27 km, Lanshi had an X much higher than that of Ma'an. Since Ma'an was located at the downstream of a node where the flood from both Makou and Sanshui joined, the variance in the upstream flood change exerted little impact. The distance between the station and the income station was the main cause.

4. Conclusions

Here, the response of flood level in a dense river network to nonstationary inflow floods and high tidal level was studied. Pearl River Delta (PRD) in South China was selected as the study area. Non-stationarity in inflow floods and tidal level was first evaluated by statistical methods such as that of Mann–Kendall, and the design value was further calculated by the time-varying moment method. The flood level estimated with nonstationary input was calculated under six scenarios in a one-dimensional hydrodynamic network model. The flood level estimated with nonstationary inputs was compared with that with stationary inputs, and the response of the flood level to non-stationarity was studied. The methodology utilized in this research for estimating the flood level response to non-stationarity can be used as a reference to investigate the design flood level change in other deltas with a dense river network. It was found that the flood level response was highly affected by nonstationary inflow floods and the high tidal level induced by environmental changes.

In PRD, not considering the non-stationarity in incoming flood flow would have resulted in an 18% underestimation of the 100-year flood level, and an underestimation up to 14% (0.37 m) in the 100-year tidal levels among the multiple outlets. The underestimation of incoming flood flow or downstream high tidal level would have further caused underestimation of flood level along river channels. This situation will not only challenge the existing standard for the local flood protection but also lead to a higher degree of loss in both economy and life. For deltas that have endured intense environmental changes just like PRD, the design flood level needs to be updated by taking the non-stationarity of hydrological inputs into account.

In this paper, the spatial pattern of the flood level response was discovered. The flood level in places located 25–42 km away from the outlets in the West River sub-delta responded the most to the non-stationarity in both inflow floods and high tidal level. Meanwhile, in the North River sub-delta, places lying 15–33 km away from delta outlets had the minimum response to the nonstationary inflow floods and the high tidal level. Due to the change in the geometries of upstream river channel, the North River sub-delta had a greater distribution ratio of inflow floods. This resulted in a greater increase in the flood level in the North River sub-delta, with a rise in the nonstationary flood and tidal level of a smaller return period. The site in the West River sub-delta situated 60 km away from the outlet was an anomalous spot since the at-site flood level merely responded to the non-stationarity.

Based on the leading factor that the flood level responded to, PRD was divided into three regions. In general, regions within 22 km of delta outlets responded the most to the increasing tidal level induced by sea level rise, and spots that were located more than 45 km away from outlets were more affected by increasing inflow floods than by increasing high tidal levels. Stations that were located 22–45 km away from the delta outlets were influenced by both the sea level rise and the inflow increase. As for these stations, distance to the income stations, river network density, and the directions of river channels could all affect the extent of flood level response to the mutual impact of the nonstationary inflow floods and high tidal level. How these factors affect the water levels and their extents at different stations remains to be studied further. Meanwhile, the boundary of the incoming flood impact or that of the tidal level in other deltas can be detected in the same way.

Acknowledgments: The research was financially supported by the National Natural Science Foundation of China (Grant No. 91547202, 51210013, 51479216, 51379223, 51479217), the Chinese Academy of Engineering Consulting Project (2015-DA-07-04-03), the Public Welfare Project of Ministry of Water Resources (Grant No. 200901043-03), and the Project for Creative Research from Guangdong Water Resources Department (Grant No. 2016-07, 2016-01).

Author Contributions: Yihan Tang and Xiaohong Chen conceived and designed the experiments; Chengjia Su performed the experiments; Yihan Tang analyzed the data with the help of Qizhong Guo; Yihan Tang, Xiaohong Chen and Chengjia Su contributed data and analysis tools; Yihan Tang wrote the paper with the help of Qizhong Guo and Xiaohong Chen.

Conflicts of Interest: The authors declare no conflict of interest.

References

1. Trenberth, K.E. Conceptual Framework for Changes of Extremes of the Hydrological Cycle with Climate Change. In *Weather and Climate Extremes*; Springer: Dordrecht, The Netherlands, 1999; pp. 327–339.
2. Groisman, P.Y.; Knight, R.W.; Karl, T.R.; Easterling, D.R.; Sun, B.; Lawrimore, J.H. Contemporary Changes of the Hydrological Cycle over the Contiguous United States: Trends Derived from In Situ Observations. *J. Hydrometeorol.* **2004**, *5*, 64–85. [CrossRef]
3. Foley, J.A.; DeFries, R.; Asner, G.P.; Barford, C.; Bonan, G.; Carpenter, S.R.; Chapin, F.S.; Coe, M.T.; Daily, G.C.; Gibbs, H.K.; et al. Global Consequences of Land Use. *Science* **2005**, *309*, 570–574. [CrossRef] [PubMed]
4. Piao, S.; Ciais, P.; Huang, Y.; Shen, Z.; Peng, S.; Li, J.; Zhou, L.; Liu, H.; Ma, Y.; Ding, Y.; et al. The impacts of climate change on water resources and agriculture in China. *Nature* **2010**, *467*, 43–51. [CrossRef] [PubMed]
5. Schaller, N.; Kay, A.L.; Lamb, R.; Massey, N.R.; Van Oldenborgh, G.J.; Otto, F.E.; Sparrow, S.N.; Vautard, R.; Yiou, P.; Ashpole, I.; et al. Human Influence on Climate in the 2014 Southern England Winter Floods and Their Impacts. *Nat. Clim. Chang.* **2016**, *6*, 627–634. [CrossRef]
6. Nicholls, R.J.; Cazenave, A. Sea-Level Rise and Its Impact on Coastal Zones. *Science* **2010**, *328*, 1517–1520. [CrossRef] [PubMed]
7. Neumann, B.; Vafeidis, A.T.; Zimmermann, J.; Nicholls, R.J. Future Coastal Population Growth and Exposure to Sea-Level Rise and Coastal Flooding—A Global Assessment. *PLoS ONE* **2015**, *10*, 1–34. [CrossRef] [PubMed]
8. Menéndez, M.; Woodworth, P.L. Changes in extreme high water levels based on a quasi-globaltide-gauge data set. *J. Geophys. Res.* **2010**, *115*, 234–244. [CrossRef]
9. Xu, S.; Huang, W. Frequency Analysis for Predicting 1% Annual Maximum Water Levels in Florida Coast. *Hydrol. Process.* **2008**, *22*, 4507–4518. [CrossRef]
10. Xu, S.; Huang, W. Effects of sea level rise on frequency analysis of 1% annual maximum water levels in the coast of Florida. *Ocean Eng.* **2013**, *71*, 96–102. [CrossRef]
11. Chen, Y.M.; Huang, W.R.; Xu, S.D. Frequency Analysis of Extreme Water Levels Affected by Sea-Level Rise in East and Southeast Coasts of China. *J. Coast. Res.* **2014**, *68*, 105–112. [CrossRef]
12. Zhong, H.; van Gelder, P.H.A.J.M.; van Overloop, P.J.A.T.M.; Wang, W. Application of a fast stochastic storm surge model on estimating the high water level frequency in the Lower Rhine Delta. *Nat. Hazards* **2014**, *73*, 743–759. [CrossRef]
13. Méndez, F.J.; Menéndez, M.; Luceño, A.; Losada, I.J. Analyzing monthly extreme sea levels with a time-dependent GEV model. *J. Atmos. Ocean. Technol.* **2007**, *24*, 894–911. [CrossRef]
14. Zheng, F.; Leonard, M.; Westra, S. Application of the design variable method to estimate coastal flood risk. *J. Flood Risk Manag.* **2015**. [CrossRef]
15. U.S. Army Corps of Engineers. *Flood Risk Management Planning Workshop*; Ocean County Community College: Toms River, NJ, USA, 2016.
16. Hydrology Bureau of Guangdong Province. *Report of Flood Control Planning in Pearl River Basin*; Zhujiang River Commission: Guangzhou, China, 2007.
17. Vidal, J. UN Report: World's Biggest Cities Merging into 'Mega-Regions'. Available online: guardian.co.uk (accessed on 13 March 2010).
18. United Nations, Department of Economic and Social Affairs, Population Division. *World Urbanization Prospects: The 2014 Revision*; United Nations: New York, NY, USA, 2014.
19. Cox, W. The World's Ten Largest Megacities. The Huffington Post, 2016. Available online: http://www.huffingtonpost.com/wendell-cox/the-worlds-ten-largest-me_b_6684694.html (accessed on 19 Feburary 2016).
20. Khaliq, M.N.; Ouarda, T.B.M.J.; Ondo, J.C.; Gachon, P.; Bobée, B. Frequency analysis of a sequence of dependentand/ornon-stationary hydro-meteorological observations: A review. *J. Hydrol.* **2006**, *329*, 534–552. [CrossRef]
21. Strupczewski, W.G.; Singh, V.P.; Mitosek, H.T. Non-stationary approach to at-site flood frequency modelling. III. Flood analysis of Polish rivers. *J. Hydrol.* **2001**, *248*, 152–167. [CrossRef]
22. Strupczewski, W.G.; Singh, V.P.; Feluch, W. Non-stationary approach to at-site flood frequency modelling. I: Maximum likelihood estimation. *J. Hydrol.* **2001**, *248*, 123–142.
23. Hydrology Bureau of Guangdong Province (HBGP). *Report on the Flood Control Plan of Pearl River Basin*; HBGP: Guangzhou, China, 1998.

24. Li, K.; Li, G.S. Calculation of return period for storm surge in the Pearl River Delta Region. *Prog. Geogr.* **2010**, *29*, 433–438.
25. Chen, Z.S.; Liu, Z.M.; Lu, J.F. Flood Joint Probability Distribution of the Xijiang River and Beijiang River in Guangdong Province. *Acta Sci. Nat. Univ. Sunyatseni* **2011**, *50*, 110–115. (In Chinese)
26. Wu, Z.Y.; Lu, G.H.; Liu, Z.Y. Trends of Extreme Flood Events in the Pearl River Basin under Climate Change. *Prog. Inquisitiones Mutat. Clim.* **2012**, *8*, 403–408. (In Chinese)
27. Liu, Z.M.; Qin, G.H.; Chen, Z.S. Study on the correlation of the water level of the tidal river with upstream flood and estuary tide level. *Shuili Xuebao* **2013**, *44*, 1278–1285. (In Chinese)
28. Liu, Y.; Guan, S. Study on the Characteristic of Multiply Events of Drought and Flood Probability in the Pearl River Basin based on Copula Function. *Pearl River* **2017**, *38*, 12–17. (In Chinese)
29. Liu, J.; Chen, H.; Wang, J.X. Comparison among Theoretical Frequency Distributions of P-III, Log P-III and GL. *J. China Hydrol.* **2013**, *33*, 1–4. (In Chinese)
30. Guan, S.; Zha, X.N.; Ding, B. Wetness-Dryness Encountering of Runoff of the Pearl River Basin Based on Copula Functions. *Trop. Geogr.* **2015**, *35*, 208–217. (In Chinese)
31. Degond, P.; Markowich, P.A. On a one-dimensional steady-state hydrodynamic model for semiconductors. *Appl. Math. Lett.* **1990**, *3*, 25–29. [CrossRef]
32. Preissmann, A. Propagation of translatory waves in channels and rivers (original in French "Propagation des intume scences dans les canaux et rivières"). In Proceedings of the first Congress of the French Association for Computation (AFCALTI) September, Grenoble, France, 14–16 September 1961; pp. 433–442.
33. Center for Water Resources and Environment. *One-Dimensional Hydrodynamic Model of the Pearl River Delta, Project Report (Part B)*; Sun Yat-sen University: Guangzhou, China, 2015; pp. 20–25.
34. Kendall, M.G. (Ed.) *Rank Correlation Methods*; Hafner: New York, NY, USA, 1975.
35. Mann, H.B. Non-parametric test against trend. *Econometrica* **1945**, *13*, 245–259. [CrossRef]
36. Tang, Y.H.; Xi, S.F.; Chen, X.H.; Lian, Y.Q. Quantification of Multiple Climate Change and Human Activity Impact Factors on Flood Regimes in the Pearl River Delta of China. *Adv. Meteorol.* **2016**. [CrossRef]
37. Zhang, Q.; Gu, X.H.; Singh, V.P.; Xiao, M.Z. Flood frequency analysis with consideration of hydrological alterations: Changing properties, causes and implications. *J. Hydrol.* **2014**, *519*, 803–813. [CrossRef]
38. Zhao, Y.F.; Zou, X.Q.; Cao, L.G.; Xu, X. Changes in precipitation extremes over the Pearl River Basin, southern China, during 1960–2012. *Quat. Int.* **2014**, *333*, 26–39. [CrossRef]
39. Gemmer, M.; Fischer, T.; Jiang, T.; Su, B.; Liu, L.L. Trends in Precipitation Extremes in the Zhujiang River Basin, South China. *J. Clim.* **2010**, *24*, 750–761. [CrossRef]
40. Zhang, S.R.; Lu, X.X. Hydrological responses to precipitation variation and diverse human activities in a mountainous tributary of the lower Xijiang, China. *Catena* **2009**, *77*, 130–142. [CrossRef]
41. Wu, X.M.; Deng, J.Q.; Cheng, R.L.; Wu, T.S. A super-large tidal physical model for the Pearl River Estuary. Presented at International Conference on Estuaries and Coasts, Hangzhou, China, 9–11 November 2003.
42. He, H.J. Analysis of the Storms and Typhoons in the Pearl River Estuary. *Ren Min Zhujiang* **1981**, *4*, 34–51. (In Chinese)
43. Ni, P. *Project Report: Water and Sediment Characteristics and the Riverbed Evolution in the Pearl River Delta, 2.1.2*; Institute of Water Resources and Hydropower Research of Guangdong Province (IWRHG): Guangzhou, China, 2014. (In Chinese)
44. Chen, X.H.; Zhang, L.; Shi, Z. Study on spatial variability of water levels in rivernet of Pearl River delta. *J. Hydraul. Eng.* **2004**, *10*, 36–42. (In Chinese)

Article

Variability of Spatially Grid-Distributed Precipitation over the Huaihe River Basin in China

Zhi-Lei Yu [1,2,3], Deng-Hua Yan [2,3,*], Guang-Heng Ni [1], Pierre Do [1] , Deng-Ming Yan [2,3,4], Si-Yu Cai [1,2,3], Tian-Ling Qin [2,3], Bai-Sha Weng [2,3] and Mei-Jian Yang [5]

1 Department of Hydraulic Engineering, Tsinghua University (THU), Beijing 100084, China; yzl16@mails.tsinghua.edu.cn (Z.-L.Y.); ghni@tsinghua.edu.cn (G.-H.N.); dodmp10@mails.tsinghua.edu.cn (P.D.); caisy@iwhr.com (S.-Y.C.)

2 State Key Laboratory of Simulation and Regulation of Water Cycle in River Basin, China Institute of Water Resources and Hydropower Research (IWHR), Beijing 100038, China; 18655058842@163.com (D.-M.Y.); tianling406@126.com (T.-L.Q.); baishaweng@126.com (B.-S.W.)

3 Department of Water Resources, IWHR, Beijing 100038, China

4 College of Environmental Science and Engineering, Donghua University, Shanghai 201620, China

5 Department of Civil and Environmental Engineering, University of Connecticut, Storrs, CT 06269, USA; meijian.yang@uconn.edu

* Correspondence: yandh1006@sina.com; Tel.: +86-10-6878-1656

Received: 8 March 2017; Accepted: 22 June 2017; Published: 5 July 2017

Abstract: This study investigates spatial characteristics of annual and decadal precipitation in the Huaihe River basin. Daily precipitation data, obtained from meteorological gauges, are analyzed for a 51-year period, from 1961 to 2011. Precipitation is analyzed in grids (5 km^2) with respect to temporal variability. The spatial distribution and intensity of annual rainfall (mm/10 year), determined by the linear regression method, reveals a slight increase of 3 mm/10 year over the basin. However, the trend did not present a significant change at 95% significance level in the most of basin. Precipitation is mostly increasing for each ten-year periods during the total 51 years. The annual precipitation randomicity was calculated from the non-uniform coefficient *Cv* (coefficient of variation) test and showed a significant non-uniform spatial distribution, indicating that randomicity of annual rainfall was the moderate variability. The Pettitt test determined that the abrupt change points occurred mainly in 1965, 1975 and 2002. Wavelet analysis showed that cyclic variations appeared almost every 5 to 10 years, accounting for 36% of the basin area. Meanwhile, these cycles tended to be delimited by the abrupt change points. This study aims to provide insights for water resources management, mitigation of climate change effects and water supply in the Huaihe River basin and surrounding watersheds.

Keywords: wtershed hydrology; grid-distributed precipitation; Huaihe River basin; linear regression method; coefficient of variation; Pettitt test; wavelet analysis

---

1. Introduction

Precipitation is the essential feature of the climate change and a key element of the hydrological processes [1,2]. Change in precipitation has greatly affected climate stability, hydrological processes, and water availability [3,4]. The spatial and temporal distribution changes of precipitation affect the frequent occurrence of extreme events, especially droughts and floods [4,5]. Therefore, exploring the characteristic changes of precipitation is of profound significance. It is essential for agricultural production, the planning and management of sustainable water resources, and overall economic development and livelihood of a country [6–8].

Many authors have documented the spatial and temporal variation characteristics of the precipitation throughout the world [9–12]. For example, scholars described precipitation trends in Nigeria using the monthly accumulations of precipitation for period 1961–2000 [2]. They observed that the spatial distribution of precipitation highly depended on the latitude and its linear relationship with longitude was not clear. Tabati and Talaee found a decreasing trend in annual precipitation in most of the 41 stations they observed in Iran from 1966 to 2005 [13]. Tian analyzed the variation characteristics of precipitation in the main rivers of China during 1957–2013 [14]. Their results showed the annual precipitation and seasonal mean precipitation changed little. Scholars described the spatio-temporal variations of the frequency of extreme precipitation using a high quality precipitation dataset of 599 stations in China for 1961–2001 [15]. Liu studied the spatio-temporal patterns of trends of the precipitation in the Yellow River basin during 1960–2006 and indicated that the precipitation possessed longitude zonality and had no obvious linear relations with the latitude [16]. Another scholar simulated the precipitation change over the past 50 years and the next 30 years in various regions of China, and the research results showed a clear warm-in trend [17].

Most previous studies were focused on the temporal scales, such as interannual [8], interdecadal [1], decadal [9], annual [18], seasonal [6,7], monthly [3], daily [19,20], and even finer temporal scales [21]. However, few investigations have been conducted on continuous spatial distribution of rainfall on a basin scale [16] and in particular the spatial variability [22]. Most studies emphasized discrete variations in the study area [6,23]. The annual precipitation describes a high spatial and temporal variability [24].

The spatial characteristics of precipitation was the major control factor of rainfall-runoff simulation and a series of other hydrological problems [25]. The uncertainty of flood forecasting mainly derived from the rainfall uncertainty and the most sensitive factor of flood peaks prediction was the spatial variation of rainfall [26]. We can obtain the precise spatial distribution of rainfall by establishing rainfall or meteorological stations with extremely high density. Although there are a lots of meteorological and rainfall stations in China, the limited observation data and scientific literature cannot meet the needs of the distributed hydrological model. Precipitation is not uniform and its change is considerably in space and time, including small scales, therefore, it is particularly hard to gauge these changes [27]. We need to extrapolate the data to spatial grids across the watershed to compensate for the unevenly distributed of meteorological stations.

In previous studies, due to the lack of scientific literature to document the spatial and temporal extent of precipitation across the basin or watershed, most researches used single point data to obtain the rain changes in different basins. Based on single point data, our study primarily extrapolated the data across the watershed and extended the findings by quantitatively analyzing the spatial variability of annual and decadal rainfall in the Huaihe River basin for a long-term period (from 1961 to 2011). This investigation used data from 45 meteorological stations distributed over the study area (Figure 1). This research aims to provide basic climate information for river basin management.

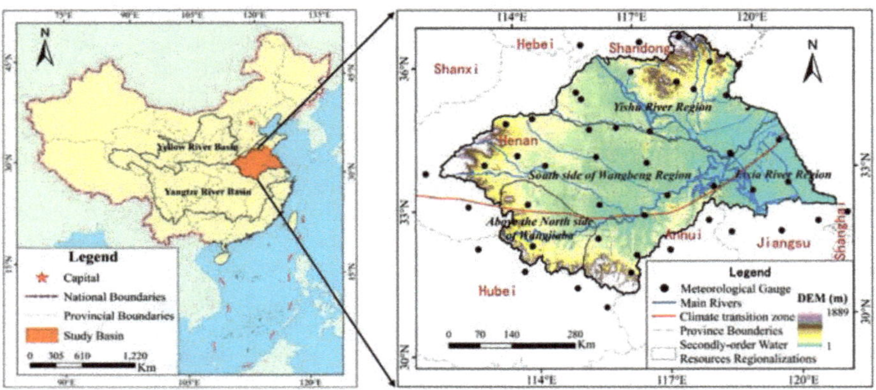

Figure 1. The inset map displays the location of the Huaihe River basin (red shading) in China. The main map shows the location of the Huaihe River basin where the bold black line reveals the boundary of the Huaihe River. The location of the meteorological stations showed by the black circle.

2. Data and Methods

2.1. Study Area

The study area is the whole Huaihe River basin, which lies approximately between 30°57′ N–36° N and 112° E–121° E in eastern China (see Figure 1). It lies in the region from the Yangtze River basin in the south to the Yellow River basin in the north. The total geographical area of the basin is approximately 259,700 km^2. The whole basin is divided into the Huaihe River catchment and the Yishusi River catchment by the paleo-channel of the Yellow River and is composed of four secondly-order water resource regionalizations. Its landscape is characterized by the near level plain and hill, among which the plain occupies more than 80% of the total basin area. The spatio-temporal distribution of annual precipitation is uneven in the basin. The mean annual precipitation in the Huaihe River basin during 1961 to 2011 is 884 mm. While the average annual maximum precipitation is 2.5 times bigger than the minimum for the period of 1961 to 2011, and the precipitation amount notably decreases from southeast to northwest. Mean annual temperature is approximately 14.47 °C, and average maximum temperature goes up to 15.79 °C, whereas the minimum temperature drops to 7.38 °C for period from 1961 to 2011. With the dramatic effects of the global climate changes and human activities, underlying surfaces and water resources changes of the Huaihe River basin have attracted more and more attention. In addition to its special geographical conditions from sub-tropical zone to warm temperate zone and from wet zone to semi-arid and arid zone and complex the climate factors, the Huaihe River basin is characterized by frequent or continuous drought or flood events, floods in the south region, and droughts in the north, and drought-flood shifting [28–30].

2.2. Data

Based on Thiesssen polygon [31], 24 h observed precipitation data were collected from 45 meteorological stations during the 1961 to 2011 period. These information are available in the National Meteorological Information Centre (NMIC) online database provided by the China Meteorological Administration [32]. For each meteorological station, discrete daily precipitation data were converted into 5 km^2 grid-distributed yearly precipitation. This interpolation used the Arc-GIS framework by inverse-distance-weighted (IDW) method. Each grid was considered as the basic dataset used to analyze the precipitation characteristics for the 1961 to 2011 period in the Huaihe River basin. Yang revealed the trend, the abrupt change, the periodicity, and the randomicity of precipitation time series for the main flood period in the Huaihe River [33]. Hence, we studied the long-term (from 1961 to 2011) spatial variability of annual and decadal precipitation in the Huaihe River basin using statistical methods through four aspects: the trend, the abrupt change, the periodicity, and the randomicity.

2.3. Methods

We use the Matlab R2012b to calculate the precipitation abrupt change and periodicity, and the ArcGis 10.2 to obtain the rainfall trend and randomicity. The ArcGis 10.2 is also used to generate the rainfall characteristics spatial interpolation maps. The study performed the spatial trend analysis of precipitation in the Huaihe River basin over the period of 1961 to 2011 using the linear regression method. The randomicity of precipitation changes was detected by the coefficient of variation (Cv) and abrupt change points of observed precipitation data were caught by the Pettitt test. The wavelet analysis determined the precipitation period. Then the spatial distribution of annual and decadal precipitation were obtained. The statistical test and methods used in this study are described below:

2.3.1. Linear Regression Method

The linear regression method is employed to investigate and analyze the long-term trends of precipitation in the time series. The main statistical parameter, the slope, is used to indicate the temporal change of the studied variable on the spatial scale. This method intuitively reflects the trend of the rainfall time series, and the slope of the linear equation represents the average change rate of the trend (to define the *slope* × 10 as the precipitation tendency rate; unit: mm/decade) [34]. The following formula is proposed by Li to computer the *slope* [35]:

$$Slope = \frac{n \times \sum_{i=1}^{n} (i \times P_i) - \sum_{i=1}^{n} i \sum_{i=1}^{n} P_i}{n \times \sum_{i=1}^{n} i^2 - \left(\sum_{i=1}^{n} i\right)^2} \tag{1}$$

where *slope* represents the estimated linear trend of precipitation during the period of 1961 to 2011. Positive values of the *slope* indicate increasing trends, while negative values of the *slope* denote decreasing trends [13]; i is the number of years in the time series (this study is from 1 to 51); and P_i is the annual precipitation amount. We used the MK test to detect whether a trend in the rainfall time series is statistically significant at a 95% confidence level [14].

2.3.2. Coefficients of Variation

The coefficient of variation (Cv) is a unit-free and effective normalized measure of dispersion [36]. Some scholars used the non-uniform coefficient Cv to quantify the evenness of monthly precipitation [15]. Where rainfall is more uneven, the randomicity is much greater. We applied the Cv for detecting and analyzing the randomicity of annual and decadal precipitation. In this study, Cv was estimated with the standard deviation and mean values [37–40]. It can be expressed as:

$$Cv = \sqrt{\frac{\sum_{i=1}^{n} (K_i - 1)^2}{n-1}}, \text{ and } K_i = \frac{P_i}{\overline{P}}, \tag{2}$$

where Cv is the randomicity of precipitation amount, namely the discreteness of precipitation. The larger the Cv value is, the greater the randomicity of precipitation changes is considered to be. In other words, the Cv is high, which indicates the uneven distribution of the annual precipitation in the basin [41]. The K_i is the coefficient of modulus. When the Cv is greater than 1, the randomicity of precipitation is defined as the strong variability; $0.1 \leq Cv \leq 1$ means the moderate variability; Cv less than 0.1 means the weak variability [42].

2.3.3. Pettitt Test

The non-parametric Pettitt statistical test is frequently applied to explore the abrupt change point in a long-term trend analysis [43]. The Pettitt test can not only judge the location and number of the abrupt change, but also estimate the significance of the abrupt change [18,44]. The Pettitt test is considered as a powerful and useful method for obtaining the abrupt change points in characterizing the trends of climate data [23]. The test statistic $U_{t,n}$ is given by:

$$U_{t,n} = \sum_{i=1}^{t} \sum_{j=t+1}^{n} sgn(\theta), \theta = P_i - P_j, 1 \leq t \leq n \tag{3}$$

where n is the time length of precipitation data set, and P is the annual precipitation amount at time i and j, and:

$$sgn(\theta) = \begin{cases} +1 & for\ \theta > 0 \\ 0 & for\ \theta = 0 \\ -1 & for\ \theta < 0 \end{cases} \tag{4}$$

The most probable change point is obtained where its value is as follows:

$$k_t = \max_{1 \leq t \leq n} |U_{t,n}| \tag{5}$$

and the significance probability ρ associated with k_i is expressed by:

$$\rho = 2exp\left(\frac{-6k_t^2}{n^3 + n^2}\right) \tag{6}$$

If ρ is less than the specific significance level, the null hypothesis is rejected. That is, if a significant change point subsists, the time series is segregated into parts at the abrupt change point t. The approximate significance probability for an abrupt change point is expressed as p = 1 − ρ. In our study, the specific significance level is at 95% confidence level.

2.3.4. Wavelet Analysis

Wavelet analysis is a powerful method to research multi-scale, non-stationary signals in finite spatial and temporal domains [11,45–47]. It has been applied in atmospheric sciences and geophysics [47,48]. The Morlet wavelet is a very accurate method for obtaining the periodicity in a few similar studies but is applied little for studying the periodicity in a database [49,50]. Therefore, we use a continuous wavelet transform (CWT) to detect the existence of oscillations and their period [46]. The Morlet wavelet is widely used to identify periodic oscillations of the real life signals and possesses an obvious strength that allows detection of time-dependent amplitude and phase for different frequencies due to its complex nature [46,51]. The Matlab R2012b software is utilized here to seek the rainfall periodicity with the above discussed Morlet wavelet.

3. Results

3.1. Spatial Variations of Annual Precipitation

The spatial distribution of the average annual precipitation presents a multi-step increase along the northwest-southeast direction throughout the basin (Figure 2a). That is, precipitation occurs more often in the southeastern basin than in the northwest. Dry regions are located in the northwest (inland) while wet ones are in the southeast (coast). The maximum mean annual rainfall concentrates in the Dabie Mountain region (Figure 2b).

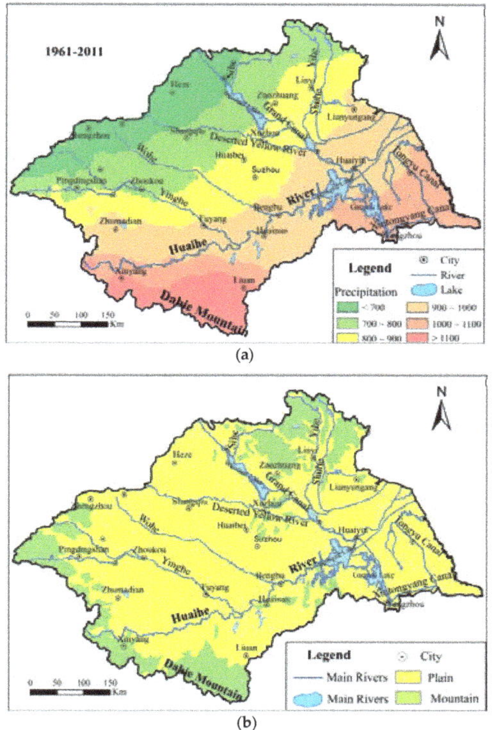

Figure 2. The spatial distribution of the average annual precipitation during 1961 to 2011 (**a**) and the main physiognomy (**b**) in the Huaihe River basin.

3.1.1. Trends of Annual Precipitation

Based on the grid annual rainfall, the spatial trend distribution of annual precipitation and their magnitudes (mm/decade) calculated by the linear regression method are given in Figure 3a. The annual precipitation in the basin presents a slight increasing trend and its mean tendency rate is 3 mm/10 year.

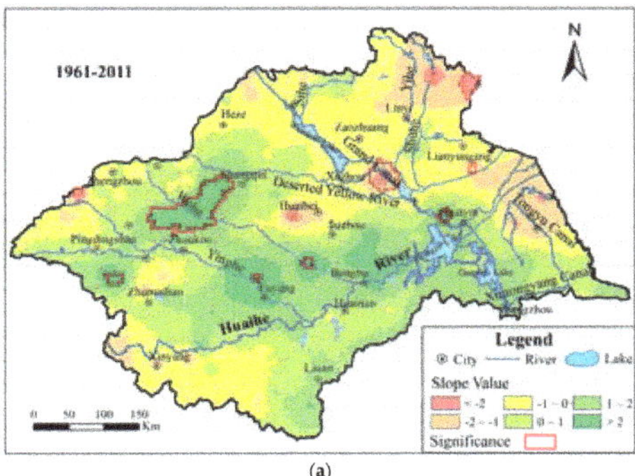

Figure 3. *Cont.*

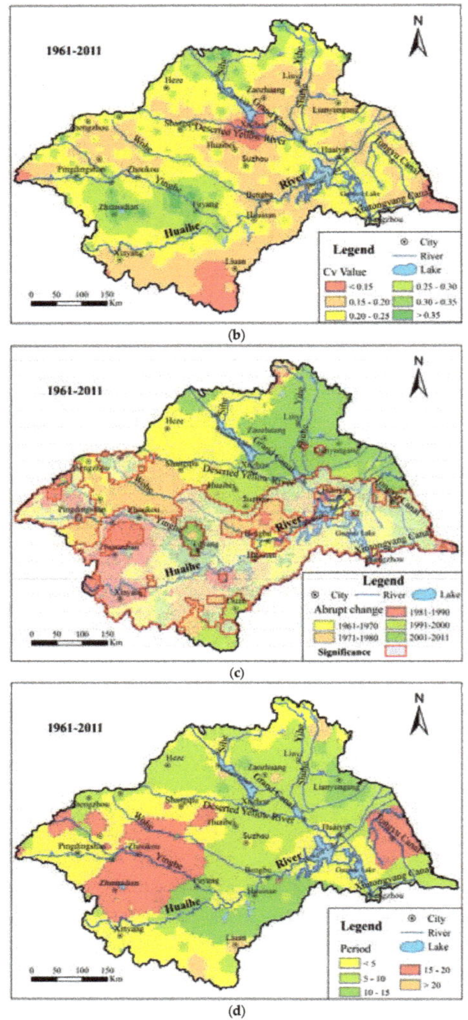

Figure 3. Spatital distribution for trend (**a**); randomicity (**b**); abrupt change (**c**) and periodicity (**d**) of annual precipitation during 1961 to 2011 in the Huaihe River basin.

About 58% of the areas had a positive trend and the rest showed a negative trend. The positive trends were found mostly in southeastern, middle, and northwestern zones along the mainstream of the Huaihe River and branches of the Guohe and Yinghe Rivers. In contrast, negative trends were discovered in Yihe and Shuhe Rivers in the northeastern region, and also in the southwest. In addition, we studied the spatial distribution of the temporal trend magnitude of annual precipitation (from 1961 to 2011) in the Huaihe River basin (Figure 3a). The positive trend magnitude varied between 0 and 40 mm/10 year, in general. The proportion of magnitude ranging between 0 and 10 mm/10 year and between 10 and 20 mm/10 year was approximately 27% and 28%, respectively. These parts were mainly grouped in the area from the southeastern basin to the northwest. The negative trend magnitude changed was from about −40 to 0 mm/10 year. The range of the magnitude between

−10 and 0 mm/10 year accounted for about 31% and was mostly distributed in the northeastern zones. In fewer regions (less than 10%), the trend magnitude was greater or less than 20 or −20 mm/10 year. These zones were scattered loosely all around the basin. Via the significance test, we found that the trend changes of annual rainfall were not significant in most of the basin.

3.1.2. Randomicity of Annual Precipitation

Based on the *Cv* test, Figure 3b indicates annual precipitation had the different spatial variations between 1961 and 2011 in the Huaihe River basin. The randomicity was greater in the part of the west and north regions than in the middle and southwest. The overall range of *Cvs* changed between 0.18 and 0.31 and the randomicity of annual rainfall was determined as the moderate variability. The *Cvs*, ranging between 0.22 and 0.26, were detected at the rate of 89% and distributed nearly in the whole area; the *Cvs* greater than 0.26 were primarily around Zhumadian and Fuyang, which accounted for about 5% of the total area analyzed. The *Cvs* were less than 0.22, and its proportion of the region area was approximately 5% in the vicinity of Xuzhou, the southwest of Liuan, and the southeast corner. The *Cvs* were less than 0.2 in the rest of the area. The annual rainfall uneven distribution was obvious in the north, upstream of the Huaihe River and downstream of the Yihe and Sihe Rivers. The geomorphology of these regions was mainly the plain (Figure 2b). In contrast, the inhomogeneous changes of the annual precipitation were slight in the mountain regions. These implicated that the annual precipitation distribution was more heterogeneous in mountain areas than in the plain.

Across the basin, the spatial distribution of the randomicity showed uneven precipitation patterns over the last 50 years. The rainfall homogeneity was greater upstream than downstream of the Huaihe River. In particular, the *Cv* was much greater in the Shahe River and Yinghe River regions. It also showed that variability of rainfall in these regions was stronger than in other areas.

3.1.3. Abrupt Change of Annual Precipitation

The Pettitt test was used to obtain the abrupt change in the precipitation time series during the study period. The different abrupt points were detected in various 10-year periods (Figure 3c) and the distribution proportion of abrupt changes varied by decade. Except from 1981 to 1990, the abrupt changes nearly increased with time. The abrupt changes occurred with a probability of ca. 16% from 1961 to 1970. The abrupt changes was almost 22% between 1971 and 1980. The 10% abrupt change occurred in the southwestern basin during 1981–1990. About 21% of the total region showed an abrupt change in rainfall from 1991 to 2000. Abrupt change points (more than 30%) emerged in the north and southeast basin, and parts of the southwest and west basin during 2001–2011. In conclusion, during the 51-year period, in all abrupt change points, 1965, 1975, and 2002 were the major abrupt change points, and their proportion of distribution accounted for 21%, 13%, and 11%, respectively.

The abrupt points were distinct during the very 10-year period. The dominant abrupt point was detected in 1965 (about 70% of the total) during 1961–1970. From 1971 to 1980, the primary abrupt changes occurred in 1975 and 1978, and accounted for 58% of the land area during these periods. There was an obvious abrupt point in 1984 and its proportion was ca. 60% from 1981 to 1990. The three abrupt change points between 1990 and 2000 occurred in 1991, 1995, and 1997, and accounted for ca. 78% in this duration. During the period of 2001–2011, 2001 and 2002 were the abrupt change points, accounting for 87%.

The abrupt changes were found at a 95% confidence level. There were significant abrupt changes in 47.11% of the regions in the trunk stream of the Huaihe River (Figure 3c). These areas were distributed on both sides of the Huaihe River. The years 1965, 1978, 1984, 1991, 1995, 1997, 2001, and 2002 presented abrupt changes at a 95% significance level.

3.1.4. Periodicity of Annual Precipitation

The periodicity variations of the annual precipitation were calculated by wavelet analysis in the basin during the 51 years. The time-frequency distributions were shown in the various spatial region

(Figure 3d). For the annual precipitation all year round, the six scales of periodic cycle existed: <5, 5–10, 10–15, 15–20, and 20–25 years, almost throughout the whole time sequence. As a whole, the different cyclic length of annual precipitation distributed irregularly in the basin. During the <5-year period (about 32% of the total area) precipitation was chiefly distributed in the north and southwestern regions of the basin. A nearly and less than 10-year cycle (about 36% of the total area) were mainly observed in northeastern, central, and parts of the southwestern basin. Approximately 30% of the regions where there was a 10–20-year period in the southwest region were identified. More than 20-year cycles accounted for ca. 2%.

3.2. *Spatial Variations of Inter-Decadal Precipitation*

We investigated the character distribution of the precipitation on the different time scales. The precipitation trends and the *Cv* in basin were analyzed by using nonparametric tests based on the grid data of decadal rainfall.

3.2.1. Trends of Inter-Decadal Precipitation

There were five 10-year periods, including the 1960s, 1970s, 1980s, 1990s, and beginning of the 21st century. The inter-decadal precipitation showed a slightly increasing trend during the various periods (from the 1960s to the beginning of the 21st century). The mean magnitude of the trend was successively about −12, −10, 3, 3, and 6 mm/year during the various periods. The magnitude changes were different in the basin's various orientations (Figure 4). The negative trends were detected in the north of the downstream of Huaihe River and the positive ones were in the south of the upstream of Huaihe River in the 1960s (Figure 4a), where the magnitude ranged from −58 to 41 mm/year. Additionally, the rainfall trends decreased sharply in parts of the northeastern basin and increased slightly in the western basin during the 1970s (Figure 4b). The tendency magnitude was between −57 and 23 mm/year. Furthermore, the tendency magnitude range was from −43 to 42 mm/year in the 1980s (Figure 4c). The majority of the northeastern basin showed an increasing trend while a decrease occurred in the parts of the southwestern basin. Conversely, the precipitation magnitude increased in most parts of the northwestern basin and decreased in the southeast of the Huaihe River in the 1990s (Figure 4d). The magnitude ranged from −48 to 57 mm/year. Finally, the rainfall magnitude changed from −28 to 38 mm/year during the beginning of the 21st century (Figure 4e). The precipitation decreased in the parts of the central and western basin and increased in most areas. It was remarkable that all trend changes had no distinguishability by the significance test.

3.2.2. Randomicity of Inter-Decadal Precipitation

The decadal precipitation heterogeneity was investigated by using the *Cvs* (Figure 5). We demonstrated that there were notable changes between the minimum and maximum *Cv*. The *Cv* vaule ranged variously (from 0.07 to 0.41) in different decades. The difference between the minimum and maximum *Cv* was 0.34. However, the randomicity was also the moderate variability. It ranged between 0.14 and 0.41 in 1960s (Figure 5a). The uneven precipitation was geographically concentrated in the northern edges of the basin, where the *Cv* value was greater than 0.25. The minimum *Cv* value was 0.07 and the maximum was 0.33 in the 1970s (Figure 5b). The precipitation was relatively even across the most areas of the basin during this period. The *Cv* ranged from 0.10 to 0.37 during the 1980s period (Figure 5c). The precipitation variability was obvious in the downstream of the Yinghe River, Yihe River and Sihe River. The *Cv* value ranged from 0.11 to 0.41 during the 1990s (Figure 5d) and 0.11 to 0.38 at the beginning of the 21st century (Figure 5e). The precipitation was uneven in the Yinghe River basin and the south parts of the middle and lower reaches of the Huaihe River in 1990s. The dissimilar precipitation was chiefly concentrated in the north parts of the Huaihe River at beginning of the 21st century.

The mean *Cvs* were heterogeneous and changed from 0.20 to 0.27. The precipitation randomicity increased during the 1960s and 1970s, and then decreased in the 1980s, and increased again between

1990s and the beginning of the 21st century. The *Cv* was 0.23 in the 1960s, which indicated that precipitation was relatively even across most of the basin. The values reached 0.27, 0.20, 0.23, and 0.23 in the 1970s, 1980s, 1990s, and early 21st century, respectively. Most of the uneven precipitation occurred in the north of the Huaihe River and changed with climate change.

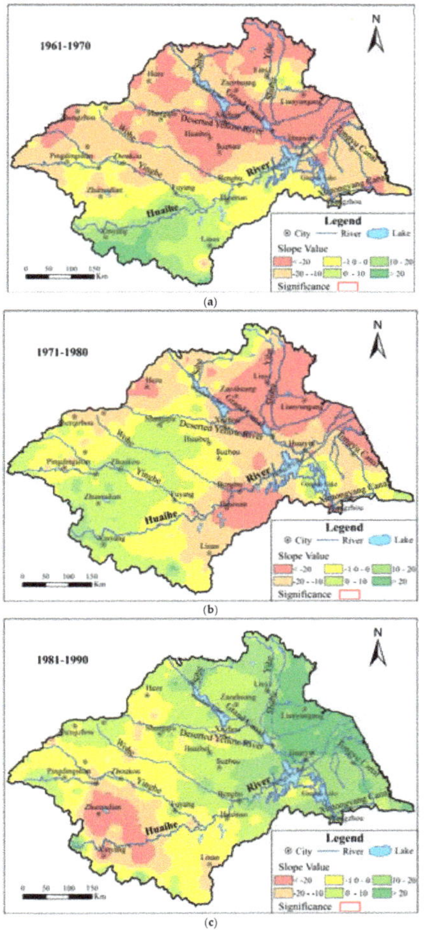

Figure 4. *Cont.*

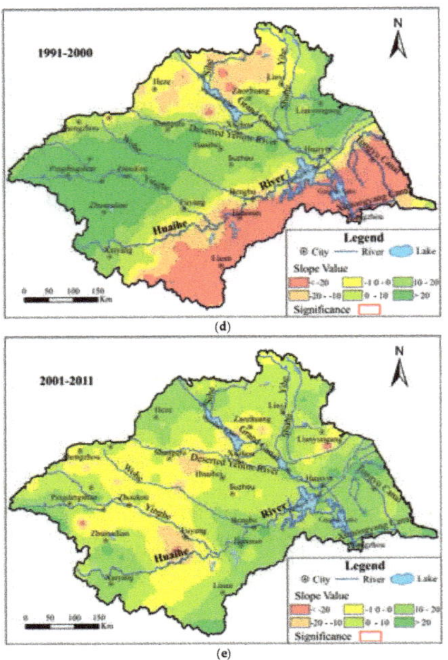

Figure 4. The spatial distribution of the tendency magnitude in the different decadal precipitation ((**a**) from 1961 to 1970; (**b**) from 1971 to 1980; (**c**) from 1981 to 1990; (**d**) from 1991 to 2000 and (**e**) from 2001 to 2011).

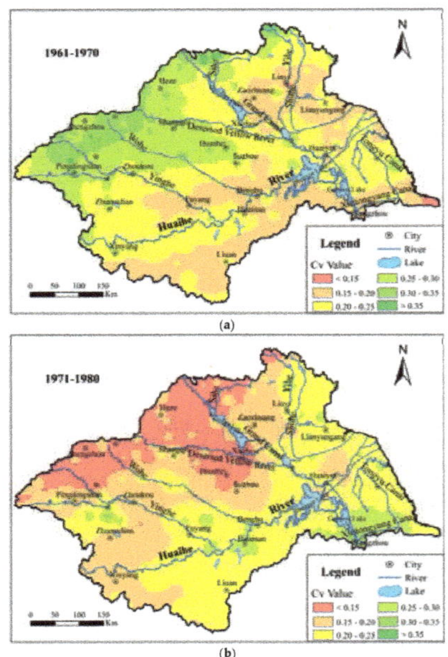

Figure 5. *Cont.*

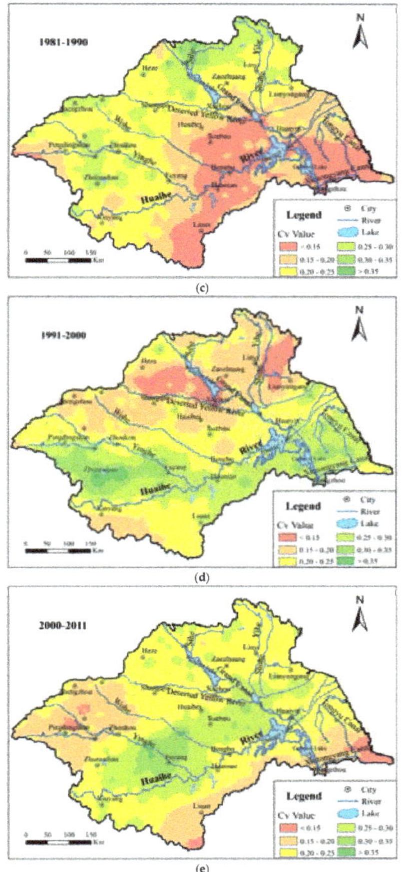

Figure 5. The spatial distribution of randomicity in the different decadal precipitation ((**a**) from 1961 to 1970; (**b**) from 1971 to 1980; (**c**) from 1981 to 1990; (**d**) from 1991 to 2000; (**e**) from 2001 to 2011).

4. Discussion

Based on the analysis of annual precipitation variability across the Huaihe River basin, we found the spatial and temporal dynamic characteristics changes, including the trends, the randomicity, abrupt change, and the periodicity. Different methods were adopted to derive these characteristic changes.

4.1. Trends of the Precipitation

Our results were consistent with scholars' findings but was different from others' [41,52–56]. The difference was derived from their various study area's boundaries. Most studies suggested that the Huaihe River basin did not include the Shandong peninsula, while others thought the opposite. The difference possibly stemmed from the fact that our study area did not include the Shandong Peninsula. Additionally, precipitation changes in China were mainly caused by the East Asian Monsoons and the precipitation pattern in China may change due to a reduction in the summer monsoon cycle [57]. The weakening of summer and winter monsoon contributed to increasing precipitation over Southeastern China [57,58]. However, there as a weakening relationship between East Asian winter monsoon and ENSO after mid-1970s [59]. The precipitation amount presented an advanced tendency after 1990s and increased obviously after 2000 [52,60]. Therefore, the summer and winter monsoon variations, especially around the mid-1970s, may be a possible factor for the precipitation changes in the Huaihe River basin.

4.2. Randomicity of Precipitation

Our results were similar to existing research findings [53,61]. The monsoon climate contributed to the uneven distribution of annual rainfall [61]. The annual rainfall uneven distribution also reflected that the Huaihe River basin was the climate transition zone (between 32°30′ N~33°55′ N and 104°31′ E~120°25′ E) (Figure 1). The Huaihe River is known as the transitional river in China [54]. It is the important north-south geographical demarcation boundary and a climate transition zone in China: while the south of the river is subject to a warm, humid temperates and the north is a subtropical, semi-arid climate [28,62,63]. The climate of the basin is mainly a warm, temperate monsoon and sub-humid climate. The annual rainfall heterogeneity is not only related to the latitude, but also to the typical physiognomy of the Huaihe River basin [64]. The rainfall evenness could be affected by other factors (e.g., temperature, topography, atmospheric circulation) and these attribution analyses might be effective targets for the future.

4.3. Abrupt Change of Precipitation

Our outcomes are close to the results obtained by Yang, Lu, and Wang [33,52,65]. In 1965 and 2002 abrupt changes occurred [66], this was consistent with our findings. In addition, the results were almost identical with the previous studies of the heavy droughts that occurred in 2001 and 2002 [67,68]. The anomalous northeast cold vortex could lead to the abrupt changes of rainfall [69]. The rainfall abrupt change points were similar with the anomalous occurrence time in our study. In general, the abrupt changes mainly occurred in 1970s and after the 1990s, which was close to the results detected by Lu and Wang [52,70]. However, the strong rainfall occurred in 2003, 2005, and 2007 [71]. The Pettitt test did not capture these abrupt change points in our study, since the data used in our study were yearly rainfall rather than the seasonal data. Additionally, there are many reservoirs in the basin, and the rainfall abrupt changes may be controlled by the distribution of these reservoirs.

4.4. Periodicity of Precipitation

The study showed that there were about 5-year, 10-year, 15-year, and 30-year cycles in the Huaihe River [72]. This was close to our findings. However, due to the difference of time series of rainfall data used, the 30-year cycle does not exist in our study. However, the other studies obtained 7–9 years as the major cycle and two years as the second cycle [66], which was different from our findings. WANG used the annual day and night rainfall as the basic data to study day/night precipitation cycle change, respectively [66]. Otherwise, we used the annual rainfall to get the cycle period rather than the only day or night rainfall. Another study demonstrated that the Yangtze-Huaihe meiyu period were 6–7 year and quasi-20-year periodical oscillations [73], which are due to the same reasons as above.

The Huaihe River basin is the climate transition zone between the north subtropical zone and the south extratropical zone of China [74]. Precipitation in this watershed is influenced by the variation of the transition zone position and ENSO [74,75]. There is an evident correlation between ENOS events and the abnormity of rainfall in the Huaihe River [76]. Some studies deemed that the rainfall periodicity was 20–25 year, and there was a positive correlation between summer precipitation and solar activities in the Huaihe River.

4.5. Advice for Management Implications in a Basin

Rainfall is the primary water source in the Huaihe River basin. According to the above analysis, the change of rainfall intensity, quantity, and pattern could cause the extreme events, such as drought and flood [77]. A shift in the storm track makes some regions wetter and the nearby-regions drier [78].

Thus, the spatio-temporal variability of rain across the watershed affects the flood control planning and comprehensive management. The precipitation increased significantly and changed unevenly remarkably in the Guohe River basin. Floods may be more likely to occur and we should strengthen the control measures and formulate the corresponding water resources utilization policy in the basin.

4.6. Future Challenges

Latitude, atmospheric circulation, sea-land distribution, ocean current, and topography are the main factors that affect the climate change. Local and regional variations in the feature of rainfall depend on the variability patterns of atmospheric circulation. Human activities, surface evaporation density, and latent energy also affect the precipitation changes [78]. However, we have not discussed and quantified the specific factors (i.e., temperatures, evaporations, land use changes and human activities) that are controlling the trends of rainfall. In the coming months, combined with the climate change, land use, and water conservancy facilities, our investigation needs to determine the factors that affect the dynamic changes of rain across the basin. Meanwhile, we could use the multi-factors regression analysis to obtain the primary and secondary factors. We could also study the data extrapolated in other watersheds to confirm the scientific merit of the study. Some scholars and researchers may worry about the accuracy and precision of extrapolation, so they may not have included that in their research. Indeed, it is a limitation. Thus, we could study the data extrapolated in other watersheds to confirm the scientific merit of the study. Additionally, we could use remote sensing data or more reference materials to prove the accuracy of the extrapolated data.

5. Conclusions

In this study, the precipitation series were used to detect the spatial distribution of temporal variation characteristics for the period 1961 to 2011 in the Huaihe River basin. Although the trend changes are not significant, the precipitation varied markedly. Its spatial distribution characteristics presented the evident difference as well. The entire basin was dominated by the slight increasing trends of inter-annual and inter-decadal precipitation, especially in the upper and middle reaches of the Huaihe River. But for the inter-annual rainfall, the change trend did not present at 95% confidence level in most of basin. For the inter-decadal, all trend changes are insignificant. The abrupt points were various in different periods. According to the proportion of abrupt area in the entire basin, we obtained nine chiefly abrupt points and three principal periods. The Cv value indicated that the spatial distributions of precipitation were heterogeneous and the randomicity were all of moderate variability.

Most previous studies concentrated on the trend, periodicity, and abrupt change of precipitation based on point data of a single meteorological station. However, our study extrapolated the single data across the watershed grid data. We obtained the rain variability characteristics from four sections: the trend, the abrupt change, the periodicity, and the randomicity. Our study could add needed value to the scientific community, especially to the basin where there is a lack of scientific literature to document. Meanwhile, our results could provide the scientific support for water resources management in river basins.

Acknowledgments: The authors extend sincere thanks for the supports of the Representative Achievements and Cultivation Project of State Key Laboratory of Simulation and Regulation of Water Cycle in River Basin (No. 2016CG02) and the National Key Research and Development Project (No. 2016YFA0601503).

Author Contributions: Z.-L.Y. completed the statistical analysis and wrote the paper; D.-H.Y. provided the writing ideas and supervised the study; G.-H.N. guided the writing and finalized the paper; P.D. was responsible for study designing and modifying the language; D.-M.Y. was responsible for data processing and diagraming; T.-L.Q. took charge of diagraming and data analysis; S.-Y.C., B.-S.W. and M.-J.Y. (in the order of initial letter in surnames) have equal contributions, mainly in charge of literature retrieval, data collection and paper translation.

Conflicts of Interest: The authors declares that there is no conflict of interests regarding the publication of this paper.

References

1. Limsakul, A.; Singhruck, P. Long-term trends and variability of total and extreme precipitation in Thailand. *Atmos. Res.* **2016**, *169*, 301–317. [CrossRef]
2. Oguntunde, P.G.; Abiodun, B.J.; Lischeid, G. Precipitation trends in Nigeria, 1901–2000. *J. Hydrol.* **2011**, *411*, 207–218. [CrossRef]
3. Sayemuzzaman, M.; Jha, M.K. Seasonal and annual precipitation time series trend analysis in North Carolina, United States. *Atmos. Res.* **2014**, *137*, 183–194. [CrossRef]
4. Wang, X.; Cui, G.; Wu, F.; Li, C.H. Analysis of temporal-spatial precipitation variations during the crop growth period in the Lancang River basin, southwestern China. *Ecol. Eng.* **2015**, *76*, 47–56. [CrossRef]
5. Xia, J.; She, D.X.; Zhang, Y.Y.; Du, H. Spatio-temporal trend and statistical distribution of extreme precipitation events in Huaihe River Basin during 1960–2009. *J. Geogr. Sci.* **2012**, *22*, 195–208. [CrossRef]
6. Chang, H.; Kwon, W.-T. Spatial variations of summer precipitation trends in South Korea, 1973–2005. *Environ. Res. Lett.* **2007**, *2*, 45012–45019. [CrossRef]
7. Chatterjee, S.; Khan, A.; Akbari, H.; Wang, Y.P. Monotonic trends in spatio-temporal distribution and concentration of monsoon precipitation (1901–2002), West Bengal, India. *Atmos. Res.* **2016**, *182*, 54–75. [CrossRef]
8. Irannezhad, M.; Marttila, H.; Chen, D.; Kløve, B. Century-long variability and trends in daily precipitation characteristics at three Finnish stations. *Adv. Clim. Chang. Res.* **2016**, *7*, 54–59. [CrossRef]
9. Zhao, C.Y.; Wang, Y.; Zhou, X.Y.; Cui, Y.; Liu, Y.L.; Shi, D.M.; Yu, H.M.; Liu, Y.Y. Changes in climatic factors and extreme climate events in Northeast China during 1961–2010. *Adv. Clim. Chang. Res.* **2013**, *4*, 92–102.
10. Tabari, H.; Abghari, H.; Hosseinzadeh Talaee, P. Temporal trends and spatial characteristics of drought and precipitation in arid and semiarid regions of Iran. *Hydrol. Process.* **2012**, *26*, 3351–3361. [CrossRef]
11. Zhang, W.; Pan, S.M.; Cao, L.G.; Cai, X.; Zhang, K.X.; Xu, Y.H.; Xu, W. Changes in extreme climate events in eastern China during 1960–2013: A case study of the Huaihe River Basin. *Quat. Int.* **2015**, *380*, 22–34. [CrossRef]
12. Chen, Y.N.; Xu, C.C.; Hao, X.M.; Ye, Z.X. Fifty-year climate change and its effect on annual runoff in the Tarim River Basin, China. *Quat. Int.* **2009**, *208*, 53–61.
13. Tabari, H.; Talaee, P.H. Temporal variability of precipitation over Iran: 1966–2005. *J. Hydrol.* **2011**, *396*, 313–320. [CrossRef]
14. Tian, Q.; Prange, M.; Merkel, U. Precipitation and temperature changes in the major Chinese river basins during 1957–2013 and links to sea surface temperature. *J. Hydrol.* **2016**, *536*, 208–221. [CrossRef]
15. Fu, G.B.; Yu, J.J.; Yu, X.B.; Ouyang, R.L.; Zhang, Y.C.; Wang, P.; Liu, W.B.; Min, L.L. Temporal variation of extreme precipitation events in China, 1961–2009. *J. Hydrol.* **2013**, *487*, 48–59. [CrossRef]
16. Liu, Q.; Yang, Z.; Cui, B. Spatial and temporal variability of annual precipitation during 1961–2006 in Yellow River Basin, China. *J. Hydrol.* **2008**, *361*, 330–338. [CrossRef]
17. Liu, Y.; Li, X.; Zhang, Q.; Guo, Y.F.; Gao, G.; Wang, J.P. Simulation of regional temperature and precipitation in the past 50 years and the next 30 years over China. *Quat. Int.* **2010**, *212*, 57–63. [CrossRef]
18. Liu, H.Y.; Gao, Q. Contribution rate of driving factors on the rainfall-sediment relationship of Longhe River watershed in the Three Gorges Reservoir region. *Sci. Soil Water Conserv.* **2015**, *13*, 1–8.
19. Portmann, R.W.; Solomon, S.; Hegerl, G.C. Spatial and seasonal patterns in climate change, temperatures, and precipitation across the United States. *Proc. Natl. Acad. Sci. USA* **2009**, *106*, 7324–7329. [CrossRef] [PubMed]
20. Keggenhoff, I.; Elizbarashvili, M.; Amiri-Farahani, A.; King, L. Trends in daily temperature and precipitation extremes over Georgia, 1971–2010. *Weather Clim. Extrem.* **2014**, *4*, 75–85. [CrossRef]
21. Beecham, S.; Chowdhury, R.K. Temporal characteristics and variability of point precipitation: A statistical and wavelet analysis. *Int. J. Climatol.* **2010**, *30*, 458–473.
22. Sang, Y.F.; Singh, V.P.; Gong, T.; Xu, K.; Sun, F.; Liu, C.; Liu, W.; Chen, R. Precipitation variability and response to changing climatic condition in the Yarlung Tsangpo River basin, China. *J. Geophys. Res. Atmos.* **2016**, *121*, 8820–8831. [CrossRef]
23. Wang, R.; Li, C. Spatiotemporal analysis of precipitation trends during 1961–2010 in Hubei province, central China. *Theor. Appl. Clim.* **2016**, *124*, 385–399. [CrossRef]

24. Karpouzos, D.K.; Kavalieratou, S.; Babajimopoulos, C. Trend analysis of precipitation data in Pieria Region (Greece). *Eur. Water* **2010**, *30*, 31–40.
25. Shi, P.; Rui, X.F. Comparison and improvement of spatial rainfall interpolation methods. *J. Hohai Univ. (Nat. Sci.)* **2005**, *33*, 361–365. (In Chinese)
26. Mohamed, A.S. *Reliabilty Estimation of Rainfall-Runoff Models*; State University of New York: New York, NY, USA, 1999.
27. Shi, P.; Ma, X.X.; Chen, X.; Qu, S.M.; Zhang, Z.C. Analysis of variation trends in precipitation in an upstream catchment of Huai River. *Math. Probl. Eng.* **2013**, *2013*, 1262–1268. [CrossRef]
28. Zhou, Y.K.; Wang, L.C.; Peng, X.Y.; Zhang, J. Chaotic dynamics of the flood series in the Huaihe river basin. *J. Hydrol.* **2002**, *258*, 100–110. [CrossRef]
29. Song, X.S.; Yan, D.H.; Wang, Y.H.; Wang, Y. Analysis on the evolution of drought and flood class in the east-central Huang-Huai-Hai plain over the last 540 years basing on Markov model. *J. Hydrol. Eng.* **2013**, *12*, 1425–1432. (In Chinese)
30. Yang, Z.Y.; Yuan, Z.; Yan, D.H.; Weng, B. Study of spatial and temporal distribution and multiple characteristics of drought and flood in Huang-Huai-Hai River basin. *Adv. Water Sci.* **2013**, *24*, 617–625. (In Chinese)
31. Croley, T.E., II; Hartmann, H.C. Resolving Thiessen polygons. *J. Hydrol.* **1985**, *76*, 363–379. [CrossRef]
32. China Meteorological Data Service Center (CMDC). Available online: http://data.cma.cn/data/cdcdetail/dataCode/SURF_CLI_CHN_MUL_DAY.html (accessed on 20 March 2012).
33. Yang, M.M.; Zhong, P.A.; Wei, P. Study on precipitation evolution rule of Huaihe River Basin in main flood period. *Water Resour. Power* **2012**, *30*, 37–40.
34. Wei, F.Y. *Xiandai Qiuhou Tongji Zhenduan Yu Yuce Jishu*; China Meteorological Press: Beijing, China, 2007; pp. 99–104.
35. Li, M.; Xia, J.; Chen, S.M.; Meng, D.J. Wavelet analysis on annual precipitation around 300 Years in Beijing Area. *J. Nat. Resour.* **2011**, *6*, 1001–1011. (In Chinese)
36. Teoh, W.L.; Khoo, M.B.C.; Castagliola, P.; Yeong, W.C.; Teh, S.Y. Run-sum control charts for monitoring the coefficient of variation. *Eur. J. Oper. Res.* **2017**, *257*, 144–158. [CrossRef]
37. Aerts, S.; Haesbroeck, G.; Ruwet, C. Multivariate coefficients of variation: Comparison and influence functions. *J. Multivar. Anal.* **2015**, *142*, 183–198. [CrossRef]
38. Dong, X.G.; Gu, W.Z.; Meng, X.X.; Liu, H.B. Change features of precipitation events in Shandong Province from 1961 to 2010. *Acta Geogr. Sin.* **2014**, *69*, 661–671. (In Chinese) [CrossRef]
39. Xia, J.; Ou, C.P.; Huang, G.H.; Wang, Z.G. The analysis of Haihe River Basin hydro-meteorological spatio-temporal variability based on GIS and information difference measure. *J. Nat. Resour.* **2007**, *22*, 409–415. (In Chinese)
40. Wang, X.J.; He, M.R.; Shang, M.T. Evolution law of precipitation in Yulin. *J. Arid Land Resour. Environ.* **2011**, *25*, 103–108. (In Chinese)
41. Yan, D.H.; Han, D.M.; Wang, G.; Yuan, Y.; Hu, Y.; Fang, H.Y. The evolution analysis of flood and drought in Huai River Basin of China based on monthly precipitation characteristics. *Nat. Hazards* **2014**, *73*, 849–858. [CrossRef]
42. Jin, Y.; Yan, A.; Jiang, P.A.; Wang, Z.; Wang, X.J. Spatial variability of filed soil water in Manas River Basin. *Xinjiang Agric. Sci.* **2013**, *50*, 1554–1559. (In Chinese)
43. Pettitt, A.N. A non-parametric approach to the change-point problem. *J. R. Stat. Soc.* **1979**, *28*, 126–135. [CrossRef]
44. Zhang, Y.M.; Tian, Y.; Lei, X.H.; Song, W.Z.; Jiang, Y.Z. Change characteristics of water resources in Sanchahe Upstream over past 50 years. *J. China Hydrol.* **2016**, *36*, 79–84.
45. Yi, H.; Shu, H. The improvement of the Morlet wavelet for multi-period analysis of climate data. *Comptes Rendus Geosci.* **2012**, *344*, 483–497. [CrossRef]
46. Hermida, L.; López, L.; Merino, A.; Berthet, C.; Gercía-Ortega, E.; Sánchez, J.L.; Dessens, J. Hailfall in southwest France: Relationship with precipitation, trends and wavelet analysis. *Atmos. Res.* **2015**, *156*, 174–188. [CrossRef]
47. Zhang, S.Y.; Wang, J.H.; Zhai, J.Q.; Li, H.H.; Zhao, Y.; Wang, Q.M.; Zhang, W. Characteristics analysis of time serial of rainfall in the Northern part of Haihe River Basin from 1956 to 2012. *South-to-North Water Transf. Water Sci. Technol.* **2016**, *14*, 36–42. (In Chinese)

48. Domingues, M.O.; Mendes, O.; da Costa, A.M. On wavelet techniques in atmospheric sciences. *Adv. Space Res.* **2005**, *35*, 831–842. [CrossRef]
49. Li, D.; Guo, Z. Some aspects of ecological modeling developments in China. *Ecol. Model.* **2000**, *132*, 3–10. [CrossRef]
50. Kovács, J.; Hatvani, I.G.; Korponai, J.; Kovács, I.S. Morlet wavelet and autocorrelation analysis of long-term data series of the Kis-Balaton water protection system (KBWPS). *Ecol. Eng.* **2010**, *36*, 1469–1477. [CrossRef]
51. Labat, D. Recent advances in wavelet analyses: Part 1. A review of concepts. *J. Hydrol.* **2005**, *314*, 275–288. [CrossRef]
52. Lu, Z.G.; Zhang, X.H.; Huo, J.L.; Wang, K.Q.; Xie, X.P. The evolution characteristics of the extreme precipitation in Huaihe river basin during 1960–2008. *J. Meteorol. Sci.* **2011**, *31*, 74–80.
53. Xing, W.Q.; Wang, W.Q.; Wu, Y.Q.; An, G.Y. Change properties of precipitation concentration in Huaihe River Basin. *Water Resour. Power* **2011**, *29*, 1–5.
54. Zhang, D.D.; Yan, D.H.; Wang, Y.C.; Lu, F.; Wu, D. Changes in extreme precipitation in the Huang-Huai-Hai River basin of China during 1960–2010. *Theor. Appl. Clim.* **2015**, *120*, 195–209. [CrossRef]
55. Zhou, L.G.; Dai, S.B. Spatial and temporal variation characteristics of heavy rainfall in the Huaihe River Basin in recent 60 years. *South-to-North Water Transf. Water Sci. Technol.* **2015**, *5*, 847–852. (In Chinese)
56. Yuan, Z.; Yang, Z.Y.; Zheng, X.D.; Yuan, Y. Spatial and temporal variations of precipitation in Huaihe River Basin in resent 50 years. *South-to-North Water Transf. Water Sci. Technol.* **2012**, *10*, 98–103. (In Chinese)
57. Xu, M.; Chang, C.P.; Fu, C.; Qi, Y.; Robock, A.; Robinson, D.; Zhang, H.M. Steady decline of East Asian monsoon winds, 1969–2000: Evidence from direct ground measurements of wind speed. *J. Geophys. Res. Atmos.* **2006**, *111*, 906–910. [CrossRef]
58. Zhou, L.T. Impact of East Asian winter monsoon on precipitation over southeastern China and its dynamical process. *Int. J. Clim.* **2011**, *31*, 677–686. [CrossRef]
59. Wang, H.J.; He, S.P. Weakening relationship between East Asian winter monsoon and ENSO after mid-1970s. *Chin. Sci. Bull.* **2012**, *57*, 3535–3540. [CrossRef]
60. Wang, Y.; Cao, M.K.; Tao, B.; Li, K.R. The characteristics of spatio-temporal patterns in precipitation in China under the background of global climate change. *Geogr. Res.* **2006**, *25*, 1031–1040. (In Chinese)
61. Gu, W.L.; Wang, J.J.; Zhu, Y.Y.; Sun, C.R. Annual distribution of precipitation over the Huaihe River Basin. *Resour. Environ. Yangtze Basin* **2010**, *19*, 429–434. (In Chinese)
62. Wang, K.; Chu, D.; Yang, Z. Flood control and management for the transitional Huaihe River in China. *Procedia Eng.* **2016**, *154*, 703–709.
63. Wang, J.; Liu, G.J.; Lu, L.L.; Zhang, J.M.; Liu, H.Q. Geochemical normalization and assessment of heavy metals (Cu, Pb, Zn, and Ni) in sediments from the Huaihe River, Anhui, China. *Catena* **2015**, *129*, 30–38. [CrossRef]
64. Ye, J.Y.; Huang, Y.; Zhang, C.L.; Li, Z.J. Characteristics of precipitation days and intensity over the Huaihe River basin in flood season during recent 50 years. *J. Lake Sci.* **2013**, *25*, 583–592. (In Chinese)
65. Wang, S.; Tian, H.; Ding, X.J.; Xie, W.S.; Tao, Y. Analysis of extreme precipitation events in rainy season over Huaihe River Basin from 1961 to 2008. *Meteorol. Sci. Technol.* **2012**, *40*, 87–91. (In Chinese)
66. Wang, S.; Xie, W.S.; Tang, W.A.; Tao, Y.; Ding, X. Change characteristics of day and night precipitation in Huaihe River Basin in 1961–2009. *Chin. J. Ecol.* **2011**, *30*, 2881–2887. (In Chinese)
67. Dai, X.G.; Wang, P.; Chou, J.F. Multi-scale characteristics of precipitation in rainy season and summer monsoon decadal decay in North China. *Chin. Sci. Bull.* **2003**, *48*, 2483–2487. (In Chinese) [CrossRef]
68. Wang, J.T.; Liang, S.X.; Yu, H. Analysis of water potential in Huaihe River Basin during 2000 to 2009. *China Flood Drought Manag.* **2011**, *21*, 21–24. (In Chinese)
69. Li, C.; Han, G.R.; Sun, Y. Anomalous features of Northeast cold vortex in late 50 a and its correlation with rainfall in Huaihe river valley. *J. Meteorol. Sci.* **2015**, *35*, 216–222. (In Chinese) [CrossRef]
70. Wang, S.; Tian, H.; Ding, X.J.; Xie, W.S.; Tao, Y. Climate characteristics of precipitation and phenomenon of drought-flood abrupt alternation during main floods season in Huaihe River Basin. *Chin. J. Agrometeorol.* **2009**, *30*, 31–34. (In Chinese)
71. Gao, C.; Jiang, T.; Zhai, J.Q. Analysis and precipitation of climate in the Huaihe River Basin. *Chin. J. Agrometeorol.* **2012**, *33*, 8–17. (In Chinese) [CrossRef]

72. Wang, J.C.; Guo, J.X.; Xu, J.; Li, F. Multi-time scales change characteristics and relationship of meteorological variables in the upper and middle regions of the Huaihe River Basin in recent 55 years. *Sci. Geogr. Sin.* **2017**, *37*, 611–619. [CrossRef]
73. Hu, Y.M.; Ding, Y.H.; Liao, F. A classification of the precipitation patterns during the Yangtze-Huaihe meiyu period for recent 52 years. *Acta Meteorol. Sin.* **2010**, *68*, 235–247. (In Chinese)
74. Liu, F.H.; Chen, X.; Chen, X.W.; Song, S. Relations hip between temperature change in climate boundary and summer precipitation over the Huaihe River basin. *Clim. Environ. Res.* **2010**, *15*, 169–178. (In Chinese)
75. Wang, Y.; Zhang, Q.; Zhang, S.; Chen, X.H. Spatial and temporal characteristics of precipitation in the Huaihe River Basin and its response to ENSO events. *Sci. Geogr. Sin.* **2016**, *36*, 128–134. [CrossRef]
76. Xin, Z.B.; Xie, Z.R. The impact of ENSO events on Huaihe River Basin's precipitation. *Sci. Meteorol. Sin.* **2005**, *25*, 346–354.
77. Shi, P.; Qiao, X.Y.; Chen, X.; Zhou, M.; Qu, X.X.; Zhang, Z.C. Spatial distribution and temporal trends in daily and monthly precipitation concentration indices in the upper reaches of the Huai River, China. *Stoch. Environ. Res. Risk Assess.* **2014**, *28*, 201–212. [CrossRef]
78. Trenberth, K.E. Changes in precipitation with climate change. *Clim. Res.* **2011**, *47*, 123–138. [CrossRef]

Article

Debris Flow Susceptibility Assessment in the Wudongde Dam Area, China Based on Rock Engineering System and Fuzzy *C*-Means Algorithm

Yanyan Li [1,*], Honggang Wang [2], Jianping Chen [3] and Yanjun Shang [1]

1 Key Laboratory of Shale Gas and Geoengineering, Institute of Geology and Geophysics, Chinese Academy of Sciences, Beijing 100029, China; jun94@mail.iggcas.ac.cn
2 Central Southern China Electric Power Design Institute Co., Ltd of CPECC, Wuhan 430071, China; whg5621@csepdi.com
3 College of Construction Engineering, Jilin University, Changchun 130026, China; chenjpwq@126.com
* Correspondence: lee_xandy@126.com; Tel.: +86-10-8299-8632

Received: 29 March 2017; Accepted: 17 June 2017; Published: 4 September 2017

Abstract: Debris flows in the Wudongde dam area, China could pose a huge threat to the running of the power station. Therefore, it is of great significance to carry out a susceptibility analysis for this area. This paper presents an application of the rock engineering system and fuzzy *C*-means algorithm (RES_FCM) for debris flow susceptibility assessment. The watershed of the Jinsha River close to the Wudongde dam site in southwest China was taken as the study area, where a total of 22 channelized debris flow gullies were mapped by field investigations. Eight environmental parameters were selected for debris flow susceptibility assessment, namely, lithology, watershed area, slope angle, stream density, length of the main stream, curvature of the main stream, distance from fault and vegetation cover ratio. The interactions among these parameters and their weightings were determined using the RES method. A debris flow susceptibility map was produced by dividing the gullies into three categories of debris flow susceptibility based on the susceptibility index (*SI*) using the FCM algorithm. The results show that the susceptibility levels for nine of the debris flow gullies are high, nine are moderate and four are low, respectively. The RES based *K*-means algorithm (RES_KM) was used for comparison. The results suggest that the RES_FCM method and the RES_KM method provide very close evaluation results for most of the debris flow gullies, which also agree well with field investigations. The prediction accuracy of the new method is 90.9%, larger than that obtained by the RES_KM method (86.4%). Therefore, the RES_FCM method performs better than the RES_KM method for assessing the susceptibility of debris flows.

Keywords: debris flow; susceptibility; rock engineering system; fuzzy *C*-means algorithm; interaction

---

1. Introduction

Debris flow is a sudden natural process that frequently occurs in mountainous areas. It has high mobility [1,2] and is able to carry meter-size boulders [3]. Consequently, debris flows have greatly destructive potential and could pose a huge threat to human lives and properties. For the management and reduction of risk posed by debris flows, susceptibility analysis aimed at delineating the potential threatened areas plays an important role.

Various approaches have been developed for debris flow susceptibility analysis by employing a specific set of environmental parameters, such as empirical models, statistical analyses and artificial intelligence. Empirical models [4,5] often need to be calibrated through small areas where past events exist before they could be used for a whole region. In fact, to establish a practical empirical model, large datasets are necessary. Artificial intelligence models, such as genetic algorithm [6], artificial neural network [7–10] and support vector machine [11] have been applied for debris flow prediction. Most of these models have been created using regional debris flow inventories derived from remotely sensed data. Statistical analyses, including logistic regression [12–15], discriminant analysis [16,17], and Bayes learning [18], are deemed to be suitable for susceptibility assessment in large and complex areas [19–21]. Using Bayes learning and logistic regression to predict debris flows in southwest Sichuan, China, Xu et al. [22] pointed out that both methods have disadvantages: Bayes requires some variable assumptions that are difficult to be completely met in practice, whereas logistic regression needs large samples for the iterative calculation to obtain stable model parameters. Other methods such as weight of evidence [23,24] and analytic hierarchy process [25,26] have also been used for susceptibility analysis. The spatial results of these approaches are generally appealing, and they give rise to qualitative and quantitative mapping of the threatened areas [27].

The occurrence of debris flows can be attributed to complex interactions among geology, topography and meteorology [22]. This paper proposes a new model for debris flow susceptibility evaluation based on spatial variables that are considered to be potential controls of debris flows in the watershed of the Jinsha River close to the Wudongde dam site in southwest China. Based on the rock engineering system (RES), which was first introduced by Hudson [28] to deal with complex engineering problems, the interactions among environmental parameters and their weightings were determined. A debris flow susceptibility zone map was created using the fuzzy C-means (FCM) algorithm, which is a powerful method in data mining and knowledge discovery proposed by Bezdek [29], according to the results obtained by RES. This work also tests the suitability of FCM to discriminate different levels of susceptibility. The novelty of this work is the integration of RES and FCM methods for the debris flow susceptibility assessment.

2. Study Area

The study area (Figure 1) lies along the lower reaches of the Jinsha River and is the reservoir region of the Wudongde hydropower station, which is located in the mountains separating the Sichuan and Yunnan provinces. The Wudongde hydropower station is one of the four largest power plants in the lower reaches of the Jinsha River. The station controls a basin area of 406,100 km^2, which occupies 86% of the Jinsha River. The studied section of Jinsha River is about 210 km long. The area of investigation along the Jinsha River was extended from the alluvial plain to the crest. Based on field investigations, 22 channelized debris flow gullies distributed on both sides of the Jinsha River were identified, as shown in Figure 2. Considering that loose materials from the debris flow gullies could enter the Jinsha River and affect the running of the power station, it is of great significance to carry out a susceptibility analysis for this area.

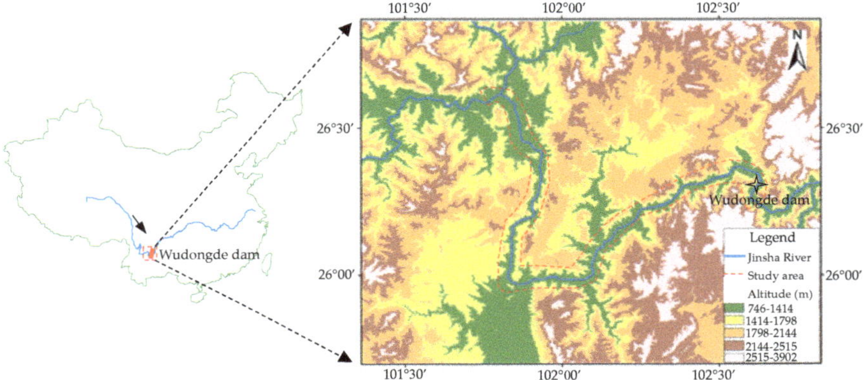

Figure 1. Location map of the study area [30].

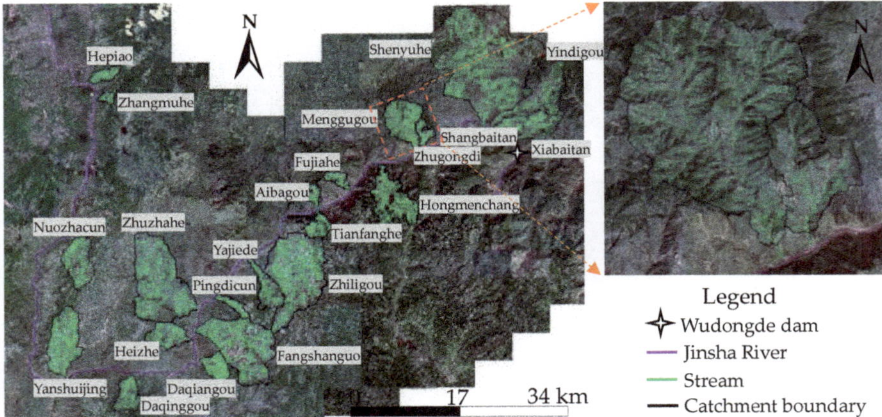

Figure 2. Location of the debris flow gullies.

2.1. Geological and Tectonic Setting

The geology comprises two major components: a pre-Sinian crystalline basement and a Sinian-Cretaceous sedimentary cover. The former mainly consists of a series of metamorphic rocks (phyllite, slate and schist), which widely outcrops along the Jinsha River. The latter includes magmatic rocks (granite and quartz diorite) and sedimentary rocks (limestone, sandstone, mudstone and shale). Figure 3 shows the lithology along the Jinsha River. According to field investigations, three types of potential source materials for debris flows outcrop in the study area [31]: (1) the Longjie silt layer from the Late Pleistocene period, mainly composed of clayey silt, silt and sand; (2) the sediment of the Madianhe Group from the Holocene period, which is mainly composed of silt and gravel; (3) and the red-bed, Triassic and the Cretaceous argillaceous evaporites.

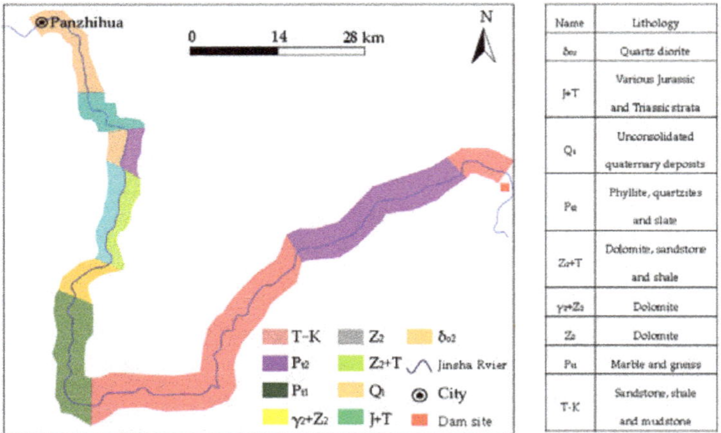

Figure 3. Lithology in the study area [31].

The study area is located in the eastern section of the Tethys-Himalaya tectonic domain, one of the tectonic zones of the Himalaya characterized by intense compressing and folding. The predominant structures are regional-scale faults constituting the famous Chuan-Dian N-S tectonic belt [32]. A total of

13 regional-scale faults that dominantly trend approximately N-S are situated in this region (Figure 4). Several strong earthquakes with magnitude greater than 6.0 have been triggered by these faults since 1955 [33], such as the Puduhe earthquake (Magnitude 6.3, 1985) triggered by the Puduhe fault, the Wozhangshan earthquake (Magnitude 6.5, 1995) triggered by the Tanglang-Yimeng fault, and the Panzhihua earthquake (Magnitude 6.1, 2008) triggered by the Mopanshan-Lvzhijiang fault. These earthquakes triggered a lot of rock falls and landslides and produced a large quantity of loose materials sufficient to potentially trigger debris flows in the drainages.

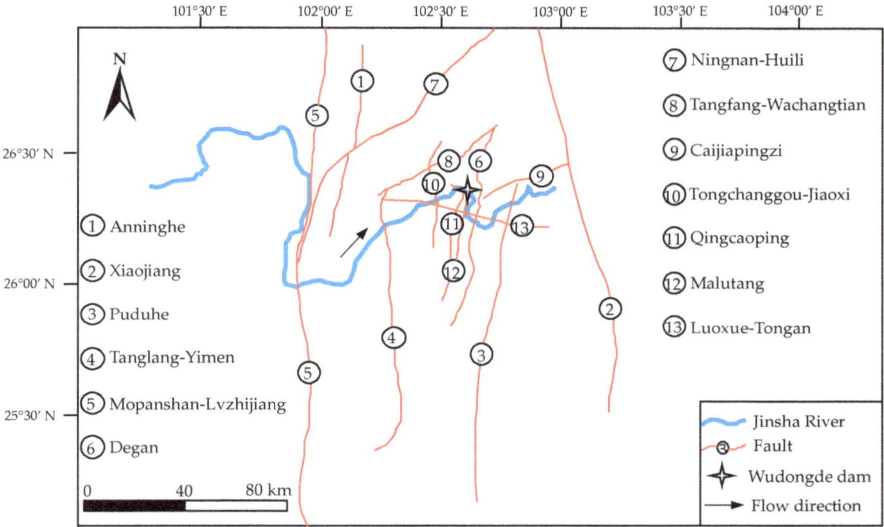

Figure 4. Regional tectonic framework map [33].

2.2. Geomorphological Setting

This area exhibits a mountain canyon geomorphology, with elevations ranging from 800 to 3600 m, as shown in Figure 1. Observed geomorphic features include cliffs, rocky slopes, ridges and Quaternary deposits along the river valleys. The average slope angles of the hillsides in the region vary from 30 to 45°. The slopes are rocky and poorly vegetated; the dominant species are grasses on the soils. The distributions of the river network and ridges are controlled by structures in some extent. The effect of high relief and structural control is also well reflected by deep gorges and narrow valleys carved by numerous streams.

2.3. Meteorological Setting

The study area experiences a low-latitude plateau subtropical monsoon meteorology, characterized by concentrated rainfall and distinct wet and dry seasons. The mean annual temperature is 20.9 °C, with the 32 year (from 1972 to 2003) mean annual precipitation being 1058 mm [34]. The rainy season concentrates from May to October, with a peak in July. The maximum 10-min, 1-h and 24-h rainfall rates recorded in 32 years are 21.7, 77.2 and 111.5 mm, respectively.

3. Influencing Parameters

The occurrence of debris flows is a complicated process that requires favorable terrain conditions, source conditions and hydrodynamic conditions [6,35]. This research selected the triggering area of the debris flow as the base spatial unit. Based on field investigations and previous studies in the study area [31,33,34], eight factors were selected as the environmental predictors for debris flows. They are lithology (P1), watershed area (P2), slope angle (P3), stream density (P4), length of the main stream (P5), curvature of the main stream (P6), distance from fault (P7), and vegetation cover ratio (P8). Note that rainfall is deemed as a triggering factor for debris flows in our study area. However, it was not considered in this work because it is relatively uniform throughout the area.

3.1. Lithology

The lithology is one of the main parameters influencing the occurrence of debris flows in the study area [31]. It controls the stability of slopes and thus affects the debris supply for drainages. Filed surveys suggest that Quaternary deposits are most prone to the initiation of debris flows, whereas magmatic rocks and limestones have the lowest susceptibility for the occurrence of debris flows.

3.2. Watershed Area

The watershed area is referenced to the rainfall that can be collected and to the volume of loose materials [7]. Debris flows in the study area primarily occurred in catchments with relatively larger areas, most probably because a larger watershed area could collect more rainfall and a larger volume of loose materials.

3.3. Slope Angle

The slope angle strongly influences the initiation and transportation of debris flows [7,36]. Previous research [5,37] shows that most debris flows occur in the areas with slope gradients higher than 15°. This value was considered as the initiation threshold for debris flows in our research.

3.4. Stream Density

The stream density is expressed as the total length of all the streams in a catchment divided by the total area [33,38]. This factor reflects the interactions among lithology, geological structures, and weathering degree of rocks in a catchment because drainages often develop in weak area [33]. In addition, it can affect the shape of a river's hydrograph during a rainstorm [38]. Investigations suggest that catchments carved by numerous streams are prone to the initiation of debris flows.

3.5. Length of the Main Stream

The length of the main stream is also an important factor influencing the occurrence of debris flows in the study region. The longer the main stream is, the more deposits a debris flow could gather together and transport to the runout zone [7].

3.6. Curvature of the Main Stream

The curvature of the main stream is related to the ratio of the main stream' curve length to its straight length [33]. It reflects the discharge capacity of debris flows. Field surveys indicate that a catchment with an abroad and straight channel often retains limited loose materials, which is not prone to the debris flow occurrence.

3.7. Distance from Fault

Faults play a crucial role in the initiation of debris flows in our study area. They trigger earthquakes, produce discontinuities in rocks, and as a consequence, furnish debris that can be mobilized. Field investigations suggest that catchments near faults are prone to experiencing debris flow events.

3.8. Vegetation Cover Ratio

The vegetation cover ratio is described as the ratio of the vegetation area to the watershed area [10]. The dominant species in the study area are grasses on the soils. The natural vegetation in the study area has been damaged because of irrational deforestation and reclamation, suggestive of the highly erosive capability of the flows, able to increase their volume as they move [31,34] A poor vegetation cover indicates a high chance for a catchment to suffer from rainfall and rock weathering, and therefore facilitates debris production.

The watershed area, slope angle, stream density, length of the main stream, curvature of the main stream, and distance from fault were derived from the digital elevation model (DEM) with a resolution of 2.5 m. The lithology was obtained from a 1:50,000 scale geological map. The vegetation cover ratio was derived from the SPOT5 remote sensing image. The environmental parameters were subdivided into classes (Table 1) based on previous studies in the study area [31,34]. A standardization method was adopted to rescale the data to a common numerical basis according to their influence on the debris flow occurrence, which was carried out by transforming raw data to scores [39]. The ratings of the influencing factors (P_i) are shown in Table 1.

Table 1. Classification of influencing factors and their rating values.

Description	Rating	Description	Rating
1. Lithology		5. Length of the main stream (km)	
Magmatic rocks, and limestones	0	<1	0
Phyllite, slate and schist	1	1–5	1
Sandstones, mudstones, and shale	2	5–10	2
Quaternary deposits	3	>10	3
2. Watershed area (km^2)		6. Curvature of the main stream	
<0.5 or >50	0	<1.1	0
0.5–10	1	1.1–1.25	1
10–35	2	1.25–1.4	2
35–50	3	>1.4	3
3. Slope angle (°)		7. Distance from fault (km)	
<15	0	>0.6	0
15–25	1	0.4–0.6	1
25–32	2	0.2–0.4	2
>32	3	<0.2	3
4. Stream density (km/km^2)		8. Vegetation cover ratio	
<5	0	>0.75	0
5–10	1	0.5–0.75	1
10–20	2	0.25–0.5	2
>20	3	<0.25	3

4. Method

4.1. Rock Engineering System

The implementation of the rock engineering system (RES) method can be achieved through an interaction matrix, which is the basic analytical device used in RES for characterizing the influencing parameters and their interaction mechanisms relevant to a particular engineering problem [40,41]. In RES, all selected parameters associated with a problem are arranged along the leading diagonal of the interaction matrix. The influence of each parameter on any other parameter, which is called interaction, is placed in the corresponding off-diagonal cells. Traditionally, the off-diagonal cells are assigned numerical values to quantify the degree of the influence of one factor on the other factors, named "coding the matrix". Various approaches have been proposed for coding the interaction matrix [42], such as the 0–1 binary, expert semi-quantitative (ESQ), and the continuous quantitative coding (CQC) methods. Among these methods, the ESQ coding is the most commonly used, whereby the interaction between the parameters is ranked based on a numerical scale. Typically, a scale from 0 to 4 is employed (Table 2) [28].

Table 2. ESQ interaction matrix coding [28].

Coding	Description
0	No interaction
1	Weak interaction
2	Medium interaction
3	Strong interaction
4	Critical interaction

Figure 5 presents an example for the simplest interaction matrix with two factors. Note that the influence of *i* on *j* is often not the same as the influence of *j* on *i*, indicating that the interaction matrix is not symmetric. Generally, the interaction matrix can contain any number of variables, depending on the engineering objective and the level of analysis required [43]. Figure 6 shows the coding of a multiple-dimensional interaction matrix. A problem that contains *N* factors will have an interaction matrix of *N* rows by *N* columns. The column passing through P_i represents the influence of other parameters on P_i, while the row through P_i represents the influence of P_i on the remaining parameters. For example, the (*i*, *j*)-th element in the matrix represents the influence of parameter *i* on parameter *j*.

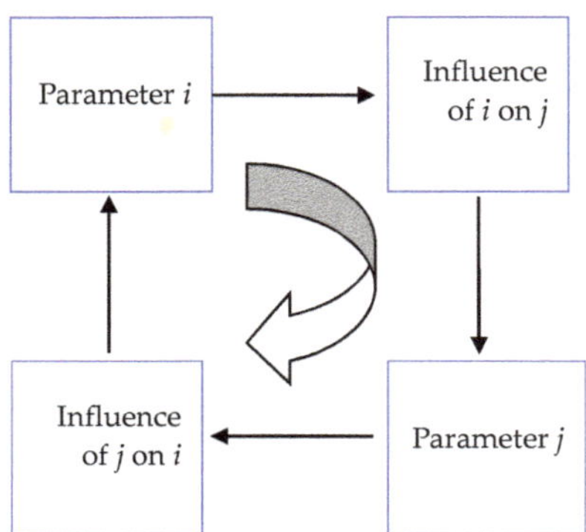

Figure 5. General illustration of the interaction matrix with two factors [28].

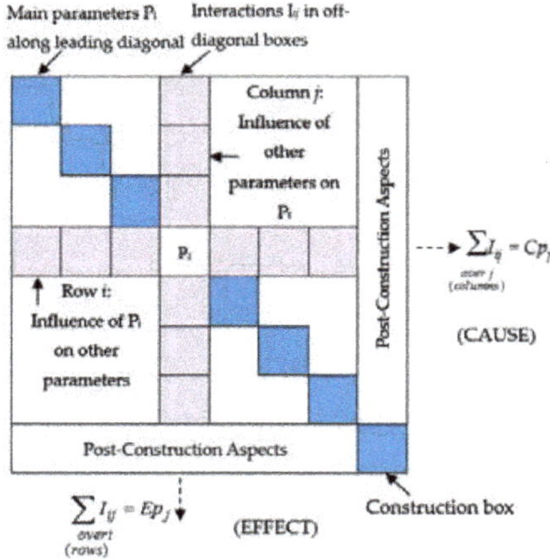

Figure 6. Summation of coding values in the row and column through each parameter to establish the cause and effect coordinates [28].

After coding the interaction matrix, the sum of each row and that of each column can be computed. For each parameter i, the sum of its row values and that of its column values are called the "cause" value (C_i) and the "effect" value (E_i), respectively. The coordinate values (C_i, E_i) for each parameter can be plotted in cause and effect space, forming the so-called cause–effect plot, which can help to understand the relative importance of each parameter within the system [43]. The percentage value of (C + E) can be used as the weighting of each parameter, which is given by:

$$w_i = \frac{(C_i + E_i)}{\sum_i C_i + \sum_i E_i} \times 100 \tag{1}$$

4.2. Fuzzy C-Means Algorithm

Fuzzy C-means (FCM) clustering method proposed by Bezdek [29] is a well-known and powerful method in data mining and knowledge discovery. It generates a fuzzy partition based on the idea of partial membership expressed by the degree of membership of each object in a given cluster. In fuzzy clustering, each object has a degree of belonging to clusters rather than belonging completely to just one cluster [44].

Given a data set of N observations obtained from N regions, each represented by a vector of P attributes, $\mathbf{X}_j = (\mathbf{X}_{j1}, \mathbf{X}_{j2}, \ldots, \mathbf{X}_{jP})$, the algorithm is designed to partition the data set into C clusters (i.e., structural domains) by iteratively minimizing the fuzzy objective function which is expressed as follows [29]:

$$J = \sum_{j=1}^{N}\sum_{i=1}^{C} (u_{ij})^m d^2(\mathbf{X}_j, \mathbf{V}_i)(C \leq N), \tag{2}$$

where u_{ij} represents the degree of membership of observation $\mathbf{X}_j$ in cluster i, m is the fuzziness index, which controls the fuzziness of the memberships, and $d(\mathbf{X}_j, \mathbf{V}_i)$ is the distance between observation $\mathbf{X}_j$ and the ith cluster center $\mathbf{V}_i$. $m = 2$ is deemed to be the best for most applications [29]. In this research, the value of P is 6 since there are six parameters that were used for structural domain determination.

The distance measure $d(\mathbf{X}_j, \mathbf{V}_i)$ is expressed as [45]:

$$d^2(\mathbf{X}_j, \mathbf{V}_i) = \sum_{p=1}^{P} (\mathbf{X}_{jp} - \mathbf{V}_{ip})^2. \tag{3}$$

u_{ij} can be calculated from [45]:

$$(u_{ij})^m = \left[\left(\frac{1}{d^2(\mathbf{X}_j, \mathbf{V}_i)}\right)^{1/(m-1)}\right]\left[\sum_{i=1}^{C}\left(\frac{1}{d^2(\mathbf{X}_j, \mathbf{V}_i)}\right)^{1/(m-1)}\right]^{-1}. \tag{4}$$

The cluster center $\mathbf{V}_i$ is computed by [45]:

$$\mathbf{V}_i = \sum_{j=1}^{N} u_{ij}^m \mathbf{X}_j / \sum_{j=1}^{N} u_{ij}^m. \tag{5}$$

5. Results and Discussion

5.1. RES Model for Debris Flow Susceptibility Assessment

With the selected eight parameters, an 8 by 8 interaction matrix was built according to Table 2, as shown in Table 3. For instance, considering that the lithology can be eroded and produce a different slope depending on its rheology, a value of 4 is assigned to the cell of the 1st row and 3th column in the matrix, suggesting that the lithology (P1) has a critical influence on the slope angle (P3). In addition, a value of 0 is assigned to the cell of the 3rd row and 1st column in the matrix, suggesting that lithology is not influenced by slope angles.

Table 3. Interaction matrix.

P1	2	4	3	3	3	2	3
0	P2	2	1	2	1	2	1
0	1	P3	2	3	2	0	2
1	0	3	P4	2	3	0	2
1	0	2	2	P5	2	0	1
1	0	1	2	4	P6	0	1
3	3	3	3	2	3	P7	2
2	0	1	3	2	2	0	P8

Based on the iteration matrix, the coordinates (C_i, E_i) of each parameter were calculated (Table 4). A cause–effect plot was drawn with the (C_i, E_i) coordinates, as shown in Figure 7. Each point in the plot represents a particular factor P_i. The cause–effect plot can help to distinguish between "less interactive" and "more interactive" parameters: the "more interactive" parameters are plotted in the upper left region, whereas the "less interactive" parameters are plotted in the lower right region [28]. Figure 7 indicates that P5 (length of the main stream) is more interactive than the other parameters, and it is greatly affected by the system. On the other hand, P1 (lithology) and P7 (distance from fault) have the maximum effect on the system.

Table 4. Coordinates and weightings of influencing parameters.

Parameter	C_i	E_i	w_i (%)
Lithology	20	8	14.58
Watershed area	9	6	7.81
Slope angle	10	16	13.54
Stream density	11	16	14.06
Length of the main stream	8	18	13.54
Curvature of the main stream	9	16	13.02
Distance from fault	19	4	11.98
Vegetation cover	10	12	11.46

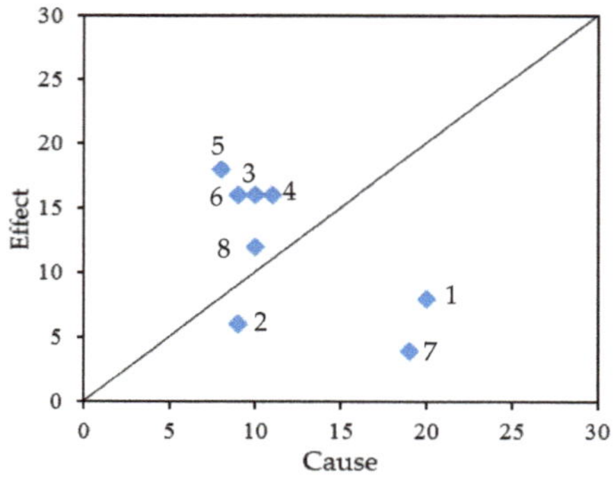

Figure 7. Cause–effect diagram.

The cause–effect plot also helps us to graphically compute the parameter interaction intensity and the parameter dominance. The interaction intensity of each parameter is represented by $(C + E)/\sqrt{2}$, and can be measured along the $C = E$ line; the parameter dominance depends on the perpendicular distance from the parameter's point representation to this line, which is calculated with $(C - E)/\sqrt{2}$ [28]. Figure 8 shows a histogram of the interaction intensity of each parameter. The histogram reveals that little changes in P2, P3, P4 and P5 will have great influence on the behavior of the system.

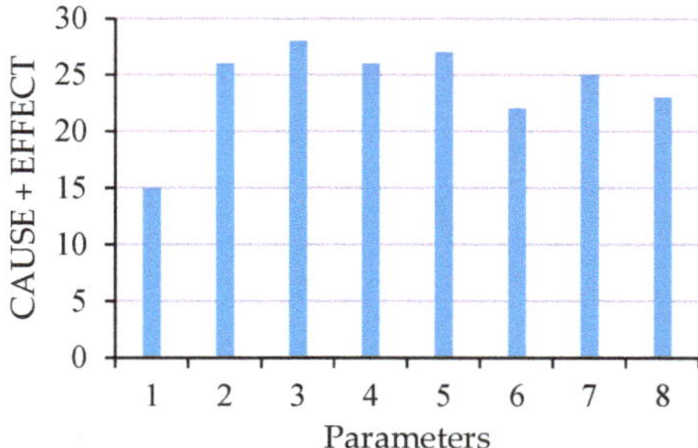

Figure 8. Histogram of interactive intensity.

Table 4 lists the weightings of the influencing parameters computed by Equation (1), which follow the order of lithology > stream density > slope angle = length of the main stream > curvature of the main stream > distance from fault > vegetation cover ratio > watershed area.

The debris flow susceptibility index (*SI*) can be calculated by

$$SI = \sum_{i=1}^{n} w_i p_i, \tag{6}$$

where w_i is the weighting of the ith parameter obtained from Table 4, p_i is the rating value of the ith parameter obtained from Table 1, and n is the total number of parameters.

5.2. Debris Flow Susceptibility Assessment

The RES model was applied to the 22 channelized debris flow gullies. The debris flow susceptibility index (*SI*) of each site was calculated by Equation (6), as listed in Table 5. In this work, the fuzzy C-means algorithm was adopted to divide the studied sites into three categories of debris flow susceptibility based on the RES model (RES_FCM). A debris flow susceptibility map was created, as shown in Figure 9. The classification results show that the susceptibility levels for nine of the debris flow gullies are high, nine are moderate and four are low, respectively.

Table 5. Susceptibility assessment results of the 22 debris flow gullies.

Gullies	Influencing Parameters								*SI*	RES_KM	RES_FCM	Actual Condition
	P1	P2	P3	P4	P5	P6	P7	P8				
Xiabaitan	T–K	3.1	36.1	5.51	3.08	1.19	0	10	188.53	High	High	High
Shangbaitan	T–K	0.91	28.5	10.29	1.87	1.08	0	10	176.03	Moderate	Moderate	Moderate
Menggugou	P_{t2}	37.1	41.37	6.73	10.52	1.13	0	40	205.19	High	High	High
Aibagou	P_{t2}	6.66	42.13	8.43	5.09	1.19	0	20	187.49	High	High	High
Nuozhacun	$\gamma_2 + Z_2$	32.61	40	4.96	10.5	1.17	0	10	194.78	High	High	High
Zhugongdi	T–K	6.5	41.8	6.24	4.98	1.15	0	15	176.55	Moderate	Moderate	Moderate
Yindigou	T–K	60.5	43.26	5.08	20.17	1.23	166	18	207.8	High	High	Moderate
Fujiahe	P_{t2}	8.62	42.7	6.34	5.16	1.26	0	17	176.55	Moderate	Moderate	Moderate
Zhangmuhe	Pt_2	4.62	29.1	9.7	5.39	1.42	0	10	199.99	High	High	Moderate
Hepiao	J + K	9.1	29.6	9.9	6.83	1.32	0	30	175.51	Moderate	Moderate	Moderate
Hongmenchang	P_{t2}	46.9	30	6.6	12.9	1.29	0	15	216.13	High	High	High
Tianfanghe	P_{t2}	13.1	34	9.3	5.6	1.17	0	16	195.3	High	High	High
Zhiligou	T–K	120.6	24	6.3	15.8	1.28	0	25	181.76	Moderate	Moderate	Moderate
Pingdicun	T–K	24.2	17	5.9	9.9	1.14	3000	40	171.34	Moderate	Moderate	Moderate
Fangshanguo	T–K	98	28	4.63	20.2	1.38	6662	10	193.22	High	High	High
Daqiangou	T–K	18.9	29	10.95	5.1	1.11	18	17	174.46	Moderate	Moderate	Moderate
Shenyuhe	T–K	256	21	2.26	29.63	1.47	0	50	169.26	Low	Moderate	Moderate
Zhuzhahe	T–K	152.6	26.6	4.32	26.3	1.7	378	20	170.3	Moderate	Moderate	Moderate

Table 5. *Cont.*

Gullies	Influencing Parameters								SI	RES_KM	RES_FCM	Actual Condition
	P1	P2	P3	P4	P5	P6	P7	P8				
Heizhe	T–K	51.7	13.5	5.12	13.9	1.15	3485	20	167.18	Low	Low	Low
Yanshuijing	P_{t1}	48.58	22.6	9.25	14.43	1.22	0	5	153.63	Low	Low	Low
Yajiede	T–K	22.3	12	4.7	9.3	1.31	0	70	145.3	Low	Low	Low
Daqinggou	T–K	31.8	32	6.02	7.32	1.1	378	15	147.38	Low	Low	Low

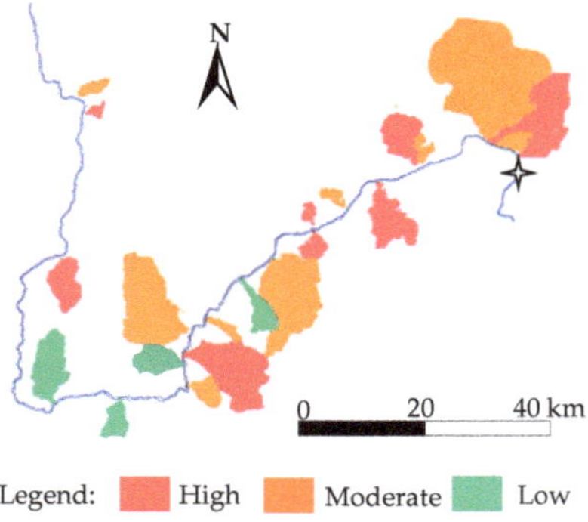

Figure 9. Debris flow susceptibility map.

5.3. Validation of the Model

The new model was validated by the field survey data. The actual conditions of the debris flow gullies were determined according to the principles listed in Table 6 based on the geological and environmental conditions, as shown in Table 5. The results show that among the 22 debris flow gullies, there are only two gullies (i.e., Yindigou and Zhangmuhe) that were assigned to different susceptibility groups compared with the actual conditions of the two gullies. The prediction accuracy of the new method is 90.9%, which is deemed satisfactory.

Table 6. Principles for assessing the actual condition of debris flow gullies.

Level	Susceptibility Degree	Description
1	High	Abundance of loose materials accumulated on slopes, steep channels, inventory of debris flows
2	Moderate	Between levels 1 and 3
3	Low	Absence of loose materials, smooth terrains , no debris flow record

In this research, the RES based *K*-means algorithm (RES_KM) was used for comparison. The results show that the RES_FCM method and the RES_KM method provide very close evaluation results for most of the debris flow gullies. The difference between the results of the two methods lies in that the susceptibility level for the Shenyuhe debris flow gully was calculated to be moderate using the RES_FCM method, whereas the result obtained by the RES_KM method shows that the gully has low susceptibility level for debris flows. In fact, field investigation for the Shenyuhe debris flow gully shows that (1) a large amount of loose materials provided by landslides were deposited along the main channel (Figure 10a); (2) some slopes composed of sediments of the Madianhe Group are unstable; and (3) a small mudflow was observed to occur along a smooth-narrow gully (Figure 10b). Combined with the field investigations, it is more reasonable to partition the Shenyuhe gully into the moderate susceptibility group. According to Table 5, the prediction accuracy of the RES_KM method is 86.4%, which is smaller than that obtained by the RES_FCM method. In addition, the susceptibility levels for Yindigou and Zhangmuhe debris flow gullies obtained by the RES_FCM method are both higher than their actual conditions, indicating that the RES_FCM method does a slight overestimation of hazard; this goes towards security. However, the RES_KM method does a slight underestimation of hazard for the Shenyuhe debris flow gully; this does not go towards general safety. Therefore, the RES_FCM method performs better than the RES_KM method for assessing the susceptibility of debris flows.

Figure 10. Geological and environmental conditions of the Shenyuhe debris flow gully. (**a**) loose materials deposited along the main channel; (**b**) unstable slopes composed of sediments of the Madianhe Group and a small mudflow.

Figure 9 illustrates that the gullies categorized into high susceptibility zone are predominantly situated in the eastern part of the study area, signifying that this part is more prone to debris flows than anywhere else. In fact, field investigations suggest that the Quaternary deposits and the fine sand layer, which are not stable and prone to slope failures (Figure 11), are mainly distributed in the eastern part of the study area. Furthermore, the strong earthquakes triggered by faults have resulted in widespread landslides or rockfalls on these unstable rock formations. Failed slopes provide sufficient loose materials for drainages and hence facilitate the initiation of debris flows. Moreover, the distribution of high susceptibility zones is in accordance with the distribution of the regional faults (Figure 4). Therefore, the debris flow susceptibility map agrees well with the environmental features favorable for the initiation of debris flows in the study area.

Figure 11. Landslides occurred in the study area. (**a**) landslides on the Quaternary deposits; (**b**) landslides on the fine sand layer.

6. Conclusions

This paper presents an application of the rock engineering system and fuzzy *C*-means algorithm (RES_FCM) for debris flow susceptibility assessment. A total of 22 channelized debris flow gullies located in the watershed of the Jinsha River close to the Wudongde dam site were investigated. The debris flow susceptibility of these gullies was assessed by introducing the concept of a susceptibility index (*SI*) based on the principles of RES. Eight parameters that are considered to be potential controls of debris flows were selected. The interactions among these parameters were determined using RES. The results show that the susceptibility levels for nine of the debris flow gullies are high, nine are moderate and four are low, respectively.

The RES based *K*-means algorithm (RES_KM) was used for comparison. The results show that the RES_FCM method and the RES_KM method provide very close evaluation results for most of the debris flow gullies, which are very similar to the evaluation results from the geological conditions in the study area. The new approach could be a simple but efficient tool for analyzing parameters influencing the occurrence of debris flows, and could be useful for evaluating the debris flow susceptibility.

This work only selected 22 large-scale debris flow gullies as our research object, considering that only large-scale debris flows could affect the stability of the dam; this could be helpful for debris flow hazard prevention. Further research work on the application of the proposed method to the total of the Wudongde dam area is necessary.

Acknowledgments: This work was financially supported by the National Natural Science Foundation of China (NSFC) (Nos. 41602327 and 41372324), and China Postdoctoral Science Foundation funded project (No. 2015M580135). The authors are grateful to two anonymous reviewers for their excellent reviews that helped to improve the manuscript.

Author Contributions: All authors were responsible for different parts of this paper. Yanyan Li, Jianping Chen conducted field investigations and wrote the whole paper; Yanjun Shang and Honggang Wang provided useful advice for data analysis and revised the paper.

Conflicts of Interest: The authors declare no conflict of interest.

References

1. Rickenmann, D. Empirical relationships for debris flows. *Nat. Hazards* **1999**, *19*, 47–77. [CrossRef]
2. Conway, S.J.; Decaulne, A.; Balme, M.R.; Murray, J.B.; Towner, M.C. A new approach to estimating hazard posed by debris flows in the Westfjords of Iceland. *Geomorphology* **2010**, *114*, 556–572. [CrossRef]
3. Kanji, M.A.; Cruz, P.T.; Massad, F. Debris flow affecting the Cubatão oil refinery, Brazil. *Landslides* **2008**, *5*, 71–82. [CrossRef]
4. Kappes, M.S.; Malet, J.P.; Remaître, A.; Horton, P.; Jaboyedoff, M.; Bell, R. Assessment of debris-flow susceptibility at medium-scale in the Barcelonnette Basin, France. *Nat. Hazards Earth Syst. Sci.* **2011**, *11*, 627–641. [CrossRef]
5. Horton, P.; Jaboyedoff, M.; Rudaz, B.; Zimmermann, M. Flow-R, a model for susceptibility mapping of debris flows and other gravitational hazards at a regional scale. *Nat. Hazards Earth Syst. Sci.* **2013**, *13*, 869–885. [CrossRef]
6. Chang, T.C.; Chien, Y.H. The application of genetic algorithm in debris flow prediction. *Environ. Geol.* **2007**, *53*, 339–347. [CrossRef]
7. Chang, T.C.; Chao, R.J. Application of back-propagation networks in debris flow prediction. *Eng. Geol.* **2006**, *85*, 270–280. [CrossRef]
8. Chang, T.C. Risk degree of debris flow applying neural networks. *Nat. Hazards* **2007**, *42*, 209–224. [CrossRef]
9. Conforti, M.; Pascale, S.; Robustelli, G.; Sdao, F. Evaluation of prediction capability of the artificial neural networks for mapping landslide susceptibility in the Turbolo River catchment (northern Calabria, Italy). *Catena* **2014**, *113*, 236–250. [CrossRef]
10. Liu, C.N.; Dong, J.J.; Peng, Y.F.; Huang, H.F. Effects of strong ground motion on the susceptibility of gully type debris flows. *Eng. Geol.* **2009**, *104*, 241–253. [CrossRef]
11. Wan, S.; Lei, T.C. A knowledge-based decision support system to analyze the debris-flow problems at Chen-Yu-Lan River, Taiwan. *Knowl.-Based Syst.* **2009**, *22*, 580–588. [CrossRef]

12. Dai, F.C.; Lee, C.F.; Tham, L.G.; Ng, K.C.; Shun, W.L. Logistic regression modelling of storm-induced shallow landsliding in time and space on natural terrain of Lantau Island, Hong Kong. *Bull. Eng. Geol. Environ.* **2004**, *63*, 315–327. [CrossRef]
13. Ayalew, L.; Yamagishi, H. The application of GIS-based logistic regression for landslide susceptibility mapping in the Kakuda-Yahiko Mountains, central Japan. *Geomorphology* **2005**, *65*, 15–31. [CrossRef]
14. Greco, R.; Sorriso-Valvo, M.; Catalano, E. Logistic regression analysis in the evaluation of mass movements susceptibility: The Aspromonte case study, Calabria, Italy. *Eng. Geol.* **2007**, *89*, 47–66. [CrossRef]
15. Manzo, G.; Tofani, V.; Segoni, S.; Battistini, A.; Catani, F. GIS techniques for regional-scale landslide susceptibility assessment: The Sicily (Italy) case study. *Int. J. Geogr. Inf. Sci.* **2013**, *27*, 1433–1452. [CrossRef]
16. Dong, J.J.; Lee, C.T.; Tung, Y.H.; Liu, C.N.; Lin, K.P.; Lee, J.F. The role of the sediment budget in understanding debris flow susceptibility. *Earth Surf. Proc. Landf.* **2009**, *34*, 1612–1624. [CrossRef]
17. Bertrand, M.; Liébault, F.; Piégay, H. Debris-flow susceptibility of upland catchments. *Nat. Hazards* **2013**, *67*, 497–511. [CrossRef]
18. Song, Y.Q.; Gong, J.H.; Gao, S.; Wang, D.C.; Cui, T.J.; Li, Y.; Wei, B.Q. Susceptibility assessment of earthquake-induced landslides using Bayesian network: A case study in Beichuan, China. *Comput. Geosci.* **2012**, *42*, 189–199. [CrossRef]
19. Ahmed, B.; Dewan, A. Application of bivariate and multivariate statistical techniques in landslide susceptibility modeling in Chittagong city corporation, Bangladesh. *Remote Sens.* **2017**, *42*, 304. [CrossRef]
20. Baeza, C.; Corominas, J. Assessment of shallow landslide susceptibility by means of multivariate statistical techniques. *Earth Surf. Proc. Landf.* **2001**, *26*, 1251–1263. [CrossRef]
21. Guzzetti, F.; Reichenbach, P.; Ardizzone, F.; Cardinali, M.; Galli, M. Estimating the quality of landslide susceptibility models. *Geomorphology* **2006**, *81*, 166–184. [CrossRef]
22. Xu, W.; Jing, S.; Yu, W.; Wang, Z.; Zhang, G.; Huang, J. A comparison between Bayes discriminant analysis and logistic regression for prediction of debris flow in southwest Sichuan, China. *Geomorphology* **2013**, *201*, 45–51. [CrossRef]
23. Kayastha, P.; Dhital, M.R.; De Smedt, F. Application of the analytical hierarchy process (AHP) for landslide susceptibility mapping: A case study from the Tinau watershed, west Nepal. *Comput. Geosci.* **2013**, *52*, 398–408. [CrossRef]
24. Meyer, N.K.; Schwanghart, W.; Korup, O.; Romstad, B.; Etzelmüller, B. Estimating the topographic predictability of debris flows. *Geomorphology* **2014**, *207*, 114–125. [CrossRef]
25. Yalcin, A. GIS-based landslide susceptibility mapping using analytical hierarchy process and bivariate statistics in Ardesen (Turkey): Comparisons of results and confirmations. *Catena* **2008**, *72*, 1–12. [CrossRef]
26. Yalcin, A.; Reis, S.; Aydinoglu, A.C.; Yomralioglu, T. A GIS-based comparative study of frequency ratio, analytical hierarchy process, bivariate statistics and logistics regression methods for landslide susceptibility mapping in Trabzon, NE Turkey. *Catena* **2011**, *85*, 274–287. [CrossRef]
27. Pradhan, B. Landslide susceptibility mapping of a catchment area using frequency ratio, fuzzy logic and multivariate logistic regression approaches. *J. Indian Soc. Remote* **2010**, *38*, 301–320. [CrossRef]
28. Hudson, J.A. *Rock Engineering Systems: Theory and Practice*; Ellis Horwood Ltd.: Chichester, UK, 1992; pp. 154–185.
29. Bezdek, J.C. *Pattern Recognition with Fuzzy Objective Function Algorithms*; Plenum Press: New York, NY, USA, 1981; pp. 203–239.
30. Li, Y.Y.; Chen, J.P.; Shang, Y.J. An RVM-based model for assessing the failure probability of slopes along the Jinsha River, close to the Wudongde dam site, China. *Sustainability* **2017**, *9*, 32. [CrossRef]
31. Niu, C. Index Selection and Rating for Debris Flow Hazard Assessment. Ph.D. Thesis, Jilin University, Changchun, China, 2013; p. 143.
32. Chen, H.; Hu, J.M.; Qu, H.J.; Wu, G.L. Early Mesozoic structural deformation in the Chuandian N-S Tectonic Belt, China. *Sci. China Ser. D* **2011**, *54*, 1651–1664. [CrossRef]
33. Zhang, W.; Chen, J.P.; Wang, Q.; An, Y.; Qian, X.; Xiang, L.; He, L. Susceptibility analysis of large-scale debris flows based on combination weighting and extension methods. *Nat. Hazards* **2013**, *66*, 1073–1100. [CrossRef]
34. Zhang, W.; Li, H.Z.; Chen, J.P.; Zhang, C.; Xu, L.M.; Sang, W.F. Comprehensive hazard assessment and protection of debris flows along Jinsha River close to the Wudongde dam site in China. *Nat. Hazards* **2011**, *58*, 459–477. [CrossRef]

35. Hu, W.; Xu, Q.; Rui, C.; Huang, R.Q.; van Asch, T.W.J.; Zhu, X.; Xu, Q.Q. An instrumented flume to investigate the initiation mechanism of the post-earthquake huge debris flow in the southwest of China. *Bull. Eng. Geol. Environ.* **2015**, *74*, 393–404. [CrossRef]
36. Lin, P.S.; Lin, J.Y.; Hung, J.C.; Yang, M.D. Assessing debris-flow hazard in a watershed in Taiwan. *Eng. Geol.* **2002**, *66*, 295–313. [CrossRef]
37. Rickenmann, D.; Zimmermann, M. The 1987 debris flows in Switzerland: Documentation and analysis. *Geomorphology* **1993**, *8*, 175–189. [CrossRef]
38. Lei, T.C.; Wan, S.; Chou, T.Y.; Pai, H.C. The knowledge expression on debris flow potential analysis through PCA + LDA and rough sets theory: A case study of Chen-Yu-Lan watershed, Nantou, Taiwan. *Environ. Earth Sci.* **2011**, *63*, 981–997. [CrossRef]
39. Mousavi, S.M.; Omidvar, B.; Ghazban, F.; Feyzi, R. Quantitative risk analysis for earthquake-induced landslides—Emamzadeh Ali, Iran. *Eng. Geol.* **2011**, *122*, 191–203. [CrossRef]
40. Hudson, J.A.; Harrison, J.P. A new approach to studying complete rock engineering problems. *Q. J. Eng. Geol. Hydrogeol.* **1992**, *25*, 93–105. [CrossRef]
41. Huang, R.; Huang, J.; Ju, N.; Li, Y. Automated tunnel rock classification using rock engineering systems. *Eng. Geol.* **2013**, *156*, 20–27. [CrossRef]
42. Faramarzi, F.; Mansouri, H.; Ebrahimi Farsangi, M.A. A rock engineering systems based model to predict rock fragmentation by blasting. *Int. J. Rock Mech. Min. Sci.* **2013**, *60*, 82–94. [CrossRef]
43. Jiao, Y.; Hudson, J.A. The fully-coupled model for rock engineering systems. *Int. J. Rock Mech. Min. Sci.* **1995**, *32*, 491–512. [CrossRef]
44. Zhang, B.; Qin, S.; Wang, W.; Wang, D.; Xue, L. Data stream clustering based on Fuzzy C-Mean algorithm and entropy theory. *Signal Process.* **2016**, *126*, 111–116. [CrossRef]
45. Hammah, R.E.; Curran, J.H. Fuzzy cluster algorithm for the automatic identification of joint sets. *Int. J. Rock Mech. Min. Sci.* **1998**, *35*, 889–905. [CrossRef]

Article

Evaluating the Effects of Low Impact Development Practices on Urban Flooding under Different Rainfall Intensities

Zhihua Zhu [1,2,3] and Xiaohong Chen [1,2,3,*]

1 Center for Water Resources and Environment, Sun Yat-sen University, Guangzhou 510275, China; persistzhzh@163.com
2 Guangdong Engineering Technology Research Center of Water Security Regulation and Control for Southern China, Sun Yat-sen University, Guangzhou 510275, China
3 Key Laboratory of Water Cycle and Water Security in Southern China of Guangdong High Education Institute, Sun Yat-sen University, Guangzhou 510275, China
* Correspondence: eescxh@mail.sysu.edu.cn; Tel.: +86-20-84114575

Received: 30 March 2017; Accepted: 12 July 2017; Published: 24 July 2017

Abstract: Low impact development (LID) is an important control measure against extreme rainfall events and is widely applied to relieve urban flood disasters. To investigate the effects of LID practices on flooding control under different rainfall scenarios, this paper constructs a rainfall–runoff model based on the storm water management model (SWMM) for a typical residential area in Guangzhou, China. The model is calibrated by using observed rainfall and runoff data. A total of 27 rainfall scenarios are constructed to simulate the change characteristics before and after the LID practices. Also, the projection pursuit method based on a particle swarm optimization (PSO) algorithm is used to assess the flooding characteristics. The results show that the constructed rainfall–runoff model can closely reflect the relationship between rainfall and runoff, with all Nash–Sutcliffe coefficients of efficiency (NSE) exceeding 0.7. It was found from the simulation and assessment of the constructed rainfall scenarios that the changes in rainfall characteristics have a considerable impact on the constructed drainage system and that LID practices can properly control the floods. However, with an increase in rainfall peak coefficient, intensity or duration, the control effects of LID tend to reduce. Particularly in the scenario of relatively high rainfall intensity, the impact of rainfall duration and the rainfall peak coefficient on the LID practices is minor.

Keywords: urban drainage system; different rainfall scenarios; control characteristics; urban runoff control; flood mitigation; low impact development (LID); China

1. Introduction

Along with significant changes in global climate and the considerably increasing disturbance caused by human activities, flood disasters in urban areas have become more serious, substantially restricting social and economic development and posing a threat to the safety of human lives and properties [1,2]. In recent years, urban flooding has caused many casualties and property losses in Japan, Singapore, The Netherlands, Britain and other countries [3–6]. Many major cities in China have also been greatly influenced by urban flooding disasters, including Guangzhou, Wuhan, Shenzhen, Nanjing, Beijing [7]. The key issue is that the drainage systems of these areas cannot adapt to climate change and human activity. Therefore, there is an urgent need to control flood disasters from a design perspective.

In order to alleviate the problem of urban flooding disasters, many strategies of flood mitigation and disaster relief have been put forward [8–10]. The traditional efforts include the following:

(i) improving the capacity of the drainage system. For example, Guangzhou requires the improvement of the drainage system capacity in areas through new construction [11]. Hong Kong and Chicago have introduced similar policies to improve drainage capacity [12]. The main measures used to improve the drainage capacity are to expand the existing drainage system and to construct a deep drainage system. (ii) Build water diversion and storage projects. Due to the restrictions on urban land resources, Singapore, Hong Kong, and Japan have constructed drainage pipes to divert water and water storage works to reduce the burden on downstream drainage [13,14]. (iii) Improve flood prevention ability. Many countries use road blocks, raise the height of existing dams, install flood gates and use other measures to prevent urban floods [14]. (iv) Implement routine maintenance of drainage systems. In order to prevent blockages in the pipe network, a common measure is to clean out solid waste in the pipes and strengthen routine maintenance of the pipe network, so as to prevent waste from entering the pipe network. However, all of these measures are generally expensive with large scale construction or can easily lead to excessive drainage of downstream areas. In highly developed urban areas, these drainage-based control measures and end-control measures do not adequately address urban floods, so it is important to find new methods to control the source of the runoff.

Due to the lack of necessary rain and flood control measures, and the unsustainable nature of traditional flood management, developed countries have studied urban flood disasters scenarios since the 1970s based on the requirements of the flood discharge space. After decades of development, a relatively complete system of urban rain and flood management has been set up in many countries [15]. A system's flood management is based on the comprehensive and multi-level considerations of individual buildings, flood drainage, runoff control and other elements. Certain well-known management measures include low impact development (LID) and best management practices (BMPs) proposed by the United States, water sensitive urban design (WSUD) proposed by Australia, sustainable drainage systems (SUDS) proposed by the Britain, and low impact urban design and development (LIUDD) proposed by New Zealand [16]. These concepts and measures place high value on the combination of structural measures and non-structural measures, as well as on natural drainage conditions and landscape features, to effectively control urban floods.

LID has been rapidly developed in Europe, the Middle East, Asia, Africa and South America [17]. It refers to a stormwater management method based on the simulation of natural hydrological conditions, causing no changes in urban hydrological characteristics. It uses ecological measures, source control and distributed control measures to accomplish stormwater control and utilization [18]. Different LID practices have different characteristics (Table 1) [19]. It results in satisfactory effects through the combined use of reservoirs, green roofs, rain gardens, bioretention swales and bioretention ponds [20,21]. The community and street reconstruction project in Seattle and the bioretention pond system implemented in the Portland Expo Center are exemplary of this approach. Additionally, Berlin Potsdamer Platz and a certain ecological urban area in Hanover, which were designed based on LID, have obtained remarkable effects [22]. According to the estimates of Unified Facilities Criteria, the LID market may have amounted to $380 billion in the United States in recent years, greatly encouraging the research, development and promotion of LID. Also, LID is vigorously promoted in China [23–25]. However, the application of LID in China is mainly confined to small-scale areas at present [26]. Also, no attention has been paid to the role of various elements and projects in the rainwater systems, and the response characteristics of different climates to LID have not been taken into account.

In order to have a clearer idea of the performance of various LID practices, the storm water management model (SWMM) of United States Environmental Protection Agency (USEPA) [27] has been widely used for simulations. Extensive research has been conducted into the performance of the SWMM model in simulating hydrologic processes [28]. For example, Burszta–Adamiak et al. [28] simulated green roof hydrologic performance based on the SWMM model. Qin et al. [29] analyzed the effects of LID on urban flooding using the SWMM model. Alfredo et al. [30] proved that green roofs can reduce 30–78% of peak runoff and delay the runoff to some extent as well. Niu et al. [31] simulated the long-term effect of LID and indicated that LID has a remarkable effect on water balance in highly urbanized areas. Ahiablame et al. [17] probed the flood control ability of LID in an effort to find the optimal way to control floods. Bedan and Clausen [32] found that arranged LID could reduce rainfall runoff by 42% and purify water. In addition, many researchers hope to obtain the optimal combination of flood control by analyzing and comparing the performance of porous pavements, green roofs, bioretention swales and other LID components [33–35].

Table 1. Comparison of some Low Impact Development (LID) practices [19].

LID Practices	Functions			Cost		Scenic Effect
	Storage	Transportation	Reducing Runoff Peak Rate	Construction Cost	Maintenance Cost	
Rain garden	A	A	A	C	B	A
Dry bioretention swales	C	A	A	A	A	A
Wet bioretention swales	C	A	C	B	A	A
Green roof	C	C	C	C	B	A
Porous pavement	C	B	B	A	A	-
Subsided Green space	C	B	B	A	A	B
Wet pond	A	B	A	C	B	A
Vegetative filter strip	C	C	C	A	A	B

Note: A represents ideal performance; B represents average performance; C represents low performance.

Compared with traditional flood control measures, LID practices are more flexible in response to climate change, which is a very important aspect for government decision-makers [36]. Also, it is considered to be a pivotal approach to sustainable urban development. Nevertheless, most studies consider only laboratory scenarios or one particular rainfall scenario. Research into the response characteristics of LID under different rainfall scenarios is sparse. In addition, the evaluation of LID practices is mainly based on the evaluation of a single factor (such as increasing peak flow attenuation, increasing lag time, flow attenuation of stormwater runoff, etc.), and it has mainly evaluated at the multidimensional scale, which is likely to cause dimensionality when a multi-factor evaluation is performed.

A large number of studies have shown that the mastering of LID control characteristics and accurately evaluating the control of LID practices under different rainfall scenarios can effectively respond to urban flood disasters and guide the formulation of appropriate measures. Therefore, a rainfall–runoff model is constructed based on the SWMM model [27]. When a large number of scenarios is considered, a better optimization procedure is required. In this paper, 27 rainfall scenarios with different rainfall intensities, rainfall durations and rainfall peak coefficients are considered and a comprehensive assessment of the change characteristics of floods is conducted in various rainfall scenarios by using particle swarm optimization (PSO)-based projection pursuit technology [37] which includes (1) the discovery of the response law of the pipeline network to different rainfall intensity before and after the implementation of LID practices; (2) the qualitative and quantitative evaluation of the variation characteristics of the flood before and after the implementation of the LID practices; and (3) the recognition of the control characteristics of LID under different complex scenarios.

2. Study Area

Guangzhou (22°26′–23°56′ N, 112°57′–114°03′ E) has a tropical and subtropical monsoon climate with extremely significant characteristics of a monsoon climate and subtropical westerly, equatorial westerly and tropical easterly winds. The area is abundant in rainfall, with average annual rainfall of 1675.5 mm, and characterized by a long rainy season and heavy rainfall of great intensity and nonuniform spatial distribution. The rainfall is mainly concentrated in the months from April to September that are affected by warm air. The rainfall from April to September accounts for 80% of annual rainfall. The rainfall in May and June represents 32% of the annual rainfall. Since the reform and opening policies in the 1980s, the forms of land use in Guangzhou have changed substantially.

Land under construction is constantly being expanded, contributing to changes in the characteristics of runoff and to increasingly serious urban flooding.

In order to probe the flood control characteristics of LID practices under different rainfall scenarios, this paper selected a typical highly-developed residential area. It covers a total of 3.13×10^5 m^2 and is primarily used for construction with some land set aside for green space and transportation (Figure 1). The drainage system uses a rainwater and sewage separation system, and its design standard is based on a 2-year rainfall intensity developed by the Guangzhou Water Affairs Bureau in 1993 (with pipe diameter between 900 and 2000 mm and slopes between 0.001 and 0.01).

The study area is divided into 19 sub-catchments, and the drainage system is generalized into 31 drainage pipes, 31 manholes and 1 drainage outlet (Figure 1) in accordance with land use, the pipe laying situation, the terrain and other data. The rainfall–runoff model is constructed based on the SWMM model to reflect the relationships between runoff and rainfall. Point A (in the downstream area) is selected to record the runoff, and the flow and rainfall are recorded simultaneously in accordance with a flow meter and rain gauge at an interval of 10 min (Figure 1).

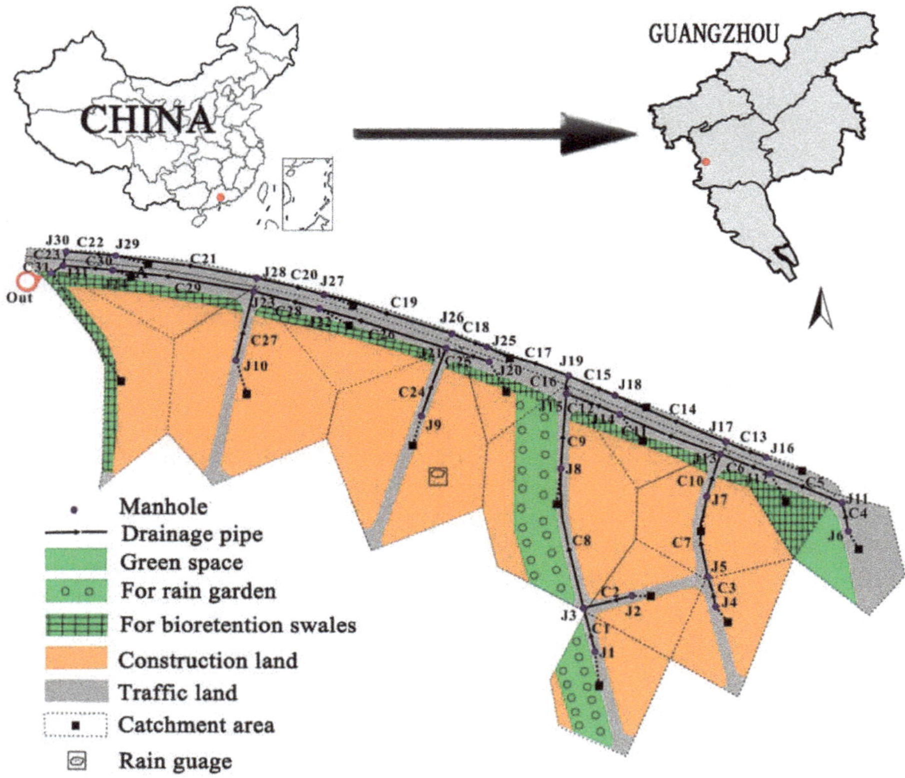

Figure 1. Generalized results of the land use and drainage system.

3. Methodology

In order to investigate the control characteristics of the highly-developed area under different rainfall scenarios after the implementation of LID practices, the observed rainfall–runoff data was employed to calibrate and verify the rainfall–runoff model based on the SWMM model. Also, different rainfall scenarios were constructed based on the rainfall intensity formula in Guangzhou [38] and the Chicago Hydrograph Model [39]. The characteristics of flood change (before and after LID) were analyzed by PSO-based projection pursuit technology.

3.1. SWMM Model

The SWMM model is a relatively flexible hydrodynamic model developed by USEPA [27], based on the physical mechanisms of mass, energy and momentum conversion, as well as a comprehensive mathematical model constructed for the mechanisms of runoff generation [27]. It can properly simulate the rainfall–runoff relationship, pollutant dispersion, flood control and other hydrological processes in urban areas under natural conditions. Also, the SWMM model is widely used to simulate the hydrologic performance of natural channels, rain and sewage diversion systems or other drainage systems, and it can evaluate the design of LID practices and BMPs. Much research on the SWMM model can be found [40–42].

The SWMM model provides different modules and algorithms for simulation as described below.

3.1.1. Surface Runoff Simulation Module

When the SWMM model is included in the calculation of surface production, the sub-basin is regarded as a non-linear reservoir, and the surface runoff refers to the rainfall depth after infiltration and surface water storage. The surface water storage modules are different for different regions. The SWMM model provides the Horton model, the Green–Ampt model and the curve number model to calculate the infiltration. This paper chooses the Horton model for calculating small watersheds. The flow is transformed into the outflow process and the process in which flow changes over time, and the two processes are calculated separately using Manning's equation and the finite difference method.

3.1.2. Transmission Calculation Module

The main method for calculating the flow of the pipe network is the Saint-Venant equation. In the SWMM model, three methods (steady flow routing, kinematic wave routing and dynamic wave routing) are used for pipe network transmission calculation. The kinematic wave routing method is used in this paper.

3.1.3. Flood Calculation Module

If the depth of the node exceeds the maximum available depth, overflow is lost from the system or water is saved at the top of the node, and it re-enters the drainage system under specific conditions. The SWMM model can simulate the approximate location of the overflow, the distribution and the corresponding overflow.

3.1.4. LID Module

The SWMM model can simulate the control effect of LID practices, and seven different types are provided: bio-retention cell, porous pavement, infiltration trench, rain barrel, vegetative swale, rain garden and green roof. In the SWMM model, different types of LID practices are divided into several vertical layers and represented by five to 23 parameters (e.g., thickness, surface roughness, offset height). More details about this can be found in [27].

The SWMM model's user manual [27] provides a range of values for sensitive parameters based on a large number of simulation studies; this provides references for calibrating the model's parameters. Thus, the recommended range of the parameters and the trial and error approach are used.

3.2. Evaluation of Model Accuracy

The Nash–Sutcliffe Efficiency (NSE) coefficient, which is often considered to be one of the most important indices for measuring the model's simulation accuracy [43], was used to evaluate the accuracy of the constructed model simulation results. Its expression is as follows:

$$NSE = 1 - \frac{\sum_{i=1}^{n} (S_i - O_i)^2}{\sum_{i=1}^{n} (O_i - \overline{O}_i)^2} \quad (1)$$

where n means the total number of time-steps; S_i represents the simulated value at time-step i; O_i represents the observed flow at time-step i; and $\overline{O}$ is the average value of the observed flow. When $NSE = 1$, the observed flow is consistent with the simulated flow. If NSE is within the range of 0 to 1, the simulated flow is acceptable. If NSE is greater than 0.5, the simulation results are satisfactory [44,45].

3.3. Projection Pursuit Method

The projection pursuit method is mainly used to project high-dimensional data to low dimensions, to avoid "dimension disaster". The characteristics of the original data are described by the optimal projection direction and projection values [46]. It has been widely applied to water resource assessment, engineering stability assessments and many other applications [47–49]. Compared with traditional methods (e.g., neural networks and principal component analysis), this method can eliminate the influence of subjective factors.

Since the flooding of an urban drainage system involves a large number of indices, it is difficult to comprehensively assess all of the flood characteristics through high-dimensional indices and data. According to the mechanisms and objectives of urban flooding analysis, the volume of flooding, the maximum flooding rate and the duration of flooding are selected as indices used to assess the characteristics of the flooding. A high-dimensional flood evaluation system may be reduced to a one-dimensional system in order to comprehensively evaluate the regional flood characteristics using the projection pursuit method. The calculation steps are as follows:

Step (1): Normalize the evaluation index. As the evaluation indices have different dimensions and lack comparability, they must be normalized. Assume an evaluation index set $x^* = \{x_{ij} | i = 1, 2, \ldots, n; j = 1, 2, 3\}$, where x_{ij} is the jth index value of the ith sample, and n is the sample size. A large number of studies have shown that the selected indices are positively related to flood evaluation. Therefore, all indices are used in Equation (2) to normalize x_{ij} to X_{ij}.

$$X_{ij} = \frac{x_{j\max} - x_{ij}}{x_{j\max} - x_{j\min}} \quad (2)$$

where $x_{j\max}$ and $x_{j\min}$ are the maximum and the minimum of evaluation index j^{th} of all samples respectively.

Step (2): Find the function of the construct projection index. The flooding evaluation index is expressed as a one-dimensional projection value $Z_i = \sum_{j=1}^{p} a_j x_{ij} \quad (i = 1, 2, \ldots, n)$ in the projection direction $\alpha = \{\alpha_j | j = 1 \sim p\}$. Therefore, the construction of the projection index function $Q(a)$ is as follows:

$$Q(a) = S_Z D_Z \quad (3)$$

where S_Z and D_Z indicate the standard deviation and local density of Z_i, respectively, i.e.,

$$S_z = \sum_{i=1}^{n} \sum_{j=1}^{p} (R - r(i,j)) \cdot u(R - r(i,j)) \quad (4)$$

$$D_z = \sqrt{\frac{\sum_{i=1}^{n} (Z_i - \overline{Z})^2}{n-1}} \quad (5)$$

where R is radius of the window, and it is usually $0.1\sqrt{\frac{\sum_{i=1}^{n}(Z_i-\overline{Z})^2}{n-1}}$; P is the number of indices; $r(i,j)$ is the distance between samples; and $u(R-r(i,j))$ is the unit step function. If $R-r(i,j)\geq 0$, then $u(R-r(i,j)=1$; if $R-r(i,j)<0$, then $u(R-r(i,j)=0$.

The change in the projection index function $Q(a)$ is determined by the projection direction. When $Q(a)$ is maximized, it is considered to be the most likely function to represent the structural features of high dimensional data, and the corresponding vector is the best projection direction α^*, where α^* is the projection direction vector involving the maximum number of high-dimensional data characteristics. The maximization of the objective function and its corresponding constraints are denoted by (6) and (7), respectively.

$$\max Q(a) = S_z D_z \tag{6}$$

$$\sum_{j=1}^{p}\alpha_j^2 = 1 \tag{7}$$

Step (3): Solving the objective function. The objective function (6) is a complex nonlinear system which needs to be solved by advanced optimization algorithms. A large number of studies have shown that the PSO algorithm has significant versatility in high-dimensional global optimization with simple and effective operation [50–53]. It is therefore used to solve the objective function in this study. The PSO algorithm is a heuristic search method that treats a solution to a problem as a "particle" of the search space. The basic idea is to initialize a group of random particles (i.e., random solutions), and then find the optimal solution by iterative calculations. The velocity, position and fitness of the particle d are V_d, S_d, f_d ($d=1,2,\ldots,N$), N is the pre-defined number of particles. During each iteration, the optimal solution (individual optimal value $p_d(t)$) will be found by the particle through the particles themselves, and the optimal solution (global optimal value $g_d(t)$) found among the entire population of particles during each calculation updates the position and velocity of the particle. Its update rules are as follows:

$$V_d(t+1) = wV_d(t) + (c_1b_1(t) + c_2b_2(t))\cdot(p_d(t) - s_d(t)) \tag{8}$$

$$S_d(t+1) = S_d(t) + V_d(t+1) \tag{9}$$

where, t is the current iteration as inertia weight; c_1 and c_2 represent individual learning factor and group learning factor, respectively; and $b_1(t)$ and $b_2(t)$ represent the individuals and groups of particles, respectively. In this paper, 150 random particles are initialized, and the maximum number of iterations is set to 1500. The maximum and minimum weighting factors are 0.9 and 0.4, respectively. $c_1=c_2=2$ gave the best overall performance [54,55]. Therefore, c_1 and c_2 are kept as constant of 2. In the iterative calculation process, the update rules of the individual optimal value and the global optimal value are shown in (10) and (11).

$$p_d(t+1) = \begin{cases} S_d(t+1) & f_d(t+1)\geq f(p_d(t)) \\ p_d(t) & f_d(t+1) < f(p_d(t)) \end{cases} \tag{10}$$

$$g_d(t+1) = S\max(t+1) \tag{11}$$

where $f_d(t+1)$ is the fitness value of the iteration of particle d in $t+1$.

To solve the coupling between PSO and the projection pursuit method, we use the following procedure:

Step (1) Put $S_d(t+1)$ into $Z_i=\sum_{j=1}^{p}a_jx_{ij}\quad(i=1,2,\ldots,n)$ to calculate the value of Z_i

Step (2) Use Equation (4) and (5) for calculating S_z and D_z

Step (3) Use Equation (3) to calculate $Q(a)$, that is $f_d(t+1)$

When $f_d(t+1) = f_d(t)$, or it achieves the maximum number of iterations, the resulting $g_d(t)$ is equal to α^*.

Step (4) Analyze the flood characteristics. α^* will be substituted into the formula $Z_i = \sum_{j=1}^{p} a_j x_{ij} \quad (i = 1, 2, \ldots, n)$ to obtain the best projection value for each manhole $Z_i{}^*$; the greater the projection value is, the larger the flood.

3.4. Setting of Rainfall Scenarios

Rainfall design is the basis and premise for planning regional drainage systems and flood control measures. To probe the impact of climate change on rainfall, the Guangzhou Water Affairs Bureau analyzed serial information about rainfall from the most recent 60 years in 2011 [38], compared the information with the Guangzhou rainstorm formula enacted in 1993 [56] and introduced a rainfall intensity formula more in line with the rainstorm characteristics of Guangzhou (Table 2). A total of 27 designed rainfall events, which are composed of different return periods (one year, five years and ten years), rainfall duration (1 h, 1.5 h and 2 h) and rainfall peak coefficients (0.375, 0.5 and 0.8), were selected to investigate the control effects of LID practices. Here, the rainfall peak coefficient means the time that the rainfall peak occurred divided by the rainfall duration.

Table 2. Design rainfall intensity i (mm/min) of Guangzhou in different rainfall return periods, P (years) [38].

P	i
$P = 2$	$\frac{5230.65(1+0.438LgP)}{167(t+14.646)^{0.815}}$
$P = 5$	$\frac{4143.327(1+0.438LgP)}{167(t+12.874)^{0.758}}$
$P = 10$	$\frac{3512.11(1+0.438LgP)}{167(t+11.61)^{0.717}}$

Note: i is design rainfall intensity, t (hour) is rainfall duration and P is rainfall return period.

4. Results and Discussion

4.1. Calibration and Verification of Model Parameters

According to the observed rainfall–runoff data, the principles of the SWMM model and the recommended parameters, five rainfalls were selected to calibrate the parameters. Based on the sensitivity analysis of the SWMM model parameters, the trial and error approach was used to calibrate the parameter values repeatedly until the simulation results agreed with the observation results. The determined model's main sensitivity parameters are shown in Table 3. The simulation runoff process was obtained by repeatedly calibrating the model's parameters, and the change process of the observed runoff was consistent with the simulated runoff (Figure 2 and Table 4). The correlation coefficient (R^2) between the simulated flow and the observed flow and NSE coefficient exceed 0.91, and the peak relative error remained within 5%. From the rainfall–runoff process, the constructed regional rainfall–runoff model accurately reflects the relationship between runoff and rainfall.

To verify the rationality and reliability of the calibrated parameters, five different rainfall scenarios were selected for parameter verification (Table 4). The results show that the NSE coefficients between the simulated runoff and observed runoff exceed 0.7, that R^2 is greater than 0.86, and that the discharge relative error and the flood peak relative error are less than 5%. Therefore, the constructed rainfall–runoff model is believed to be highly accurate.

Table 3. Values and ranges of the main sensitivity parameters of the rainfall–runoff model.

Parameter		Recommended Parameter Range [27]	Parameter Range of the Rainfall–Runoff Model
Flow Width		/	10 m~277 m
Imperviousness		0~100%	25%
The Manning roughness coefficient	Impervious area Manning's roughness	0.015	0.011~0.015
	Pervious area Manning's roughness	0.4	0.014~0.8
	Roughness	0.01~0.013	0.01~0.012
Depth of depression storage	Impervious area depression storage	1 mm	0.2 mm~2 mm
	Pervious area depression storage	10 mm~11 mm	2 mm~13 mm
Horton	Maximum infiltration volume	103.81 mm/h	50 mm/h~150 mm/h
	Minimum infiltration volume	11.44 mm/h	0~20 mm/h
	Drying Time	2~7 day	7 day

Table 4. Calibration and validation results of the rainfall–runoff model.

Event	Rainfall Duration (min)	Rainfall (mm)	*NSE*	R^2	Discharge Relative Error (%)	Flood Peak Relative Error (%)
Calibrated events	140	21.4	0.92	0.93	2.84	2.18
	260	25	0.91	0.92	2.91	1.54
	210	19.7	0.92	0.98	4.52	3.14
	320	17.9	0.90	0.94	4.01	3.38
	200	16.8	0.91	0.93	4.12	3.55
Validated events	540	43.3	0.78	0.911	4.01	3.56
	840	51.9	0.72	0.885	3.85	4.35
	110	8.3	0.74	0.889	5.82	3.78
	480	23.9	0.71	0.884	5.91	4.08
	650	34.6	0.73	0.867	6.88	7.08

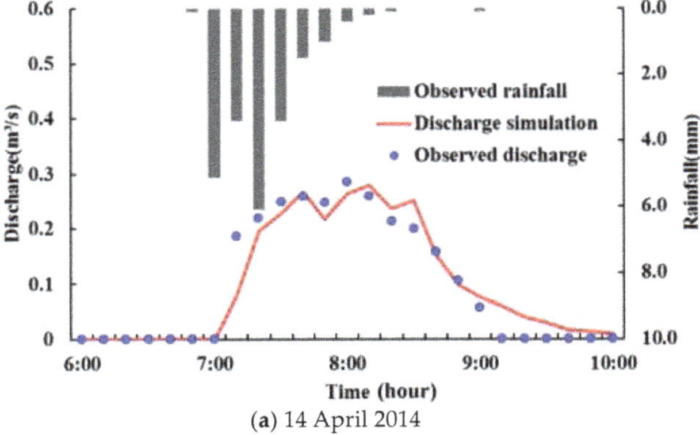

(**a**) 14 April 2014

Figure 2. *Cont.*

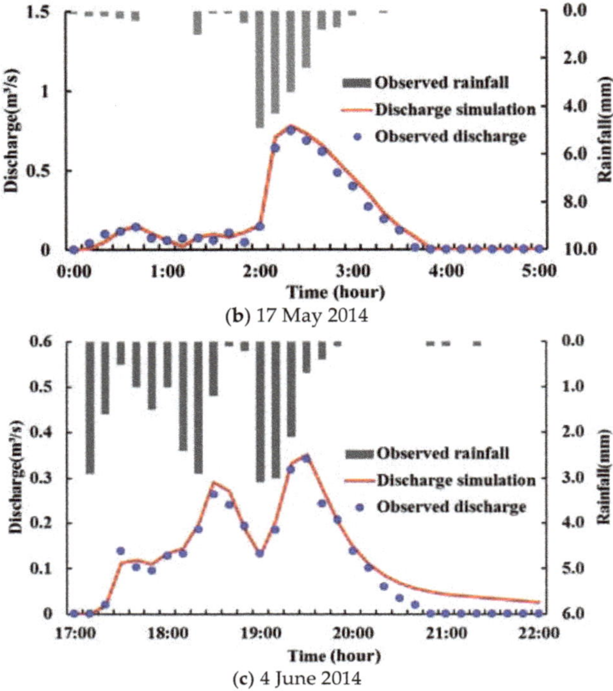

(**b**) 17 May 2014

(**c**) 4 June 2014

Figure 2. The observed and simulated hydrographs.

4.2. Analyze the Existing Situation of the Drainage System

To adapt to the changes of rainfall characteristics and land use, the constructed rainfall–runoff model simulates the functioning of the pipe network in different scenarios [38]. As can be seen from Table 5, there are seven to ten overload pipes in the two-year rainfall scenario, and the overload time is more than one hour (Table 5). The drainage system cannot satisfy the designed standard. With the increased rainfall duration or the rainfall peak coefficient, the drainage system is under increasing pressure. To ensure safety, appropriate flood controls must be implemented.

Table 5. Situation of drainage capacity under different rainfall scenarios.

Rainfall Scenarios	2y-2h-0.375	2y-2h-0.5	2y-2h-0.8	2y-1.5h-0.375	2y-1.5h-0.5	2y-1.5h-0.8	2y-1h-0.375	2y-1h-0.5	2y-1h-0.8
Number of overload pipes	7	7	10	7	7	10	7	7	9
Flood duration (hour)	1.88	1.91	1.96	1.83	1.88	1.90	1.79	1.81	1.84
Flood volume (m^3)	2651	2789	2954	2552	2733	2804	2412	2629	2693

Note: In the Rainfall Scenarios row, y indicates rainfall return period, h indicates rainfall duration; and the number below y and h indicates rainfall peak coefficient.

4.3. Control Characteristics of LID Practices under Complex Scenarios

In order to ensure safety and resource conservation, it was evaluated how well LID can contribute to flood control and it's characteristics under complex scenarios were analyzed. Since rain gardens and bioretention swales have better control ability and visual effects than other LID practices, the control characteristics of these two LID practices are analyzed under complex scenarios. The calculation is based on the volume of runoff produced by the rainfall as measured once in a two-year period [38]. The layout position is as shown in Figure 1.

4.3.1. Influence of Rainfall Intensity on Flooding and LID Practices

To investigate the comprehensive characteristics of floods with changes in rainfall intensity before and after the implementation of LID practices, the comprehensive characteristics of flood distribution under different rainfall scenarios (2y-2h-0.375, 5y-2h-0.375, 10y-2h-0.375) were analyzed. As can be seen from Figure 3, before LID practices were taken, the drainage system failed to drain the stormwater runoff produced by the rainfall intensity once in a two-year period and above. As the rainfall intensity increases, the projection values also rise. Figure 3 also illustrates that the flood-affected area increases and the flood become more serious with increasing rainfall intensity. After LID practices are implemented, the drainage system was able to promptly drain the runoff produced by the 2y-2h-0.375 rainfall scenario, and it was rendered free from flooding. However, with the increase in the intensity of the rainfall, the stormwater runoff generated by the 5y-2h-0.375 rainfall scenario caused relatively serious flooding in J9 and flooding in J15. As rainfall intensity increased to once in ten years, the flood-affected area and flood manholes increased. However, LID practices can alleviate the drainage pressure on the pipe network. It is effective in controlling the flood-affected area and the number of flood manholes, especially under the low intensity rainfall scenario.

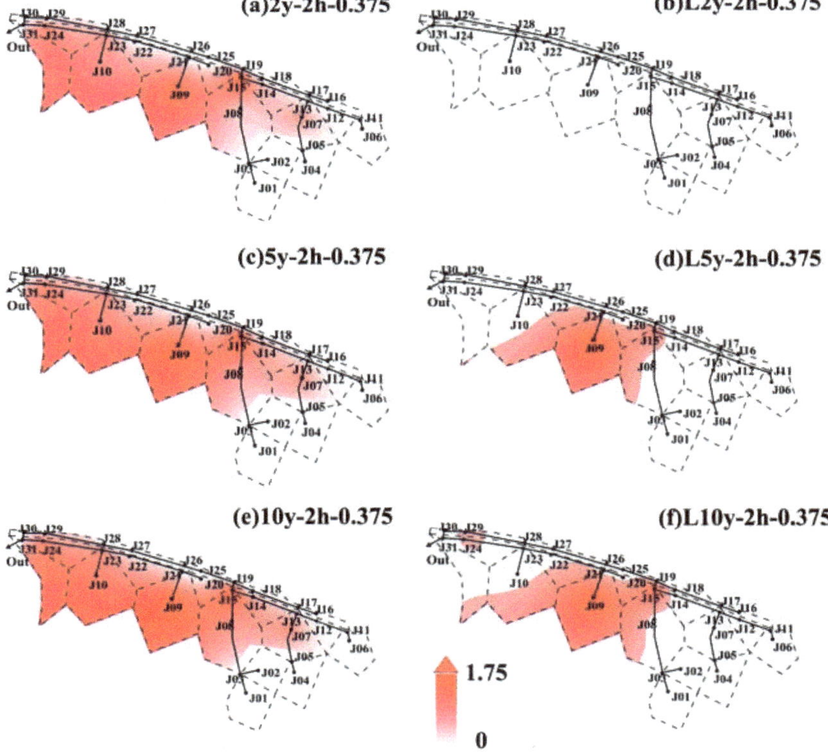

Figure 3. Impact of changes in rainfall intensity on flooding and LID practices. Characteristics of flooding spatial change are reflected by the projection value. (**a**,**c**,**e**) show where LID practices are not taken; (**b**,**d**,**f**) show where LID practices are taken.

4.3.2. Influence of Rainfall Duration on Flooding and LID Practices

In order to analyze the comprehensive characteristics of the influence of rainfall duration on flooding and LID practices, three rainfall scenarios (5y-1h-0.375, 5y-1.5h-0.375, 5y-2h-0.375) are selected. Before the implementation of LID practices, the flood-affected area and manholes with high projection values change slightly with the increase in rainfall duration (Figure 4). However, flood-affected areas tend to increase, indicating that long-duration rainfall increases flooding. After the implementation of LID practices, the flooding is significantly controlled, and the flood-affected area is drastically reduced. The shorter the rainfall duration, the smaller the flood-affected area. Yet, as rainfall duration increases, the flood-affected range gradually increases.

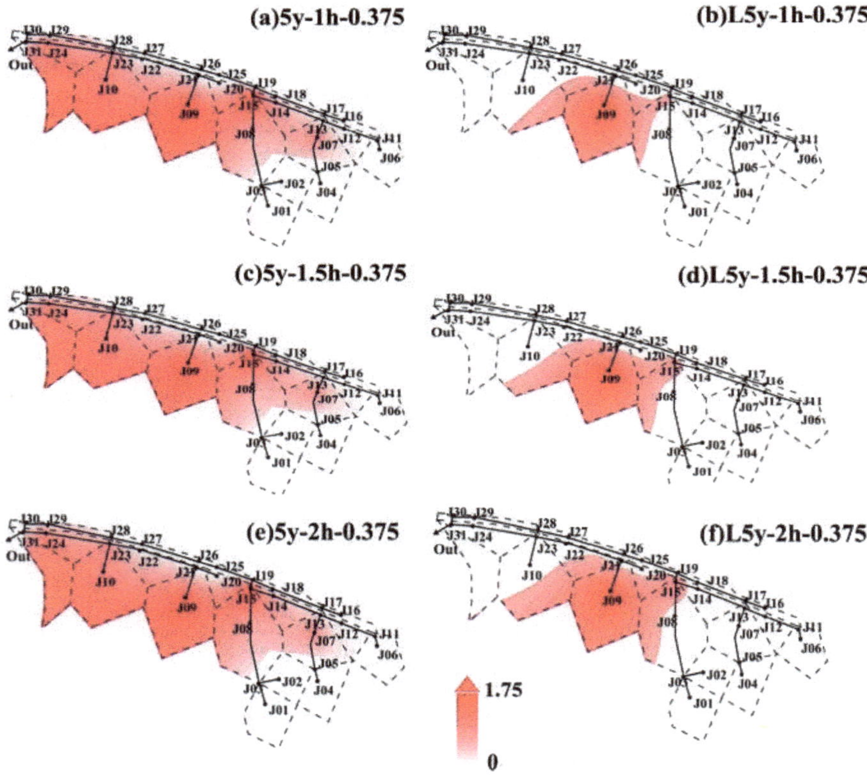

Figure 4. Influence of changes in rainfall duration on flooding and LID practices. Characteristics of flooding spatial change are reflected by the projection value. (**a**,**c**,**e**) show where LID practices are not taken; (**b**,**d**,**f**) show where LID practices are taken.

4.3.3. Influence of Rainfall Peak Coefficient on Flooding and LID Practices

The rainfall peak coefficient is one of the most important factors which influences flooding. Three rainfall scenarios with the same rainfall intensity and duration but different rainfall peak coefficients (5y-2h-0.375, 5y-2h-0.5, 5y-2h-0.8) were selected to analyze the influence of the rainfall peak coefficient on flooding and LID practices. As can be seen from Figure 5, when no LID control practices were taken, the impact of flooding is small with the changes in rainfall peak coefficient. The flood-affected area and the projection values are virtually unchanged. However, the rainfall peak coefficient has some influence on LID practices. The control effects of LID practices are diminished with the increase in the rainfall coefficient (Figure 5). The flood-affected area and the projection values of manholes J9 and J15 tended to increase (Figure 5). They are the largest under the 5y-2h-0.8 rainfall scenario.

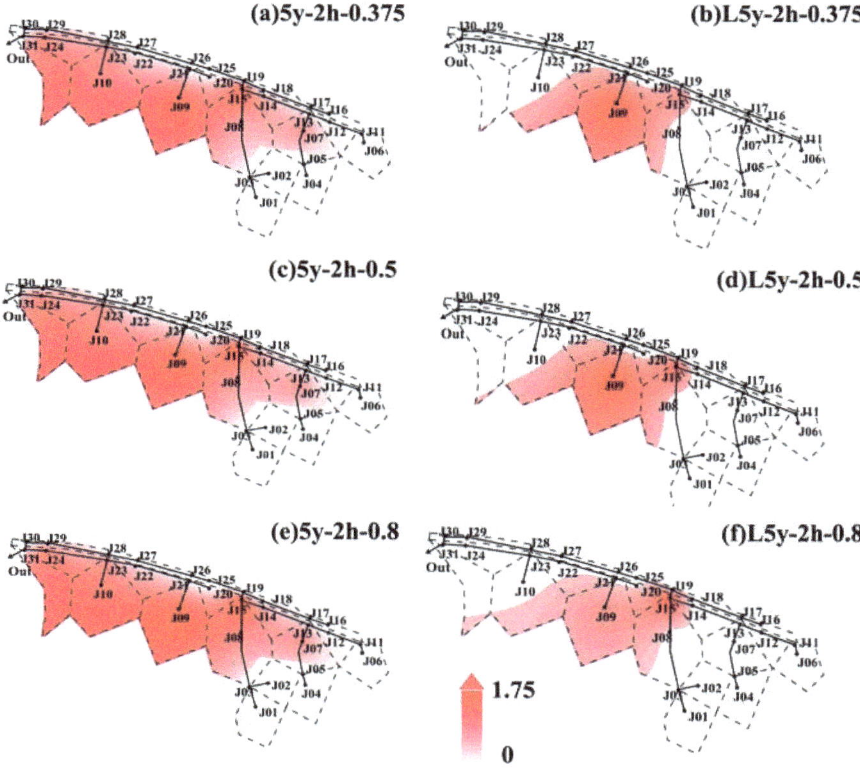

Figure 5. Influence of changes in rainfall peak coefficient on flooding and LID practices. Characteristics of flooding spatial change are reflected by the projection value. (**a**,**c**,**e**) show where LID practices are not taken; (**b**,**d**,**f**) show where LID practices are taken.

4.3.4. Changes in the Comprehensive Flood and LID Control Characteristics under Different Rainfall Scenarios

According to the comprehensive distribution characteristics of flooding, the projection values of the manholes J9 and J15 and the downstream area are high. The reasons for this are as follows: (i) for J15, it is located in the joint point of two pipes, and therefore its convergence area is large; (ii) J9 covers a large proportion of impervious surface in this area, and therefore rapid convergence speed and the severe overflow of the downstream pipe network contribute to grave flooding. The downstream flooding of the drainage system is caused by the failure of the timely drainage of stormwater, the (overly) full drainage network and the rise of the water level in the manholes.

In order to more comprehensively analyze the change of projection values and the LID control characteristics under complex scenarios, areas in J9 and J15 that have the greatest projection values were selected for analyzing the change characteristics of flooding under different complex rainfall scenarios (Figures 6 and 7). After the implementation of LID practices, the projection values of these two manholes were significantly reduced by at least 28%. However, the projection values increase with the increase in rainfall intensity, rainfall duration or rainfall peak coefficient. With the increase in rainfall intensity, there is a more obvious reduction in the impact of rainfall duration and rainfall peak coefficient on projection in the area (J9) with the greater amount of impervious areas. Some similar conclusions are obtained by Qin et al. [29].

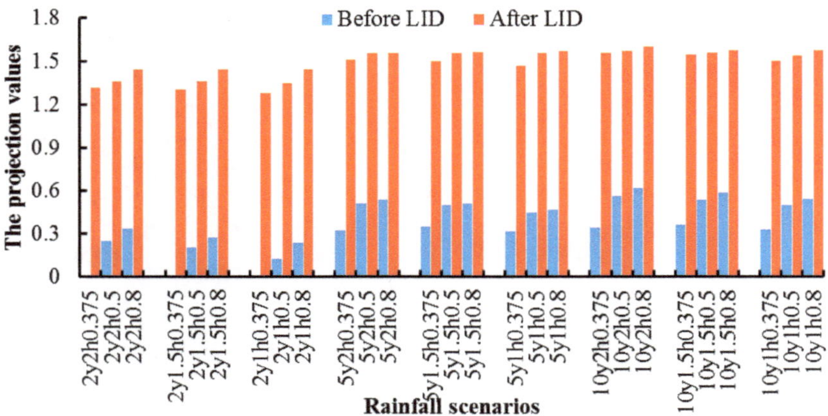

Figure 6. Changes in the projection values of J9 and LID control characteristics under complex scenarios.

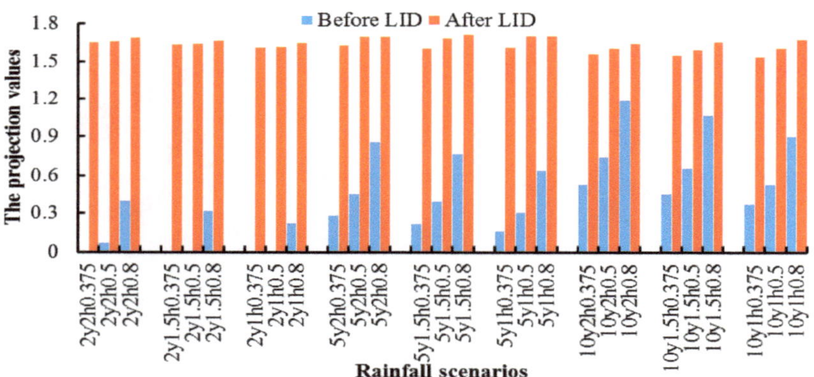

Figure 7. Changes in the projection values of J15 and LID control characteristics under complex scenarios.

It can be seen from the comparison between J9 and J15 that the influence of rainfall intensity, rainfall duration and rainfall peak coefficient on J15 are more obvious than that on J9. This is because LID implementation in J9 accounts for only a small proportion, and its effects on mitigating flooding are limited, making the control characteristics of LID practice not significant. When the rainfall duration is greater than 1.5 h or the rainfall peak coefficient is greater than 0.5, the projection value of J9 increases slightly. However, the upstream and downstream LID practices can alleviate the drainage pressure on the drainage network, thereby reducing the projection values of the areas where no LID practices are taken. On the contrary, the percentage of LID is relatively high upstream (above manhole J15), and the flood control characteristics of the LID practices are more obvious. The projection value of J15 is greater with changes in the rainfall duration and rainfall peak coefficient, suggesting that the area where many LID practices were implemented is significantly affected by rainfall duration and rainfall peak coefficient.

5. Conclusions

Rainfall has a great impact on urban flooding in the drainage system. In this study, the effect of LID practices on flood reduction in a typical residential area of Guangzhou were investigated under different rainfall scenarios. The flooding hot spots and their comprehensive characteristics were analyzed by the SWMM model with the PSO-based projection pursuit technique. The main conclusions were that: (i) the projection values of different manholes have different responses to the changes in rainfall scenarios, but the values of manholes J9 and J15 were the largest before and after the implementation of LID practices. Therefore, manholes J9 and J15 were flooding hotspots and should be paid more attention; (ii) LID practices are effective in flood reduction. They can control the flooding under the 2y-2h-0.375 rainfall scenario. However, the control effects of LID practices tend to decrease as the rainfall intensity, rainfall duration or rainfall peak coefficient increase; (iii) compared with different rainfall scenarios, the control ability of LID practices are more effective in flood reduction for shorter duration, lower intensity and smaller peak coefficient rainfall events; (iv) as the rainfall intensity increases, the impact of rainfall duration and rainfall peak coefficient on projection values is reduced, particularly in the area with the greater amount of impervious areas. Among rainfall intensity, rainfall duration and rainfall peak coefficient, the control effects of LID practices are most affected by rainfall intensity.

Acknowledgments: The research is financially supported by the National Natural Science Foundation of China (Grant No. 91547202, 51210013, 51479216), the Chinese Academy of Engineering Consulting Project (2015-ZD-07-04-03), the Public Welfare Project of Ministry of Water Resources (Grant No. 200901043-03), the Project for Creative Research from Guangdong Water Resources Department (Grant No. 2016-07, 2016-01), Research program of Guangzhou Water Authority (2017).

Author Contributions: Zhihua Zhu and Xiaohong Chen conceived and designed the experiments; Zhihua Zhu performed the experiments; Zhihua Zhu and Xiaohong Chen analyzed the data; Xiaohong Chen contributed reagents/materials/analysis tools; Zhihua Zhu and Xiaohong Chen wrote the paper.

Conflicts of Interest: The authors declare no conflict of interest.

References

1. Albano, R.; Sole, A.; Mirauda, D.; Adamowski, J. Modelling large floating bodies in urban area flash-floods via a Smoothed Particle Hydrodynamics model. *J. Hydrol.* **2016**, *541*, 344–358. [CrossRef]
2. Guinot, V.; Sanders, B.F.; Schubert, J.E. Dual integral porosity shallow water model for urban flood modelling. *Adv. Water Resour.* **2017**, *103*, 16–31. [CrossRef]
3. Duan, W.L.; He, B.; Takara, K.; Luo, P.P.; Nover, D.; Yamashiki, Y.; Huang, W.R. Anomalous atmospheric events leading to Kyushu's flash floods, 11–14 July 2012. *Nat. Hazards* **2014**, *73*, 1255–1267. [CrossRef]
4. Public Utilities Board of Singapore. *Report on Key Conclusions and Recommendations of the Expert Panel on Drainage Design and Flood Protection Measures*; PUB: Singapore, 2012.
5. Escuder-Bueno, I.; Castillo-Rodríguez, J.T.; Zechner, S.; Jöbstl, C.; Perales-Momparler, S.; Petaccia, G. A quantitative flood risk analysis methodology for urban areas with integration of social research data. *Nat. Hazards Earth Syst. Sci.* **2012**, *12*, 2843–2863. [CrossRef]
6. Parliamentary Office of Science and Technology. *Urban Flooding*; Postnote Number 289; POST: London, UK, 2007.
7. Chen, S.Y.; Xue, Z.C.; Li, M.; Zhu, X.P. Variable sets method for urban flood vulnerability assessment. *Sci. China Technol. Sci.* **2013**, *56*, 3129–3136. [CrossRef]
8. Bubeck, P.; Botzen, W.J.W.; Aerts, J.C.J.H. A review of risk perceptions and other factors that influence flood mitigation behavior. *Risk Anal.* **2012**, *32*, 1481–1495. [CrossRef] [PubMed]
9. Lee, J.Y.; Moon, H.J.; Kim, T.I.; Kim, H.W.; Han, M.Y. Quantitative analysis on the urban flood mitigation effect by the extensive green roof system. *Environ. Pollut.* **2013**, *181*, 257–261. [CrossRef] [PubMed]
10. Wang, J.; Gao, W.; Xu, S.Y.; Yu, L.Z. Evaluation of the combined risk of sea level rise, land subsidence, and storm surges on the coastal areas of Shanghai, China. *Clim. Chang.* **2012**, *115*, 537–558. [CrossRef]
11. Zhu, G.W. Ponderation over the Flood Control Functions of the Artifical Landforms in PRD. *Trop. Geogr.* **2012**, *32*, 378–384.

12. Wesselink, A.; Warner, J.; Syed, M.A.; Chan, F.; Tran, D.D.; Huq, H.; Huthoff, F.; Thuy, N.L.; Pinter, N.; Staveren, M.V.; et al. Trends in flood risk management in deltas around the world: Are we going 'soft'? *Int. J. Water Gov.* **2015**, *3*, 25–46. [CrossRef]
13. Tortajada, C.; Joshi, Y.; Biswas, A.K. *The Singapore Water Story: Sustainable Development in an Urban City State*; Routledge: London, UK; New York, NY, USA, 2013.
14. Chui, S.K.; Leung, J.K.Y.; Chu, C.K. The development of a comprehensive flood prevention strategy for Hong Kong. *Int. J. River Basin Manag.* **2006**, *4*, 5–15. [CrossRef]
15. Dietz, M.E. Low impact development practices: A review of current research and recommendations for future directions. *Water Air Soil Pollut.* **2007**, *186*, 351–363. [CrossRef]
16. Fletcher, T.D.; Shuster, W.; Hunt, W.F.; Ashley, R.; Butler, D.; Arthur, S.; Trowsdale, S.; Barraud, S.; Semadeni-Davies, A.; Bertrand-Krajewski, J.L.; et al. SUDS, LID, BMPs, WSUD and more. The evolution and application of terminology surrounding urban drainage. *Urban Water J.* **2015**, *12*, 525–542. [CrossRef]
17. Ahiablame, L.M.; Engel, B.A.; Chaubey, I. Effectiveness of low impact development practices in two urbanized watersheds: Retrofitting with rain barrel/cistern and porous pavement. *J. Environ. Manag.* **2013**, *119*, 151–161. [CrossRef] [PubMed]
18. Seo, M.; Jaber, F.; Srinivasan, R.; Jeong, J.H. Evaluating the Impact of Low Impact Development (LID) Practices on Water Quantity and Quality under Different Development Designs Using SWAT. *Water* **2017**, *9*, 193. [CrossRef]
19. PRC Ministry of Construction. *Guiding Technology on Constructing the Sponge Cities—Constructing the Rainwater System Based on Low Impact Development*; PRC Ministry of Construction: Beijing, China, 2014.
20. Lim, H.S.; Lu, X.X. Sustainable urban stormwater management in the tropics: An evaluation of Singapore's ABC Waters Program. *J. Hydrol.* **2016**, *538*, 842–862. [CrossRef]
21. Silva, M.M.; Costa, J.P. Flood adaptation measures applicable in the design of urban public spaces: Proposal for a conceptual framework. *Water* **2016**, *8*, 284. [CrossRef]
22. Xia, J.; Zhang, Y.Y.; Xiong, L.H.; He, S.; Wang, L.F.; Yu, Z.B. Opportunities and challenges of the Sponge City construction related to urban water issues in China. *Sci. China Earth Sci.* **2017**, *60*, 652–658. [CrossRef]
23. Li, T.; Shan, S.C.; Liu, J.; She, N.; Chen, B.N.; Wu, L.Y. Applying New Features of Low-Impact Development Techniques in the Master Planning of Guangzhou Educational Town. *Int. Low Impact Dev.* **2015**. [CrossRef]
24. Chang, C.G.V. Reimagining Urban Drainage in the World's Biggest Construction Sites: Three LID Stories in Eastern. *China Int. Low Impact Dev.* **2015**. [CrossRef]
25. Jia, H.; Yao, H.; Yu, S.L. Advances in LID BMPs research and practice for urban runoff control in China. *Front. Environ. Sci. Eng.* **2013**, *7*, 709–720. [CrossRef]
26. Li, P.; Liu, J.; Fu, R.; Liu, X.; Zhou, Y.Y.; Luan, M. The performance of LID (low impact development) practices at different locations with an urban drainage system: A case study of Longyan, China. *Water Pract. Technol.* **2015**, *10*, 739–746. [CrossRef]
27. Rossman, L.A. *Storm Water Management Model User's Manual, Version 5.0*; National Risk Management Research Laboratory, Office of Research and Development, US Environmental Protection Agency: Cincinnati, OH, USA, 2010.
28. Burszta-Adamiak, E.; Mrowiec, M. Modelling of green roofs' hydrologic performance using EPA's SWMM. *Water Sci. Technol.* **2013**, *68*, 36–42. [CrossRef] [PubMed]
29. Qin, H.P.; Li, Z.X.; Fu, G. The effects of low impact development on urban flooding under different rainfall characteristics. *J. Environ. Manag.* **2013**, *129*, 577–585. [CrossRef] [PubMed]
30. Alfredo, K.; Montalto, F.; Goldstein, A. Observed and modeled performances of prototype green roof test plots subjected to simulated low-and high-intensity precipitations in a laboratory experiment. *J. Hydrol. Eng.* **2010**, *15*, 444–457. [CrossRef]
31. Niu, S.; Cao, L.; Li, Y.; Huang, J.H. Long-Term Simulation of the Effect of Low Impact Development for Highly Urbanized Areas on the Hydrologic Cycle in China. *Int. J. Environ. Sci. Dev.* **2016**, *7*, 225–228. [CrossRef]
32. Bedan, E.S.; Clausen, J.C. Stormwater runoff quality and quantity from traditional and low impact development watersheds. *J. Am. Water Res. Assoc.* **2009**, *45*, 998–1008. [CrossRef]
33. Jia, H.; Lu, Y.; Yu, S.L.; Chen, Y. Planning of LID–BMPs for urban runoff control: The case of Beijing Olympic Village. *Sep. Purif. Technol.* **2012**, *84*, 112–119. [CrossRef]

34. Huang, C.L.; Hsu, N.S.; Wei, C.C.; Luo, W.J. Optimal spatial design of capacity and quantity of rainwater harvesting systems for urban flood mitigation. *Water* **2015**, *7*, 5173–5202. [CrossRef]
35. Du, S.Q.; Shi, P.J.; Van Rompaey, A.; Wen, J.H. Quantifying the impact of impervious surface location on flood peak discharge in urban areas. *Nat. Hazards* **2015**, *76*, 1457–1471. [CrossRef]
36. Zhou, Q.; Leng, G.; Huang, M. Impacts of future climate change on urban flood risks: Benefits of climate mitigation and adaptations. *Hydrol. Earth Syst. Sci. Discuss.* **2016**. [CrossRef]
37. Fang, F.; Qiao, L.L.; Cao, J.S.; Li, Y.; Xie, W.M.; Sheng, G.P.; Yu, H.Q. Quantitative evaluation of A_2O and reversed A_2O processes for biological municipal wastewater treatment using a projection pursuit method. *Sep. Purif. Technol.* **2016**, *166*, 164–170. [CrossRef]
38. Guangzhou Water Affairs Bureau. *The Calculation Formulas and Diagrams of Urban Stormy of Guangzhou*; Guangzhou Water Affairs Bureau: Guangzhou, China, 2011.
39. Chen, Z.H.; Yin, L.; Chen, X.H.; Wei, S.; Zhu, Z.H. Research on the characteristics of urban rainstorm pattern in the humid area of Southern China: A case study of Guangzhou City. *Int. J. Climatol.* **2015**, *35*, 4370–4386. [CrossRef]
40. Barco, J.; Wong, K.M.; Stenstrom, M.K. Automatic calibration of the US EPA SWMM model for a large urban catchment. *J. Hydraul. Eng.* **2008**, *134*, 466–474. [CrossRef]
41. Granata, F.; Gargano, R.; de Marinis, G. Support vector regression for rainfall-runoff modeling in urban drainage: A comparison with the EPA's Storm Water Management Model. *Water* **2016**, *8*, 69. [CrossRef]
42. Jang, S.; Cho, M.; Yoon, J.; Yoon, Y.; Kim, S.; Kim, G.; Kim, L.; Aksoy, H. Using SWMM as a tool for hydrologic impact assessment. *Desalination* **2007**, *212*, 344–356. [CrossRef]
43. Pushpalatha, R.; Perrin, C.; Le Moine, N.; Andréassian, V. A review of efficiency criteria suitable for evaluating low-flow simulations. *J. Hydrol.* **2012**, *420*, 171–182. [CrossRef]
44. Dechmi, F.; Burguete, J.; Skhiri, A. SWAT application in intensive irrigation systems: Model modification, calibration and validation. *J. Hydrol.* **2012**, *470*, 227–238. [CrossRef]
45. Yang, X.L.; Zhu, B.; Li, Y.L.; Hua, K.K. Simulation of nonpoint source nitrogen transport in two separated catchments in the hilly area of purple soil. *J. Hydraul. Eng.* **2013**, *44*, 1197–1203.
46. Zhao, J.; Jin, J.L.; Guo, Q.Z.; Liu, L.; Chen, Y.Q.; Pan, M. Dynamic risk assessment model for flood disaster on a projection pursuit cluster and its application. *Stoch. Environ. Res. Risk Assess.* **2014**, *28*, 2175–2183. [CrossRef]
47. Zhou, Y.; Guo, S.; Xu, C.Y.; Liu, D.; Chen, L.; Ye, Y. Integrated optimal allocation model for complex adaptive system of water resources management (I): Methodologies. *J. Hydrol.* **2015**, *531*, 964–976. [CrossRef]
48. Pei, W.; Fu, Q.; Liu, D.; Li, T.X.; Cheng, K. Assessing agricultural drought vulnerability in the Sanjiang Plain based on an improved projection pursuit model. *Nat. Hazards* **2016**, *82*, 683–701. [CrossRef]
49. Yang, G.; Guo, S.L.; Li, L.P.; Hong, X.J.; Wang, L. Multi-objective operating rules for Danjiangkou reservoir under climate change. *Water Res. Manag.* **2016**, *30*, 1183–1202. [CrossRef]
50. Poli, R.; Rennedy, J.; Blackwell, T. Particle swarm optimization. *Swarm Intell.* **2007**, *1*, 33–57. [CrossRef]
51. Gao, Y.; Du, W.B.; Yan, G. Selectively-informed particle swarm optimization. *Sci. Rep.* **2015**, *5*, 9295. [CrossRef] [PubMed]
52. Shi, Y. Particle swarm optimization: Developments, applications and resources. In Proceedings of the 2001 Congress on Evolutionary Computation, Seoul, South Korea, 27–30 May 2001; IEEE: New York, NY, USA; Volume 1, pp. 81–86.
53. Zhu, Z.H.; Chen, Z.H.; Chen, X.H.; He, P.Y. Approach for evaluating inundation risks in urban drainage systems. *Sci. Total Environ.* **2016**, *553*, 1–12. [CrossRef] [PubMed]
54. Eberhart, R.C.; Shi, Y. Comparing inertia weights and constriction factors in particle swarm optimization. In Proceedings of the 2000 Congress on Evolutionary Computation, La Jolla, CA, USA, 16–19 July 2000; IEEE: New York, NY, USA; Volume 1, pp. 84–88.
55. Wang, Y.C.; Lv, J.; Zhu, L.; Ma, Y.M. Crystal structure prediction via particle-swarm optimization. *Phys. Rev. B* **2010**, *82*, 094116. [CrossRef]
56. Guangzhou Water Affairs Bureau. *The Calculation Formulas and Diagrams of Urban Stormy of Guangzhou*; Guangzhou Water Affairs Bureau: Guangzhou, China, 1993.

MDPI
St. Alban-Anlage 66
4052 Basel, Switzerland
Tel. +41 61 683 77 34
Fax +41 61 302 89 18
http://www.mdpi.com

Water Editorial Office
E-mail: water@mdpi.com
http://www.mdpi.com/journal/water

www.ingramcontent.com/pod-product-compliance
Lightning Source LLC
Chambersburg PA
CBHW051727210326
41597CB00032B/5635

* 9 7 8 3 0 3 8 4 2 9 5 1 7 *